*Nonlinear Signal
Processing*

Nonlinear Signal Processing

A Statistical Approach

Gonzalo R. Arce
University of Delaware
Department of Computer
and Electrical Engineering

A JOHN WILEY & SONS, INC., PUBLICATION

Published by John Wiley & Sons, Inc., Hoboken, New Jersey.
Published simultaneously in Canada.

For general information on our other products and services please contact our Customer Care
Department within the U.S. at 877-762-2974, outside the U.S. at 317-572-3993 or fax 317-572-4002.

Wiley also publishes its books in a variety of electronic formats. Some content that appears in print,
however, may not be available in electronic format.

Library of Congress Cataloging-in-Publication Data:

Arce, Gonzalo R.
    Nonlinear signal processing : a statistical approach / Gonzalo R. Arce.
      p. cm.
    Includes bibliographical references and index.
    ISBN 0-471-67624-1 (cloth : acid-free paper)
      1. Signal processing—Mathematics.   2. Statistics.   I. Title.

   TK5102.9.A77 2004
   621.382'2—dc22   2004042240

Printed in the United States of America.

10 9 8 7 6 5 4 3 2 1

To Catherine, Andrew, Catie, and my beloved parents.

Preface

Linear filters today enjoy a rich theoretical framework based on the early and important contributions of Gauss (1795) on Least Squares, Wiener (1949) on optimal filtering, and Widrow (1970) on adaptive filtering. Linear filter theory has consistently provided the foundation upon which linear filters are used in numerous practical applications as detailed in classic treatments including that of Haykin [99], Kailath [110], and Widrow [197]. Nonlinear signal processing, however, offers significant advantages over traditional linear signal processing in applications in which the underlying random processes are nonGaussian in nature, or when the systems acting on the signals of interest are inherently nonlinear. Practice has shown that nonlinear systems and nonGaussian processes emerge in a broad range of applications including imaging, teletraffic, communications, hydrology, geology, and economics. Nonlinear signal processing methods in all of these applications aim at exploiting the system's nonlinearities or the statistical characteristics of the underlying signals to overcome many of the limitations of the traditional practices used in signal processing.

Traditional signal processing enjoys the rich and unified theory of linear systems. Nonlinear signal processing, on the other hand, lacks a unified and universal set of tools for analysis and design. Hundreds of nonlinear signal processing algorithms have been proposed in the literature. Most of the proposed methods, although well tailored for a given application, are not broadly applicable in general. While nonlinear signal processing is a dynamic and rapidly growing field, large classes of nonlinear signal processing algorithms can be grouped and studied in a unified framework. Textbooks on higher-and lower-order statistics [148], polynomial filters [141], neural-networks [100], and mathematical morphology have appeared recently with the common goal of grouping a "self-contained" class of nonlinear signal processing algorithms into a unified framework of study.

This book focuses on unifying the study of a broad and important class of nonlinear signal processing algorithms that emerge from statistical estimation principles, and where the underlying signals are nonGaussian processes. Notably, by concentrating on just two nonGaussian models, a large set of tools is developed that encompasses a large portion of the nonlinear signal processing tools proposed in the literature over the past several decades. In particular, under the generalized Gaussian distribution, signal processing algorithms based on weighted medians and their generalizations are developed. The class of stable distributions is used as the second nonGaussian model from which weighted myriads emerge as the fundamental estimate from which general signal processing tools are developed. Within these two classes of nonlinear signal processing methods, a goal of the book is to develop a unified treatment on optimal and adaptive signal processing algorithms that mirror those of Wiener and Widrow, extensively presented in the linear filtering literature.

The current manuscript has evolved over several years while the author regularly taught a nonlinear signal processing course in the graduate program at the University of Delaware. The book serves an international market and is suitable for advanced undergraduates or graduate students in engineering and the sciences, and practicing engineers and researchers. The book contains many unique features including:

- Numerous problems at the end of each chapter.

- Numerous examples and case studies provided throughout the book in a wide range of applications.

- A set of 60+ MATLAB software m-files allowing the reader to quickly design and apply any of the nonlinear signal processing algorithms described in the book to an application of interest.

- An accompanying MATLAB software guide.

- A companion PowerPoint presentation with more than 500 slides available for instruction.

The chapters in the book are grouped into three parts.

Part I provides the necessary theoretical tools that are used later in text. These include a review of nonGaussian models emphasizing the class of generalized Gaussian distributions and the class of stable distributions. The basic principles of order statistics are covered, which are of essence in the study of weighted medians. Part I closes with a chapter on maximum likelihood and robust estimation principles which are used later in the book as the foundation on which signal processing methods are build upon.

Part II comprises of three chapters focusing on signal processing tools developed under the generalized Gaussian model with an emphasis on the Laplacian model. Weighted medians, L-filters, and several generalizations are studied at length.

Part III encompasses signal processing methods that emerge from parameter estimation within the stable distribution framework.

The chapter sequence is thus assembled in a self-contained and unified framework of study.

Acknowledgments

The material in this textbook has benefited greatly from my interaction with many bright students at the University of Delaware. I am particularly indebted to my previous graduate students Juan Gonzalez, Sebastian Hoyos, Sudhakar Kalluri, Yinbo Li, David Griffith, Yeong-Taeg Kim, Edwin Heredia, Alex Flaig, Zhi Zhou, Dan Lau, Karen Bloch, Russ Foster, Russ Hardie, Tim Hall, and Michael McLoughlin. They have all contributed significantly to material throughout the book. I am very grateful to Jan Bacca and Dr. Jose-Luis Paredes for their technical and software contributions. They have generated all of the MATLAB routines included in the book as well as the accompanying software guide. Jan Bacca has provided the much needed electronic publishing support to complete this project.

I am particularly indebted to Dr. Neal C. Gallagher of the University of Central Florida for being a lifelong mentor, supporter, and friend.

It has been a pleasure working with the Non-linear Signal Processing Board: Dr. Hans Burkhardt of the Albert-Ludwigs-University, Freiburg Germany, Dr. Ed Coyle of Purdue University, Dr. Moncef Gabbouj of the Tampere University of Technology, Dr. Murat Kunt of the Swiss Federal Institute of Technology, Dr. Steve Marshall of the University of Strathclyde, Dr. John Mathews of the University of Utah, Dr. Yrjo Neuvo of Nokia, Dr. Ioannis Pitas of the Aristotle University of Thessaloniki, Dr. Jean Serra of the Center of Mathematical Morphology, Dr. Giovanni Sicuranza of the University of Trieste, Dr. Akira Taguchi of the Musashi Institute of Technology, Dr. Anastasios N. Venetsanopoulos of the University of Toronto, and Dr. Pao-Ta Yu of the National Chung Cheng University. Their contribution in the organization of the international workshop series in this field has provided the vigor required for academic excellence.

My interactions with a number of outstanding colleagues has deepened my understanding of nonlinear signal processing. Many of these collaborators have made important contributions to the theory and practice of nonlinear signal processing. I am most grateful to Dr. Ken Barner, Dr. Charles Boncelet, Dr. Xiang Xia, and Dr. Peter Warter all from the University of Delaware, Dr. Jackko Astola, Dr. Karen Egiazarian, Dr. Oli Yli-Harja, Dr I. Tăbus, all from the Tampere University of Technology, Dr. Visa Koivunen of the Helsinki University of Technology, Dr. Saleem Kassam of the University of Pennsylvania, Dr. Sanjit K. Mitra of the University of California, Santa Barbara, Dr. David Munson of the University of Michigan, Dr. Herbert David of Iowa State University, Dr. Kotroupolus of the Universtiy of Thessaloniki, Dr. Yrjo Neuvo of Nokia, Dr. Alan Bovik and Dr. Ilya Shmulevich, both of the University of Texas, Dr. Francesco Palmieri of the University of Naples, Dr. Patrick Fitch of the Lawrence Livermore National Laboratories, Dr. Thomas Nodes of TRW, Dr. Brint Cooper of Johns Hopkins University, Dr. Petros Maragos of the University of Athens, and Dr. Y. H. Lee of KAIST University.

I would like to express my appreciation for the research support I received from the National Science Foundation and the Army Research laboratories, under the Federated Laboratories and Collaborative Technical Alliance programs, for the many years of research that led to this textbook. I am particularly grateful to Dr. John Cozzens and Dr. Taieb Znati, both from NSF, and Dr. Brian Sadler, Dr. Ananthram Swami, and Jay Gowens, all from ARL. I am also grateful to the Charles Black Evans Endowment that supports my current Distinguished Professor appointment at the University of Delaware.

I would like to thank my publisher George Telecki and the staff at Wiley for their dedicated work during this project and Heather King for establishing the first link to Wiley.

G. R. A.

Contents

Acronyms

ADSL	Asymmetric digital suscriber line
BIBO	Bounded-input bounded-output
BR	Barrodale and Roberts' (algorithm)
CMA	Constant modulus algorithm
CWM	Center-weighted median
CWMY	Center-weighted myriad
DWMTM	Double window modified Trimmed mean
DWD	Discrete Wigner distribution
FIR	Finite impulse response
FLOS	Fractional lower-order statistics
FLOM	Fractiona lower-order moments
HOS	higher-order statistics
i.i.d	Independent and identically distributed
IIR	Infinite impulse response
LCWM	Linear combination of weighted medians
LS	Least squares
LAD	Least absolute deviation

LLS	Logarithmic least squares
LMS	Least mean square
LMA	Least mean absolute
LP	Linearity parameter
MSE	Mean square error
ML	Maximum likelihood
MAE	Mean absolute error
MTM	Modified trimmed mean
PAM	Phase amplitude modulation
pdf	Portable document format
PLL	Phase lock loop
PSNR	Peak signal-to-noise ratio
PBF	Positive boolean function
RTT	Round trip time
SαS	Symmetric α-stable
SSP	Sample selection probabilities
TCP/IP	Internet transfer protocol
TD	Threshold Decomposition
WM	Weighted median
WMM	Weighted multichannel median
WD	Wigner distribution
ZOS	Zero-order statistics

1

Introduction

Signal processing is a discipline embodying a large set of methods for the representation, analysis, transmission, and restoration of information-bearing signals from various sources. As such, signal processing revolves around the mathematical manipulation of signals. Perhaps the most fundamental form of signal manipulation is that of filtering, which describes a rule or procedure for processing a signal with the goal of separating or attenuating a desired component of an observed signal from either noise, interference, or simply from other components of the same signal. In many applications, such as communications, we may wish to remove noise or interference from the received signal. If the received signal was in some fashion distorted by the channel, one of the objectives of the receiver is to compensate for these disturbances. Digital picture processing is another application where we may wish to enhance or extract certain image features of interest. Perhaps image edges or regions of the image composed of a particular texture are most useful to the user. It can be seen that in all of these examples, the signal processing task calls for separating a desired component of the observed waveform from any noise, interference, or undesired component. This segregation is often done in frequency, but that is only one possibility. Filtering can thus be considered as a system with arbitrary input and output signals, and as such the filtering problem is found in a wide range of disciplines including economics, engineering, and biology.

A classic filtering example, depicted in Figure 1.1, is that of bandpass filtering a frequency rich chirp signal. The frequency components of the chirp within a selected band can be extracted through a number of linear filtering methods. Figure 1.1*b* shows the filtered chirp when a linear 120-tap finite impulse response (FIR) filter is used. This figure clearly shows that linear methods in signal processing can indeed

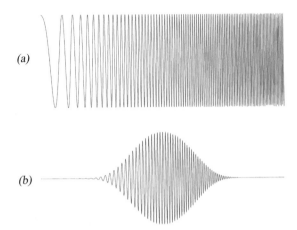

Figure 1.1 Frequency selective filtering: (*a*) chirp signal, (*b*) linear FIR filter output.

be markedly effective. In fact, linear signal processing enjoys the rich theory of linear systems, and in many applications linear signal processing algorithms prove to be optimal. Most importantly, linear filters are inherently simple to implement, perhaps the dominant reason for their widespread use.

Although linear filters will continue to play an important role in signal processing, nonlinear filters are emerging as viable alternative solutions. The major forces that motivate the implementation of nonlinear signal-processing algorithms are the growth of increasingly challenging applications and the development of more powerful computers. Emerging multimedia and communications applications are becoming significantly more complex. Consequently, they require the use of increasingly sophisticated signal-processing algorithms. At the same time, the ongoing advances of computers and digital signal processors, in terms of speed, size, and cost, makes the implementation of sophisticated algorithms practical and cost effective.

Why Nonlinear Signal Processing? Nonlinear signal processing offers advantages in applications in which the underlying random processes are nonGaussian. Practice has shown that nonGaussian processes do emerge in a broad array of applications, including wireless communications, teletraffic, hydrology, geology, economics, and imaging. The common element in these applications, and many others, is that the underlying processes of interest tend to produce more large-magnitude (outlier or impulsive) observations than those that would be predicted by a Gaussian model. That is, the underlying signal density functions have tails that decay at rates lower than the tails of a Gaussian distribution. As a result, linear methods which obey the superposition principle suffer from serious degradation upon the arrival of samples corrupted with high-amplitude noise. Nonlinear methods, on the other hand, exploit the statistical characteristics of the noise to overcome many of the limitations of the traditional practices in signal processing.

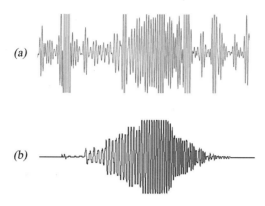

(a)

(b)

Figure 1.2 Frequency selective filtering in nonGaussian noise: (*a*) linear FIR filter output, (*b*) nonlinear filter.

To illustrate the above, consider again the classic bandpass filtering example. This time, however, the chirp signal under analysis has been degraded by nonGaussian noise during the signal acquisition stage. Due to the nonGaussian noise, the linear FIR filter output is severely degraded as depicted in Figure 1.2*a*. The advantages of an equivalent nonlinear filter are illustrated in Figure 1.2*b* where the frequency components of the chirp within the selected band have been extracted, and the ringing artifacts and the noise have been suppressed [1].

Internet traffic provides another example of signals arising in practice that are best modeled by nonGaussian models for which nonlinear signal processing offer advantages. Figure 1.3 depicts several round trip time delay series, each of which measures the time that a TCP/IP packet takes to travel between two network hosts. An RTT measures the time difference between the time when a packet is sent and the time when an acknowledgment comes back to the sender. RTTs are important in retransmission transport protocols used by TCP/IP where reliability of communications is accomplished through packet reception acknowledgments, and, when necessary, packet retransmissions. In the TCP/IP protocol, the retransmission of packets is based on the prediction of future RTTs. Figure 1.3 depicts the nonstationary characteristics of RTT processes as their mean varies dramatically with the network load. These processes are also nonGaussian indicating that nonlinear prediction of RTTs can lead to more efficient communication protocols.

Internet traffic exhibits nonGaussian statistics, not only on the RTT delay data mechanisms, but also on the data throughput. For example, the traffic data shown in Figure 1.4 corresponds to actual Gigabit (1000 Mb/s) Ethernet traffic measured on a web server of the ECE Department at the University of Delaware. It was measured using the TCPDUMP program, which is part of the Sun Solaris operating system. To

---

[1]The example uses a weighted median filter that is developed in later sections.

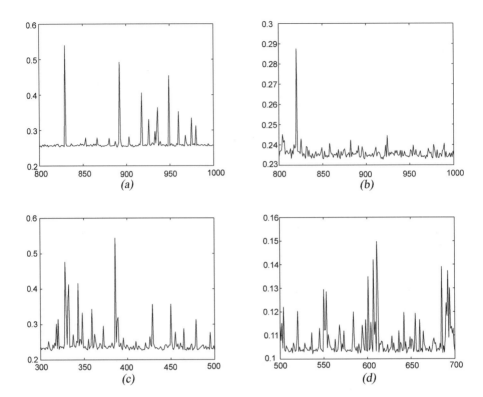

Figure 1.3 RTT time series measured in seconds between a host at the University of Delaware and hosts in (*a*) Australia (12:18 AM - 3:53 AM); (*b*) Sydney, Australia (12:30 AM - 4:03 AM); (*c*) Japan (2:52 PM - 6:33 PM); (*d*) London, UK (10:00 AM - 1:35 PM). All plots shown in 1 minute interval samples.

generate this trace, all packets coming to the server were captured and time-stamped during several hours. The figure considers byte counts (size of the transferred data) measured on 10ms intervals, which is shown in the top plot of Figure 1.4. The overall length of the recordings is approximately four hours (precisely 14,000s). The other plots in Figure 1.4 represent the "aggregated" data obtained by averaging the data counts on increasing time intervals. The notable fact in Figure 1.4 is that the aggregation does not smooth out the data. The aggregated traffic still appears bursty even in the bottom plot despite the fact that each point in it is the average of one thousand successive values of the series shown in the top plot of Figure 1.4. Similar behavior in data traffic has been observed in numerous experimental setups, including Cappé et al. (2002) [42], Beran et al. (1995) [31], Leland et al. (1994) [127], and Paxson and Floyd (1995) [159].

Another example is found in high-speed data links over telephone wires, such as Asymmetric Digital Subscriber Lines (ADSL), where noise in the communications channel exhibits impulsive characteristics. In these systems, telephone twisted pairs

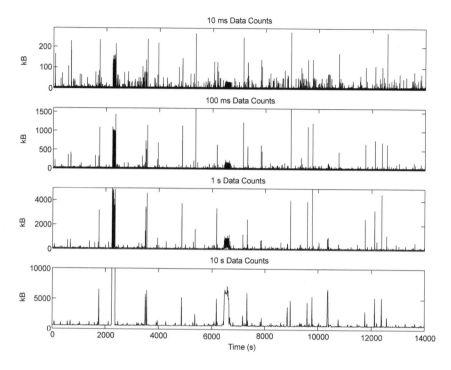

Figure 1.4 Byte counts measured over 14,000 seconds in a web server of the ECE Department at the University of Delaware viewed through different aggregation intervals: from top to bottom, 10ms, 100ms 1s, 10s.

are unshielded, and are thus susceptible to large electromagnetic interference. Potential sources of impulsive interference include light switching and home appliances, as well as natural weather phenomena. Severe interference is also generated by cross talk among multiple twisted pairs making up a telephone cable. The interference is inherently impulsive and nonstationary leading to poor service reliability. The impact of impulsive noise on ADSL systems depends on the impulse energy, duration, interarrival time, and spectral characteristics. Isolated impulses can reach magnitudes significantly larger than either additive white noise or crosstalk interference. A number of models to characterize ADSL interference have been proposed [139]. Current ADSL systems are designed conservatively under the assumption of a worst-case scenario due to severe nonstationary and nonGaussian channel interference [204]. Figure 1.5 shows three ADSL noise signals measured at a customer's premise. These signals exhibit a wide range of spectral characteristics, burstiness, and levels of impulsiveness. In addition to channel coding, linear filtering is used to combat ADSL channel interference [204]. Figure 1.5a–c depicts the use of linear and nonlinear filtering. These figures depict the improvement attained by nonlinear filtering in removing the noise and interference.

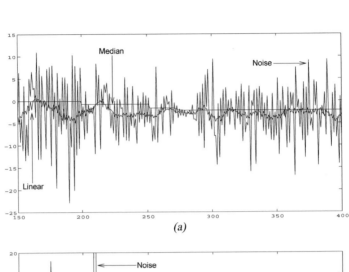

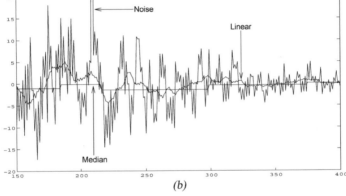

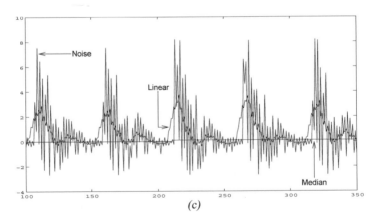

Figure 1.5 (*a–c*) Different noise and interference characteristics in ADSL lines. A linear and a nonlinear filter (recursive median filter) are used to overcome the channel limitations, both with the same window size (adapted from [204]).

The last example (Fig. 1.6), visually illustrates the advantages of nonlinear signal processing. This figure depicts an enlarged section of an image which has been JPEG compressed for storage in a Web site. Since compression reduces and often eliminates the high frequency components, compressed images contain edge artifacts and tend to look blurred. As a result, images found on the Internet are often sharpened. Figure 1.6b shows the output of a traditional sharpening algorithm equipped with linear FIR filters. The amplification of the compression artifacts are clearly seen. Figure 1.6c depicts the sharpening output when nonlinear filters are used. Nonlinear sharpeners avoid noise and artifact amplification and are as effective as linear sharpeners in highlighting the signal edges.

The examples above suggest that significant improvements in performance can be achieved by nonlinear methods of signal processing. Unlike linear signal processing, however, nonlinear signal processing lacks a unified and universal set of tools for analysis and design. Hundreds of nonlinear signal processing algorithms have been proposed [21, 160]. While nonlinear signal processing is a dynamic, rapidly growing field, a large class of nonlinear signal algorithms can be studied in a unified framework. Since signal processing focuses on the analysis and transformation of signals, nonlinear filtering emerges as the fundamental building block of nonlinear signal processing. This book develops the fundamental signal-processing tools that arise when considering the filtering of nonGaussian, rather than Gaussian, random processes. By concentrating on just two nonGaussian models, a large set of tools is developed that notably encompass a significant portion of the nonlinear signal-processing tools proposed in the literature over the past several decades.

1.1   NONGAUSSIAN RANDOM PROCESSES

In statistical signal processing, signals are modeled as random processes and many signal-processing tasks reduce to the proper statistical analysis of the observed signals. Selecting the appropriate model for the application at hand is of fundamental importance. The model, in turn, determines the signal processing approach. Classical linear signal-processing methods rely on the popular Gaussian assumption. The Gaussian model appears naturally in many applications as a result of the Central Limit Theorem first proved by De Moivre (1733) [69].

THEOREM 1.1 (CENTRAL LIMIT THEOREM) *Let $X_1, X_2, \ldots$, be a sequence of i.i.d. random variables with zero mean and variance σ^2. Then as $N \to \infty$, the normalized sum*

$$S_N = \frac{1}{\sqrt{N}} \sum_{i=1}^{N} X_i \tag{1.1}$$

converges almost surely to a zero-mean Gaussian variable with the same variance as X_i.

Conceptually, the central limit theorem explains the Gaussian nature of processes generated from the superposition of many small and independent effects. For ex-

Figure 1.6 (*a*) Enlarged section of a JPEG compressed image, (*b*) output of unsharp masking using FIR filters, (*c*) and (*d*) outputs of median sharpeners.

ample, thermal noise generated as the superposition of a large number of random independent interactions at the molecular level. The Central Limit Theorem theoretically justifies the appearance of Gaussian statistics in real life.

However, in a wide range of applications, the Gaussian model does not produce a good fit which, at first, may seem to contradict the principles behind the Central Limit Theorem. A careful revision of the conditions of the Central Limit Theorem indicates that, in order for this theorem to be valid, the variance of the superimposed random variables must be finite. If the random variables possess infinite variance, it can be shown that the series in the Central Limit Theorem converges to a non-Gaussian impulsive variable [65, 207]. This important generalization of the Central Limit Theorem explains the apparent contradictions of its "traditional" version, as well as the presence of non-Gaussian, infinite variance processes, in practical problems. In the same way as the Gaussian model owes most of its strength to the Central Limit Theorem, the Generalized Central Limit Theorem constitutes a strong theoretical argument to the development of models that capture the impulsive nature of these signals, and of signal processing tools that are adequate in these nonGaussian environments.

Perhaps the simplest approach to address the effects of nonGaussian signals is to detect outliers that may be present in the data, reject these heuristically, and subsequently use classical signal-processing algorithms. This approach, however, has many disadvantages. First, the detection of outliers is not simple, particularly when these are bundled together. Second, the efficiency of these methods is not optimal and is generally difficult to measure since the methods are based on heuristics.

The approach followed in this book is that of exploiting the rich theories of robust statistics and non-Gaussian stochastic processes, such that a link is established between them leading to signal processing with solid theoretical foundations. This book considers two model families that encompass a large class of random processes. These models described by their distributions allow the rate of tail decay to be varied: the *generalized Gaussian* distribution and the class of *stable* distributions. The tail of a distribution can be measured by the mass of the tail, or *order*, defined as $P_r(X > x)$ as $x \to \infty$. Both distribution families are general in that they encompass a wide array of distributions with different tail characteristics. Additionally, both the generalized Gaussian and stable distributions contain important special cases that lead directly to classes of nonlinear filters that are tractable and optimal for signals with heavy tail distributions.

1.1.1 Generalized Gaussian Distributions and Weighted Medians

One approach to modeling the presence of outliers is to start with the Gaussian distribution and allow the exponential rate of tail decay to be a free parameter. This results directly in the generalized Gaussian density function. Of special interest is the case of first order exponential decay, which yields the double exponential, or Laplacian, distribution. Optimal estimators for the generalized Gaussian distribution take on a particularly simple realization in the Laplacian case. It turns out that weighted median filters are optimal for samples obeying Laplacian statistics, much like linear filters are optimal for Gaussian processes. In general, weighted median filters are more efficient than linear filters in impulsive environments, which can be directly attributed to the heavy tailed characteristic of the Laplacian distribution. Part II of the book uncovers signal processing methods using median-like operations, or order statistics.

1.1.2 Stable Distributions and Weighted Myriads

Although the class of generalized Gaussian distributions includes a spectrum of impulsive processes, these are all of exponential tails. It turns out that a wide variety of processes exhibit more impulsive statistics that are characterized with algebraic tailed distributions. These impulsive processes found in signal processing applications arise as the superposition of many small independent effects. For example, radar clutter is the sum of many signal reflections from an irregular surface; the transmitters in a multiuser communication system generate relatively small independent signals, the sum of which represents the ensemble at a user's receiver; rotating electric machinery generates many impulses caused by contact between distinct parts of the machine; and standard atmospheric noise is known to be the superposition of many electrical discharges caused by lightning activity around the Earth. The theoretical justification for using stable distribution models lies in the Generalized Central Limit Theorem which includes the well known "traditional" Central Limit Theorem as a special case. Informally:

A random variable X is stable if it can be the limit of a normalized sum of i.i.d. random variables.

The generalized theorem states that if the sum of i.i.d. random variables with or without finite variance converges to a distribution, the limit distribution must belong to the family of stable laws [149, 207]. Thus, nonGaussian processes can emerge in practical applications as sums of random variables in the same way as Gaussian processes.

Stable distributions include two special cases of note: the standard Gaussian distribution and the Cauchy distribution. The Cauchy distribution is particularly important as its tails decay algebraically. Thus, the Cauchy distribution can be used to model very impulsive processes. It turns out that for a wide range of stable-distributed signals, the so-called weighted myriad filters are optimal. Thus, weighted myriad filters emerging from the stable model are the counterparts to linear and median filters related to the Gaussian and Laplacian environments, respectively. Part III of the book develops signal-processing methods derived from stable models.

1.2 STATISTICAL FOUNDATIONS

Estimation theory is a branch of statistics concerned with the problem of deriving information about the properties of random processes from a set of observed samples. As such, estimation theory lies at the heart of statistical signal processing. Given an observation waveform $\{X(n)\}$, one goal is to extract information that is embedded within the observed signal. It turns out that the embedded information can often be modeled parametrically. That is, some parameter β of the signal represents the information of interest. This parameter may be the local mean, the variance, the local range, or some other parameter associated with the received waveform. Of course, finding a good parametric model is critical.

Location Estimation   Because observed signals are inherently random, these are described by a probability density function (pdf), $f(x_1, x_2, \ldots, x_N)$. The pdf may be parameterized by an unknown parameter β. The parameter β thus defines a class of pdfs where each member is defined by a particular value of β. As an example, if our signal consists of a single point ($N = 1$) and β is the mean, the pdf of the data under the Gaussian model is

$$f(x_1; \beta) = \frac{1}{\sqrt{2\pi\sigma^2}} exp\left[-\frac{1}{2\sigma^2}(x_1 - \beta)^2\right] \qquad (1.2)$$

which is shown in Figure 1.7 for various values of β. Since the value of β affects the probability of X_1, intuitively we should be able to infer the value of β from the observed value of X_1. For example, if the observed value of X_1 is a large positive number, the parameter β is more likely to be equal to β_1 than to β_2 in Figure 1.7. Notice that β determines the location of the pdf. As such, β is referred to as the *location parameter*. Rules that infer the value of β from sample realizations of the data are known as *location estimators*. Although a number of parameters can be associated with a set of data, *location* is a parameter that plays a key role in the design of filtering algorithms. The filtering structures to be defined in later chapters have their roots in location estimation.

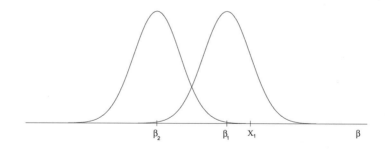

Figure 1.7   Estimation of parameter β based on the observation X_1.

Running Smoothers   Location estimation and filtering are intimately related. The *running mean* is the simplest form of filtering and is most useful in illustrating this relationship. Given the data sequence $\{\ldots, X(n-1), X(n), X(n+1), \ldots\}$, the running mean is defined as

$$Y(n) = \text{MEAN}(X(n - N), X(n - N + 1), \ldots, X(n + N)). \quad (1.3)$$

At a given point n, the output is the average of the samples within a window centered at n. The output at $n + 1$ is the average of the samples within the window centered at $n + 1$, and so on. Thus, at each point n, the running mean computes a location estimate, namely the sample mean. If the underlying signals are not Gaussian, it would be reasonable to replace the mean by a more appropriate location estimator. Tukey (1974) [189], for instance, introduced the running median as a robust alternative to the running mean.

Although running smoothers are effective in removing noise, more powerful signal processing is needed in general to adequately address the tasks at hand. To this end, the statistical foundation provided by running smoothers can be extended to define optimal filtering structures.

1.3   THE FILTERING PROBLEM

Filtering constitutes a system with arbitrary input and output signals, and consequently the filtering problem is found in a wide range of disciplines. Although filtering theory encompasses continuous-time as well as discrete-time signals, the availability of digital computer processors is causing discrete-time signal representation to become the preferred method of analysis and implementation. In this book, we thus consider signals as being defined at discrete moments in time where we assume that the sampling interval is fixed and small enough to satisfy the Nyquist sampling criterion.

Denote a random sequence as $\{X\}$ and let $\mathbf{X}(n)$ be a N-long element, real valued observation vector

$$\begin{aligned} \mathbf{X}(n) &= [X(n), \ X(n - 1), \ldots, X(n - N + 1)]^T \\ &= [X_1(n), \ X_2(n), \ldots, \ X_N(n)]^T \end{aligned} \quad (1.4)$$

where $X_i(n) = X(n - i + 1)$ and where T denotes the transposition operator. R denotes the real line. Further, assume that the observation vector $\mathbf{X}(n)$ is statistically related to some desired signal denoted as $D(n)$. The filtering problem is then formulated in terms of joint process estimation as shown in Figure 1.8. The observed vector, $\mathbf{X}(n)$, is formed by the elements of a shifting window, the output of the filter is the estimate $\hat{D}(n)$ of a desired signal $D(n)$. The optimal filtering problem thus reduces to minimizing the cost function associated with the error $e(n)$ under a given criterion, such as the mean square error (MSE).

Under Gaussian statistics, the estimation framework becomes linear and the filter structure reduces to that of FIR linear filters. The linear filter output is defined as

$$Y(n) = \text{MEAN}(W_1 \cdot X_1(n), W_2 \cdot X_2(n), \ldots, W_N \cdot X_N(n)), \quad (1.5)$$

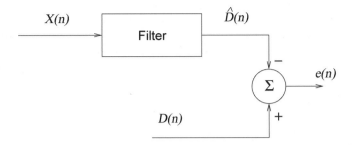

Figure 1.8 Filtering as a joint process estimation

where the W_i are real-valued weights assigned to each input sample.

Under the Laplacian model, it will be shown that the median becomes the estimate of choice and weighted medians become the filtering structure. The output of a weighted median is defined as

$$Y(n) = \text{MEDIAN}(W_1 \diamond X_1(n), W_2 \diamond X_2(n), \dots, W_N \diamond X_N(n)), \qquad (1.6)$$

where the operation $W_i \diamond X_i(n)$ replicates the sample $X_i(n)$, W_i times. Weighting in median filters thus takes on a very different meaning than traditional weighting in linear filters.

For stable processes, it will be derived shortly that the weighted myriad filter emerges as the ideal structure. In this case the filter output is defined as

$$Y(n) = \text{MYRIAD}\left(K; \ W_1 \circ X_1, W_2 \circ X_2, \dots, W_N \circ X_N\right), \qquad (1.7)$$

where $W_i \circ X_i(n)$ represents a nonlinear weighting operation to be described later, and K in (1.7) is a free tunable parameter that will play an important role in weighted myriad filtering. It is the flexibility provided by K that makes the myriad filter a more powerful filtering framework than either the linear FIR or the weighted median filter frameworks.

1.3.1 Moment Theory

Historically, signal processing has relied on second-order moments, as these are intimately related to Gaussian models. The first-order moment

$$\mu_X = E\{X(n)\} \qquad (1.8)$$

and the second-order moment characterization provided by the *autocorrelation* of stationary processes

$$R_X(k) = E\left\{X(n)X(n+k)\right\} \qquad (1.9)$$

are deeply etched into traditional signal processing practice. As it will be shown later, second-order descriptions do not provide adequate information to process non-Gaussian signals. One popular approach is to rely on *higher-order* statistics that exploit moments of order greater than two. If they exist, higher-order statistics provide information that is unaccessible to second-order moments [148]. Unfortunately, higher-order statistics become less reliable in impulsive environments to the extent that often they cease to exist.

The inadequacy of second- or higher-order moments leads to the introduction of alternate moment characterizations of impulsive processes. One approach is to use *fractional lower-order statistics* (FLOS) consisting of moments for orders less than two [136, 149]. Fractional lower-order statistics are not the only choice. Much like the Gaussian model naturally leads to second-order based methods, selecting a Laplacian model will lead to a different natural moment characterization. Likewise, adopting the stable laws will lead to a different, yet natural, moment characterization.

Part I

Statistical Foundations

2

NonGaussian Models

The Gaussian distribution model is widely accepted in signal processing practice. Theoretically justified by the Central Limit Theorem, the Gaussian model has attained a privileged place in statistics and engineering. There are, however, applications where the underlying random processes do not follow Gaussian statistics. Often, the processes encountered in practice are impulsive in nature and are not well described with conventional Gaussian distributions. Traditionally, the design emphasis has often relied on a continuity principle: optimal processing at the ideal Gaussian model should be almost optimal nearby. Unfortunately, this reliance on continuity is unfounded and in many cases one finds that optimum signal-processing methods can suffer drastic performance degradations, even for small deviations from the nominal assumptions. As an example, synchronization, detection, and equalization, basic in all communication systems, fail in impulsive noise environments whenever linear processing is used.

In order to model nonGaussian processes, a wide variety of distributions with heavier-than-Gaussian tails have been proposed as viable alternatives. This chapter reviews several of these approaches and focuses on two distribution families, namely the class of generalized Gaussian distributions and the class of stable distributions. These two distribution families are parsimonious in their characterization leading to a balanced trade-off between fidelity and complexity. On the one hand, fidelity leads to more efficient signal-processing algorithms, while the complexity issue stands for simpler models from which more tractable algorithms can be derived. The Laplacian distribution, a special case of the generalized Gaussian distribution, lays the statistical foundation for a large class of signal-processing algorithms based on the sample median. Likewise, signal processing based on the so-called sample myriad emerges from the statistical foundation laid by stable distributions.

2.1 GENERALIZED GAUSSIAN DISTRIBUTIONS

The Central Limit Theorem provides a theoretical justification for the appearance of Gaussian processes in nature. Intimately related to the Gaussian model are linear estimation methods and, to a large extent, a large section of signal-processing algorithms based on operations satisfying the linearity property. While the Central Limit Theorem has provided the key to understanding the interaction of a large number of random independent events, it has also provided the theoretical burden favoring the use of linear methods, even in circumstances where the nature of the underlying signals are decidedly non-Gaussian.

One approach used in the modeling of non-Gaussian processes is to start from the Gaussian model and slightly modify it to account for the appearance of clearly inappropriate samples or outliers. The *Gaussian mixture* or *contaminated Gaussian* model follows this approach, where the ϵ-contaminated density function takes on the form

$$f(x) = (1 - \epsilon)f_n(x) + \epsilon f_c(x) \tag{2.1}$$

where $f_n(x)$ is the *nominal* Gaussian density with variance σ_n^2, ϵ is a small positive constant determining the percentage of contamination, and $f_c(x)$ is the contaminating Gaussian density with a large relative variance, such that $\sigma_c^2 \gg \sigma_n^2$. Intuitively, one out of $1/\epsilon$ samples is allowed to be contaminated by the higher variance source. The advantage of the contaminated Gaussian distribution lies in its mathematical simplicity and ease of computer simulation. Gaussian mixtures, however, present drawbacks. First, dispersion and impulsiveness are characterized by three parameters, $\epsilon, \sigma_n, \sigma_c$, which may be considered overparameterized. The second drawback, and perhaps the most serious, is that its sum density function formulation makes it difficult to manipulate in general estimation problems.

A more accurate model for impulsive phenomena was proposed by Middleton (1977) [143]. His class A, B, and C models are perhaps the most credited statistical-physical characterization of radio noise. These models have a direct physical interpretation and have been found to provide good fits to a variety of noise and interference measurements. Contaminated Gaussian mixtures can in fact be derived as approximations to Middleton's Class A model. Much like Gaussian mixtures, however, Middleton's models are complicated and somewhat difficult to use in laying the foundation of estimation algorithms.

Among the various extensions of the Gaussian distributions, the most popular models are those characterized by the *generalized Gaussian distribution*. These have been long known, with references dating back to 1923 by Subbotin [183] and 1924 by Frèchet [74]. A special case of the generalized Gaussian distribution class is the well known Laplacian distribution, which has even older roots; Laplace introduced it more than two hundred years ago [122]. In the generalized Gaussian distribution, the presence of outlier samples can be modeled by modifying the Gaussian distribution, allowing the exponential rate of tail decay to be a free parameter. In this manner, the tail of the generalized Gaussian density function is governed by the parameter k.

DEFINITION 2.1 (GENERALIZED GAUSSIAN DISTRIBUTION) *The probability density function for the generalized Gaussian distribution is given by*

$$f(x) = \frac{k\alpha}{2\Gamma(1/k)} e^{-(\alpha|x-\beta|)^k},$$ (2.2)

where $\Gamma(\cdot)$ is the Gamma function $\Gamma(x) = \int_0^\infty t^{x-1}e^{-t}dt$, α is a constant defined as $\alpha = \sigma^{-1}\sqrt{\Gamma(3/k)\left(\Gamma(1/k)\right)^{-1}}$ and σ is the standard deviation[1].

In this representation, the scale of the distribution is determined by the parameter $\sigma > 0$ whereas the impulsiveness is related to the parameter $k > 0$. As expected, the representation in (2.2) includes the standard Gaussian distribution as a special case for $k = 2$. Conceptually, the lower the value of k, the more impulsive the distribution is. For $k < 2$, the tails decay slower than in the Gaussian case, resulting in a heavier tailed distribution. A second special case of the generalized Gaussian distribution that is of particular interest is the case $k = 1$, which yields the double exponential, or Laplacian distribution ,

$$f(x) = \frac{1}{\sqrt{2}\sigma} e^{-\frac{\sqrt{2}}{\sigma}|x-\beta|} = \frac{\lambda}{2}e^{-\lambda|x-\beta|}.$$ (2.3)

where the second representation is the most commonly used and is obtained making $\sigma = \sqrt{2}/\lambda$.

The effect of decreasing k on the tails of the distribution can be seen in Figures 2.1 and 2.2. As these figures show, the Laplacian distribution has heavier tails than the Gaussian distribution. One of the weaknesses of the generalized Gaussian distribution is the shape of these distributions around the origin for $k < 2$. The "peaky" shape of these distributions contradicts the widely accepted *Winsor's principle*, according to which, all density functions of practical appeal are bell-shaped [87, 188].

2.2 STABLE DISTRIBUTIONS

Stable distributions describe a rich class of processes that allow heavy tails and skewness. The class was characterized by Lévy in 1925 [128]. Stable distributions are described by four parameters: an *index of stability* $\alpha \in (0, 2]$, a scale parameter $\gamma > 0$, a skewness parameter $\delta \in [-1, 1]$, and a location parameter $\beta \in \mathcal{R}$. The stability

---

[1]The gamma function satisfies: $\Gamma(x) = (x-1)\Gamma(x-1)$ for $x > 1$. For positive integers it follows that $\Gamma(x) = (x-1)!$ and for a non integer $x > 0$ such that $x = i + u$ where $0 \leq u < 1$, $\Gamma(x) = (x-1)(x-2)\cdots u\Gamma(u)$. For $x = \frac{1}{2}, \Gamma(\frac{1}{2}) = \sqrt{\pi}$.

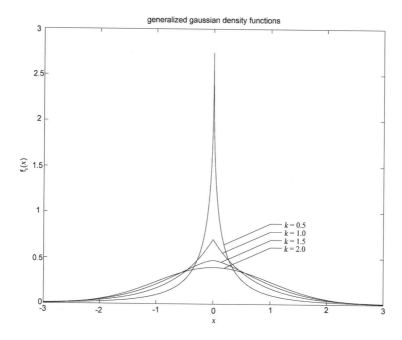

Figure 2.1   Generalized Gaussian density functions for different values of the tail constant k.

parameter α measures the thickness of the tails of the distribution and provides this model with the flexibility needed to characterize a wide range of impulsive processes. The scale parameter γ, also called the dispersion, is similar to the variance of the Gaussian distribution. The variance equals twice the square of gamma in the Gaussian case when $\alpha = 2$. When the skewness parameter is set to $\delta = 0$, the stable distribution is symmetric about the location parameter β. Symmetric stable processes are also referred to as symmetric α-stable or simply as SαS. A stable distribution with parameter α is said to be *standard* if $\beta = 0$ and $\gamma = 1$. For any stable variable X with parameters $\alpha, \beta, \gamma, \delta$, the corresponding standardized stable variable is found as $(X - \beta)/\gamma$, for $\alpha \neq 1$.

Stable distributions are rapidly becoming popular for the characterization of impulsive processes for the following reasons. Firstly, good empirical fits are often found using stable distributions on data exhibiting skewness and heavy tails. Secondly, there is solid theoretical justification that nonGaussian stable processes emerge in practice, such as multiple access interference in a Poisson-distributed communication network [179], reflection off a rotating mirror [69], and Internet traffic [127]; see Uchaikin and Zolotarev (1999) [191] and Feller (1971) [69] for additional examples. The third argument for modeling with stable distributions is perhaps the most significant and compelling. Stable distributions satisfy an important generalization

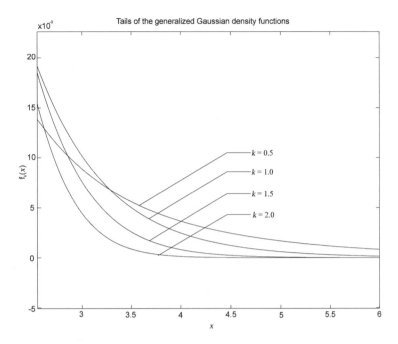

Figure 2.2 Tails of the Generalized Gaussian density functions for different values of the tail constant k.

of the Central Limit Theorem which states that the only possible limit of normalized sums of independent and identically distributed terms is stable.

A wide variety of impulsive processes found in signal processing applications arise as the superposition of many small independent effects. While Gaussian models are clearly inappropriate, stable distributions have the theoretical underpinnings to accurately model these type of impulsive processes [149, 207]. Stable models are thus appealing, since the generalization of the Central Limit Theorem explains the apparent contradictions of its "ordinary" version, which could not naturally explain the presence of heavy tailed signals.

The Generalized Central Limit Theorem and the strong empirical evidence is used by many to justify the use of stable models. Examples in finance and economics are given in Mandelbrot (1963) [138] and McCulloch (1966) [142]; in communication systems by Stuck and Kleiner (1974)[182], Nikias and Shao (1995) [149], and Ilow and Hatzinakos (1997) [106]. A number of monographs providing in-depth discussion of stable processes have recently appeared: Zolotarev (1986) [207], Samorodnitsky and Taqqu (1994) [75], Nikias and Shao (1995) [149], Uchaikin and Zolotarev (1999) [191], Adler et al. (2002) [67], and Nolan (2002) [151].

2.2.1   Definitions

Gaussian random variables obey the important property that the sum of any two Gaussian variables is itself a Gaussian random variable. Formally, for any two independent Gaussian random variables X_1 and X_2 and any positive constants a, b, c,

$$aX_1 + bX_2 \overset{d}{=} cX + d,$$

where d is a real-valued constant[2]. As their name implies, stable random variables obey this property as well.

DEFINITION 2.2 (STABLE RANDOM VARIABLES) *A random variable X is stable if for X_1 and X_2 independent copies of X and for arbitrary positive constants a and b, there are constants c and d such that*

$$aX_1 + bX_2 \overset{d}{=} cX + d. \tag{2.4}$$

A symmetric *stable random variable distributed around* 0 *satisfies* $X \overset{d}{=} -X$.

Informally, the stability property states that the shape of X is preserved under addition up to scale and shift. The stability property (2.4) for Gaussian random variables can be readily verified yielding $c^2 = a^2 + b^2$ and $d = (a + b - c)\mu$, where μ is the mean of the parent Gaussian distribution. Other well known distributions that satisfy the stable property are the Cauchy and Lévy distributions, and as such, both distributions are members of the stable class. The density function, for $X \sim$ Cauchy(γ, β) has the form

$$f(x) = \frac{1}{\pi} \frac{\gamma}{\gamma^2 + (x - \beta)^2}, \quad -\infty < x < \infty. \tag{2.5}$$

The Lévy density function, sometimes referred to as the Pearson distribution, is totally skewed concentrating on $(0, \infty)$. The density function for $X \sim$ Lévy(γ, β) has the form

$$f(x) = \sqrt{\frac{\gamma}{2\pi}} \frac{1}{(x - \beta)^{3/2}} \exp\left(-\frac{\gamma}{2(x - \beta)}\right), \quad \beta < x < \infty. \tag{2.6}$$

Figure 2.3 shows the plots of the standardized Gaussian, Cauchy, and Lévy distributions. Both Gaussian and Cauchy distributions are symmetric and bell-shaped. The main difference between these two densities is the area under their tails — the Cauchy having much larger area or heavier tails. In contrast to the Gaussian and Cauchy, the Lévy distribution is highly skewed, with even heavier tails than the Cauchy.

General stable distributions allow for varying degrees of skewness, the influence of the parameter δ in the distribution of an α-stable random variable is shown in Figure 2.4.

---

[2]The symbol $\overset{d}{=}$ defines equality in distribution

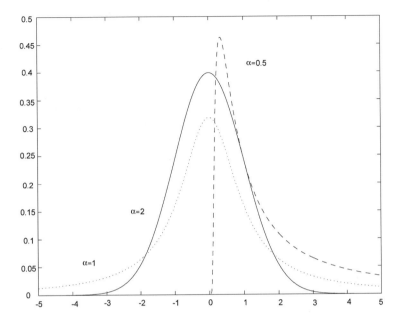

Figure 2.3 Density functions of standardized Gaussian ($\alpha = 2$), Cauchy ($\alpha = 1$), and Lévy ($\alpha = 0.5$, $\delta = 1$).

Although some practical processes might be better modeled by skewed distributions, we will focus on *symmetric* stable processes for several reasons. First, the processes found in a number of signal-processing applications are symmetric; second, asymmetric models can lead to a significant increase in the computational complexity of signal-processing algorithms; and, more important, estimating the location of an asymmetric distribution is not a well-defined problem. All of the above constitute impediments to the derivation of a general theory of nonlinear filtering.

2.2.2 Symmetric Stable Distributions

Symmetric α-stable or $S\alpha S$ distributions are defined when the skewness parameter δ is set to zero. In this case, a random variable obeying the symmetric stable distribution with scale γ is denoted as $X \sim S\alpha S(\gamma)$. Although the stability condition in Definition 2.2 is sufficient to characterize all stable distributions, a second and more practical characterization of stable random variables is through their characteristic function.

$$\phi(\omega) = E \exp(j\omega X) = \int_{-\infty}^{\infty} \exp(j\omega x) f(x) dx \qquad (2.7)$$

where $f(x)$ is the density function of the underlying random variable.

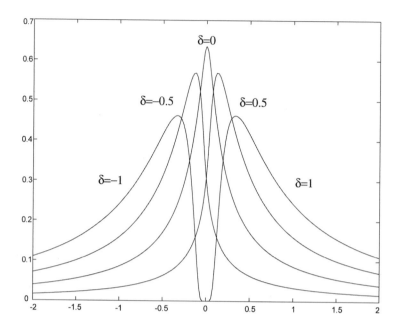

Figure 2.4 Density functions of skewed stable variables ($\alpha = 0.5$, $\gamma = 1$, $\beta = 0$).

DEFINITION 2.3 (CHARACTERISTIC FUNCTION OF SαS DISTRIBUTIONS) *A random variable X is symmetrically stable if and only if $X \stackrel{d}{=} AZ + B$ where $0 < \alpha \leq 2, A \geq 0, B \in \mathcal{R}$ and $Z = Z(\alpha)$ is a random variable with characteristic function*

$$\phi(\omega) = e^{-\gamma^{\alpha}|\omega|^{\alpha}}. \tag{2.8}$$

The dispersion parameter γ is a positive constant related to the scale of the distribution. Again, the parameter α is referred to as the index of stability. In order for (2.8) to define a characteristic function, the values of α must be restricted to the interval $(0; 2]$. Conceptually speaking, α determines the impulsiveness or tail heaviness of the distribution (smaller values of α indicate increased levels of impulsiveness). The limit case, $\alpha = 2$, corresponds to the zero-mean Gaussian distribution with variance $2\gamma^2$.[3] All other values of α correspond to heavy-tailed distributions.

Figure 2.5 shows plots of normalized unitary-dispersion stable densities. Note that lower values of α correspond to densities with heavier tails, as shown in Figure 2.6.

---

[3]The characteristic function of a Gaussian random variable with zero mean and variance σ^2 is given by: $\phi(\omega) = \exp\left(-\frac{\omega^2\sigma^2}{2}\right)$, from this equation and (2.8) with $\alpha = 2$, the relationship shown between γ and σ^2 can be obtained.

Symmetric stable densities maintain many of the features of the Gaussian density. They are smooth, unimodal, symmetric with respect to the mode, and bell-shaped.

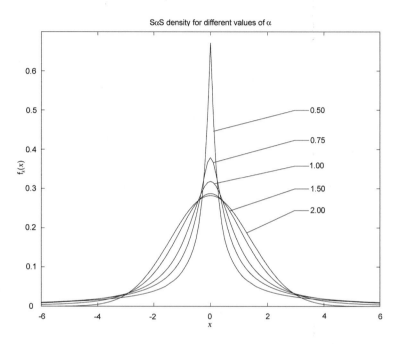

Figure 2.5 Density functions of Symmetric stable distributions for different values of the tail constant α.

A major drawback to stable distribution modeling is that with a few exceptions stable density or their corresponding cumulative distribution functions lack closed form expressions. There are three cases for which closed form expressions of stable density functions exist: the Gaussian distribution ($\alpha = 2$), the Cauchy distribution ($\alpha = 1$), and the Lévy ($\alpha = \frac{1}{2}$) distribution. For other values of α, no closed form expressions are known for the density functions, making it necessary to resort to series expansions or integral transforms to describe them.

DEFINITION 2.4 (SYMMETRIC STABLE DENSITY FUNCTIONS) *A general, "zero-centered," symmetric stable random variable with unitary dispersion can be characterized by the power series density function representation [207]:*

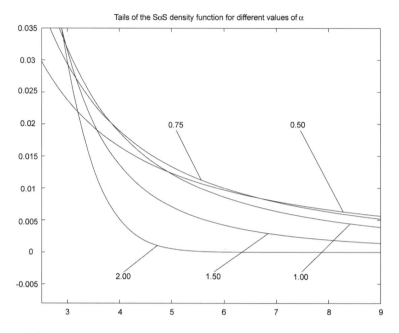

Figure 2.6   Tails of symmetric stable distributions for different values of the tail constant α.

$$f_\alpha(x) = \begin{cases} \frac{1}{\pi}\sum_{k=1}^{\infty} \frac{(-1)^{k-1}}{k!}\Gamma(k\alpha+1)\sin(\frac{\pi k \alpha}{2})|x|^{-k\alpha-1} & for\ 0 < \alpha < 1,\ x \neq 0 \\[2mm] \frac{1}{\pi(x^2+1)} & for\ \alpha = 1 \\[2mm] \frac{1}{\pi\alpha}\sum_{k=0}^{\infty} \frac{(-1)^k}{(2k)!}\Gamma(\frac{2k+1}{\alpha})x^{2k} & for\ 1 < \alpha < 2 \\[2mm] \frac{1}{2\sqrt{\pi}}\exp[-\frac{x^2}{4}] & for\ \alpha = 2. \end{cases}$$

$$(2.9)$$

DEFINITION 2.5 (CHARACTERISTIC FUNCTION OF A STABLE RV.) *[151] A random variable X is stable with characteristic exponent α, dispersion γ, location β and skewness δ if X has a characteristic function:*

$$\phi(\omega) = \begin{cases} \exp\left(-\gamma^\alpha|\omega|^\alpha\left[1 - j\delta\left(\tan\frac{\pi\alpha}{2}\right)(sgn\ \omega)\right] + j\beta\omega\right) & for\ \alpha \neq 1 \\ \exp\left(-\gamma|\omega|\left[1 + j\delta\frac{2}{\pi}(sgn\ \omega)\ln|\omega|\right] + j\beta\omega\right) & for\ \alpha = 1. \end{cases}$$

$$(2.10)$$

EXAMPLE 2.1 (STANDARD STABLE RANDOM VARIABLES)

As stated previously, if X is a stable random variable with location β and dispersion γ, the variable $X' = \frac{X-\beta}{\gamma}$ ($\alpha \neq 1$) is standard stable. This can be demonstrated with the help of the general characteristic function. Define

$$
\begin{aligned}
\phi(\omega') &= E\left[\exp\left(j\omega'X'\right)\right] = E\left[\exp\left(j\omega'\frac{X-\beta}{\gamma}\right)\right] \\
&= \exp\left(-j\frac{\omega'}{\gamma}\beta\right) E\left[\exp\left(j\frac{\omega'}{\gamma}X\right)\right] \quad \text{using (2.10)} \\
&= \exp\left(-j\frac{\omega'}{\gamma}\beta\right) \exp\left(-\gamma^{\alpha}\left|\frac{\omega'}{\gamma}\right|^{\alpha}\left[1 - j\delta\left(\tan\frac{\pi\alpha}{2}\right)\left(\operatorname{sgn}\frac{\omega'}{\gamma}\right)\right] + j\beta\frac{\omega'}{\gamma}\right)
\end{aligned}
$$

but $\gamma \geq 0$, then $|\gamma| = \gamma$ and $\operatorname{sgn}\left(\frac{\omega'}{\gamma}\right) = \operatorname{sgn}(\omega')$, then

$$
\phi(\omega') = \exp\left(-|\omega|^{\alpha}\left[1 - j\delta\left(\tan\frac{\pi\alpha}{2}\right)(\operatorname{sgn}\omega')\right]\right) \tag{2.11}
$$

is the characteristic function of a stable random variable with $\gamma = 1$ and $\beta = 0$. ∎

EXAMPLE 2.2

Let $X \sim S(\alpha, \gamma, \beta)$, a symmetric stable random variable, then for $a \neq 0$ it is shown that

$$
aX + b \sim S(\alpha, |a|\gamma, a\beta + b).
$$

Following the procedure used in the previous example, define $X' = aX + b$:

$$
\begin{aligned}
\phi(\omega') &= E\left[\exp\left(j\omega'X'\right)\right] = E\left[\exp\left(j\omega'\left(aX + b\right)\right)\right] \\
&= \exp\left(j\omega'b\right) E\left[\exp\left(j\left(\omega'a\right)X\right)\right] \quad \text{using (2.10) with } \delta = 0 \\
&= \exp\left(j\omega'b\right) \exp\left(-\gamma^{\alpha}\left|\omega'a\right|^{\alpha} + j\beta\left(\omega'a\right)\right) \\
&= \exp\left(-\left(|a|\gamma\right)^{\alpha}\left|\omega'\right|^{\alpha} + j\omega'\left(a\beta + b\right)\right), \tag{2.12}
\end{aligned}
$$

which is the characteristic function of a symmetric stable random variable with dispersion $|a|\gamma$ and location $a\beta + b$. ∎

EXAMPLE 2.3

Let $X_1 \sim S(\alpha, \gamma_1, \beta_1)$ and $X_2 \sim S(\alpha, \gamma_2, \beta_2)$ be independent symmetric stable random variables, it is shown here that $X_1 + X_2 \sim S(\alpha, \gamma, \beta)$, where $\gamma^\alpha = \gamma_1^\alpha + \gamma_2^\alpha$ and $\beta = \beta_1 + \beta_2$.

Define $X' = X_1 + X_2$ and find the characteristic function of X' as:

$$
\begin{aligned}
\phi(\omega') &= E\left[\exp\left(j\omega'X'\right)\right] = E\left[\exp\left(j\omega'\left(X_1 + X_2\right)\right)\right] \\
&= E\left[\exp\left(j\omega'X_1\right)\right] E\left[\exp\left(j\omega'X_2\right)\right] \quad \text{since the variables are independent} \\
&= \exp\left(-\gamma_1^\alpha |\omega'|^\alpha + j\beta_1\omega'\right)\exp\left(-\gamma_2^\alpha |\omega'|^\alpha + j\beta_2\omega'\right) \\
&= \exp\left(-\left(\gamma_1^\alpha + \gamma_2^\alpha\right)|\omega'|^\alpha + j(\beta_1 + \beta_2)\omega'\right),
\end{aligned}
\tag{2.13}
$$

which is the characteristic function of a symmetric stable random variable with $\gamma^\alpha = \gamma_1^\alpha + \gamma_2^\alpha$ and $\beta = \beta_1 + \beta_2$. ∎

2.2.3 Generalized Central Limit Theorem

Much like Gaussian signals, a wide variety of non-Gaussian processes found in practice arise as the superposition of many small independent effects. At first, this may point to a contradiction of the Central Limit Theorem, which states that, in the limit, the sum of such effects tends to a Gaussian process. A careful revision of the conditions of the Central Limit Theorem indicates that, in order for the Central Limit Theorem to be valid, the variance of the superimposed random variables must be finite. If the variance of the underlying random variables is infinite, an important generalization of the Central Limit Theorem emerges. This generalization explains the apparent contradictions of its "ordinary" version, as well as the presence of non-Gaussian processes in practice.

THEOREM 2.1 (GENERALIZED CENTRAL LIMIT THEOREM [75]) *Let* X_1, $X_2, \ldots$ *be an independent, identically distributed sequence of (possibly shift corrected) random variables. There exist constants* a_n *such that as* $n \to \infty$ *the sum*

$$
a_n(X_1 + X_2 + \cdots) \overset{d}{\to} Z
\tag{2.14}
$$

if and only if Z *is a stable random variable with some* $0 < \alpha \le 2$.

In the same way as the Gaussian model owes most of its strength to the Central Limit Theorem, the Generalized Central Limit Theorem constitutes a strong theoretical argument compelling the use of stable models in practical problems.

At first, the use of infinite variance in the definition of the Generalized Central Limit Theorem may lead to some skepticism as infinite variance for real data having bounded range may seem inappropriate. It should be noted, however, that the variance is but one measure of spread of a distribution, and is not appropriate for all problems. It is argued that in stable environments, γ may be more appropriate as a measure of spread. From an applied point of view, what is important is capturing the shape of a distribution. The Gaussian distribution is, for instance, routinely used to model bounded data, even though it has unbounded support. Although in some cases there are solid theoretical reasons for believing that a stable model is appropriate, in other more pragmatic cases the stable model can be used if it provides a good and parsimonious fit to the data at hand.

2.2.4 Simulation of Stable Sequences

Computer simulation of random processes is important in the design and analysis of signal processing algorithms. To this end, Chambers, Mallows, and Stuck (1976) [43] developed an algorithm for the generation of stable random variables. The algorithm is described in the following theorem.

THEOREM 2.2 (SIMULATION OF STABLE VARIABLES [151]) *Let* Θ *and* W *be independent with* Θ *uniformly distributed on* $\left(-\frac{\pi}{2}, \frac{\pi}{2}\right)$ *and* W *exponentially distributed with mean 1.* $Z \sim S(\alpha, \delta)$ *is generated as*

$$
Z = \begin{cases} c(\alpha, \delta) \dfrac{\sin \alpha(\Theta + \theta_0)}{(\cos \Theta)^{1/\alpha}} \left(\dfrac{\cos(\Theta - \alpha(\Theta + \theta_0))}{W} \right)^{(1-\alpha)/\alpha} & \alpha \neq 1 \\ \dfrac{2}{\pi} \left[\left(\dfrac{\pi}{2} + \delta\Theta \right) \tan \Theta - \delta \ln \left(\dfrac{\frac{\pi}{2} W \cos \Theta}{\frac{\pi}{2} + \delta\Theta} \right) \right] & \alpha = 1 \end{cases} \tag{2.15}
$$

where $c(\alpha, \delta) = (1 + (\delta \tan \frac{\pi\alpha}{2})^2)^{1/(2\alpha)}$ *and* $\theta_0 = \alpha^{-1} \arctan(\delta \tan \frac{\pi\alpha}{2})$. *In particular, for* $\alpha = 1, \delta = 0$ *(Cauchy),* $Z \sim \mathrm{Cauchy}(\gamma)$ *is generated as*

$$
Z = \gamma \tan(\Theta) = \gamma \tan \left(\pi \left(U - \frac{1}{2} \right) \right) \tag{2.16}
$$

where U *is a uniform random variable in* $(0, 1)$.

Figure 2.7 illustrates the impulsive behavior of symmetric stable processes as the characteristic exponent α is varied. Each one of the plots shows an independent and identically distributed (i.i.d.) "zero-centered" symmetric stable signal with unitary geometric power[4]. In order to give a better feeling of the impulsive structure of the data, the signals are plotted twice under two different scales. As it can be appreciated, the Gaussian signal ($\alpha = 2$) does not show impulsive behavior. For values of α close to 2 ($\alpha = 1.7$ in the figure), the structure of the signal is still similar to the Gaussian,

---

[4]The geometric power is introduced in the next section as a strength indicator of processes with infinite variance.

although some impulsiveness can now be observed. As the value of α is decreased, the impulsive behavior increases progressively.

2.3   LOWER-ORDER MOMENTS

Statistical signal processing relies, to a large extent, on the statistical characterization provided by second-order moments such as the variance $Var(X) = E(X^2) - (EX)^2$ with EX being the first moment. Second-order based estimation methods are sufficient whenever the underlying signals obey Gaussian statistics. The characterization of nonGaussian processes by second-order moments is no longer optimal and other moment characterizations may be required. To this end, *higher-order statistics* (HOS) exploiting third- and fourth-order moments (cummulants) have led to improved estimation algorithms in nonGaussian environments, provided that higher-order moments exist and are finite [148]. In applications where the processes are inherently impulsive, second-order and HOS may either be unreliable or may not even exist.

2.3.1   Fractional Lower-Order Moments

The different behavior of the Gaussian and nonGaussian distributions is to a large extent caused by the characteristics of their tails. The existence of second-order moments depends on the behavior of the tail of the distribution. The tail "thickness" of a distribution can be measured by its asymptotic mass $P(|X| > x)$ as $x \to \infty$. Given two functions $h(x)$ and $g(x)$, they have asymptotic similarity ($h(x) \sim g(x)$) if for $x \to \infty$: $\lim_{x \to \infty} h(x)/g(x) = 1$, the Gaussian distribution can be shown to have exponential order tails with asymptotic similarity

$$P(|X| > x) \sim \sqrt{\frac{2}{\pi}} x^{-1} e^{-x^2/2}. \tag{2.17}$$

Second order moments for the Gaussian distribution are thus well behaved due to the exponential order of the tails. The tails of the Laplacian distribution are heavier than that of the Gaussian distribution but remain of exponential order with

$$P(|X| > x) \sim e^{-x/\sigma}. \tag{2.18}$$

The tails of more impulsive nonGaussian distributions, however, behave very differently. Infinite variance processes that can appear in practice as a consequence of the Generalized Central Limit Theorem are modeled by probability distributions with algebraic tails for which

$$P(X > x) \sim cx^{-\alpha} \tag{2.19}$$

for some fixed c and $\alpha > 0$. The tail-heaviness of these distributions is determined by the tail constant α, with increased impulsiveness corresponding to small values of α. Stable random variables, for $\alpha < 2$, are examples of processes having algebraic tails as described by the following theorem.

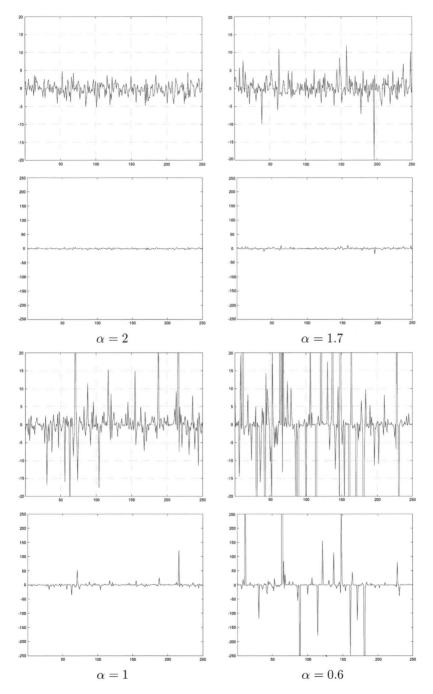

Figure 2.7 Impulsive behavior of i.i.d. α-stable signals as the tail constant α is varied. Signals are plotted twice under two different scales.

THEOREM 2.3 (STABLE DISTRIBUTION TAILS [151]) *Let $X \sim S(\alpha)$ be a symmetric, standard stable random variable with $0 < \alpha < 2$, then as $x \to \infty$,*

$$P(X > x) \sim \Gamma(\alpha) \frac{sin(\pi\alpha/2)}{\pi} x^{-\alpha}. \tag{2.20}$$

For stable and other distributions having algebraic tails, the following theorem is important having a significant impact on the statistical moments that can be used to process and analyze these signals.

THEOREM 2.4 *Algebraic-tailed random variables exhibit finite absolute moments for orders less than α*

$$E(|X|^p) < \infty, \quad \text{if } p < \alpha. \tag{2.21}$$

Conversely, if $p \geq \alpha$, the absolute moments become infinite.

Proof: The variable Y is replaced by $|X|^p$ in the first moment relationship

$$EY = \int_{-\infty}^{\infty} P(Y > y)dy \tag{2.22}$$

yielding

$$E(|X|^p) = \int_0^{\infty} P(|X|^p > t)dt \tag{2.23}$$

$$= \int_0^{\infty} pu^{p-1}P(|X| > u)du, \tag{2.24}$$

which, from (2.19), diverges for any distribution having algebraic tails.   ∎

Given that second-order, or higher-order moments, do not exist for algebraic tailed processes, the result in (2.21) points to the fact that in this case, it is better to rely on *fractional lower-order moments* (FLOMs): $E|X|^p = \int_{-\infty}^{\infty} |x|^p f(x)dx$, which exist for $0 < p < \alpha$. FLOMs for arbitrary processes can be computed from the definitions. Zolotarev (1957) [207], for instance, derived the FLOMs of $S\alpha S$ random variables as

PROPERTY 2.1 *The FLOMs for a $S\alpha S$ random variable with zero location parameter and dispersion γ is given by*

$$E(|X|^p) = C(p, \alpha)\gamma^p \quad 0 < p < \infty, \tag{2.25}$$

where

$$C(p, \alpha) = \frac{2^{p+1}\Gamma\left(\frac{p+1}{2}\right)\Gamma(-p/\alpha)}{\alpha\sqrt{\pi}\Gamma(-p/2)}. \tag{2.26}$$

Figure 2.8 depict the fractional lower-order moments for standardized $S\alpha S$ ($\gamma = 1$, $\delta = 0$) as functions of p for various values of α.

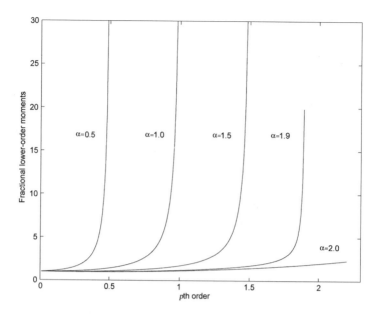

Figure 2.8   Fractional lower-order moments of the standardized $S\alpha S$ random variable.

2.3.2   Zero-Order Statistics

Fractional lower-order statistics do not provide a universal framework for the characterization of algebraic-tailed processes: for a given $p > 0$, there will always be a "remaining" class of processes (those with $\alpha \leq p$) for which the associated FLOMs do not exist. On the other hand, restricting the values of p to the valid interval $(0; \alpha)$ requires either the previous knowledge of α or a numerical procedure to estimate it. The former may not be possible in most practical applications, and the later may be inexact and/or computationally expensive. Unlike lower- or higher-order statistics, the advantage of *zero-order statistics* (ZOS) is that they provide a common ground for the analysis of basically *any* distribution of practical use [85, 48, 47, 50, 49]. In the same way as *pth*-order moments constitute the basis of FLOS and HOS techniques, zero-order statistics are based on logarithmic "moments" of the form $E \log |X|$.

THEOREM 2.5 *Let X be a random variable with algebraic or lighter tails. Then,* $E \log |X| < \infty$.

Proof: If X has algebraic or lighter tails, there exists a $p > 0$ such that $E|X|^p < \infty$. Jensen's inequality [65] guarantees that for a concave function ϕ, and a random variable Z, $E\phi(Z) \leq \phi(EZ)$. Letting $\phi(x) = \log |x|/p$ and $Z = |X|^p$ leads to

$$E \log |X| = E\left(\frac{\log |X|^p}{p}\right) \leq \frac{\log(E|X|^p)}{p} < \infty, \qquad (2.27)$$

which is the desired result.   ■

Random processes for which Theorem 2.5 applies, are referred to as being of "logarithmic order," in analogy with the term "second order" used to denote processes with finite variance. The logarithmic moment, which is finite for all logarithmic-order processes, can be used as a tool to characterize these signals. The strength of a signal is one attribute that can be characterized by logarithmic moments. For second-order processes, the *power* EX^2 is a widely accepted measure of signal strength. This measure, however, is always infinite when the processes exhibit algebraic tails, failing to provide useful information. To this end, zero-order statistics can be used to define an alternative strength measure referred to as the *geometric power*.

DEFINITION 2.6 (GEOMETRIC POWER [85]) *Let X be a logarithmic-order random variable. The* geometric power *of X is defined as*

$$S_0 = S_0(X) = e^{E \log |X|}. \tag{2.28}$$

The geometric power gives a useful strength characterization along the class of logarithmic-order processes having the advantage that it is mathematically and conceptually simple. In addition, it has a rich set of properties that can be effectively used. The geometric power is a scale parameter satisfying $S_0(X) \geq 0$ and $S_0(cX) = |c|S_0(X)$, and as such, it can be effectively used as an indicator of process strength or "power" in situations where second-order methods are inadequate. The geometric power takes on the value $S_0(X) = 0$ if and only if $P(X = 0) > 0$, which implies that zero power is only attained when there is a discrete probability mass located in zero [85].

The geometric power of any logarithmic-order process can be computed by the evaluation of (2.28). The geometric power of symmetric stable random variables, for instance, can be obtained in the closed-form expression.

PROPOSITION 2.1 (GEOMETRIC POWER OF STABLE PROCESSES) *The geometric power of a symmetric stable variable is given by*

$$S_0 = \frac{\gamma C_g^{1/\alpha}}{C_g}, \tag{2.29}$$

where $C_g = e^{C_e} \approx 1.78$, is the exponential of the Euler constant.

Proof: From [207], p. 215, the logarithmic moment of a zero-centered symmetric α-stable random variable with unitary dispersion is given by

$$E \log |X| = \left(\frac{1}{\alpha} - 1 \right) C_e, \tag{2.30}$$

where $C_e = 0.5772 \ldots$ is the Euler constant. This gives

$$S_0(X) \Big|_{\gamma=1} = e^{E \log |X|} = \left(e^{C_e} \right)^{\frac{1}{\alpha} - 1} = \frac{C_g^{1/\alpha}}{C_g}, \tag{2.31}$$

where $C_g = e^{C_e} \approx 1.78$. If X has a non-unitary dispersion γ, it is easy to see that

$$S_0(X) = \gamma \left[S_0(X)|_{\gamma=1} \right] = \frac{\gamma C_g^{1/\alpha}}{C_g}. \tag{2.32}$$

∎

The geometric power is well defined in the class of stable distributions for any value of $\alpha > 0$. Being a scale parameter, it is always multiple of γ and, more interestingly, it is a decreasing function of α. This is an intuitively pleasant property, since we should expect to observe more process strength when the levels of impulsiveness are increased.

Figure 2.9 illustrates the usefulness of the geometric power as an indicator of process strength in the α-stable framework. The scatter plot on the left side was generated from a stable distribution with $\alpha = 1.99$ and geometric power $S_0 = 1$. On the right-hand side, the scatter plot comes from a Gaussian distribution ($\alpha = 2$) also with unitary geometric power. After an intuitive inspection of Figure 2.9, it is reasonable to conclude that both of the generating processes possess the same strength, in accordance with the values of the geometric power. Contrarily, the values of the second-order power lead to the misleading conclusion that the process on the left is much stronger than the one on the right.

A similar example to the above can be constructed to depict the disadvantages of FLOS-based indicators of strength in the class of logarithmic-order processes. Fractional moments of order p present the same type of discontinuities as the one illustrated in Figure 2.9 for processes with tail constants close to $\alpha = p$. The geometric power, on the other side, is consistently continuous along all the range of values of α. This "universality" of the geometric power provides a general framework for comparing the strengths of any pair of logarithmic-order signals, in the same way as the (second-order) power is used in the classical framework.

The term zero-order statistics used to describe statistical measures using logarithmic moments is coined after the following relationship of the geometric power with fractional order statistics.

THEOREM 2.6 *Let $S_p = (E|X|^p)^{1/p}$ denote the scale parameter derived from the pth-order moment of X. If S_p exists for sufficiently small values of p, then*

$$S_0 = \lim_{p \to 0} S_p. \tag{2.33}$$

Furthermore, $S_0 \leq S_p$, for any $p > 0$.

Proof: It is enough to prove that $\lim_{p \to 0} \frac{1}{p} \log E|X|^p = E \log |X|$. Applying L'Hospital rule,

$$\lim_{p \to 0} \frac{\log E|X|^p}{p} = \lim_{p \to 0} \frac{d}{dp} \log E|X|^p \tag{2.34}$$

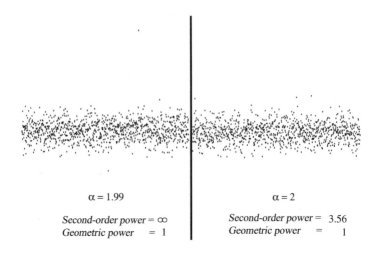

$\alpha = 1.99$ $\alpha = 2$

Second-order power = ∞ *Second-order power* = 3.56
Geometric power   = 1   *Geometric power*   = 1

Figure 2.9   Comparison of second-order power vs. geometric power for i.i.d. α-stable processes. Left: $\alpha = 1.99$. Right: $\alpha = 2$. While the values of the geometric power give an intuitive idea of the relative strengths of the signals, second-order power can be misleading.

$$= \lim_{p \to 0} \frac{E\left(\frac{d}{dp}|X|^p\right)}{E|X|^p} \tag{2.35}$$

$$= \lim_{p \to 0} \frac{E(|X|^p \log |X|)}{E|X|^p} \tag{2.36}$$

$$= E \log |X|. \tag{2.37}$$

To prove that $S_0 \leq S_p$, Jensen's inequality [65] guarantees that for a convex function ϕ and a random variable Z, $\phi(EZ) \leq E\phi(Z)$. Making $\phi(x) = e^x$ and $Z = \log |X|^p$ we get,

$$S_0^p = e^{(E \log |X|^p)} \leq E e^{\log |X|^p} = E|X|^p = S_p^p, \tag{2.38}$$

which leads to the desired result.   ∎

Theorem 2.6 indicates that techniques derived from the geometric power are the limiting *zero-order* relatives of FLOMs.

2.3.3   Parameter Estimation of Stable Distributions

The generalized central limit method and the theoretical formulation of several stochastic processes justify the use of stable distribution models. In some other cases, the approach can be more empirical where large data sets exhibit skewness and heavy tails in such fashion that stable models provide parsimonious and effective characterization. Modeling a sample set by a stable probability density function thus requires estimating the parameters of the stable distribution, namely the characteristic exponent $\alpha \in (0, 2]$; the symmetry parameter $\delta \in [-1, 1]$, which sets the skewness; the scale parameter $\gamma > 0$; and the location parameter β.

The often-preferred maximum likelihood parameter estimation approach, which offers asymptotic efficiency, is not readily available as stable distributions lack closed form analytical expressions. This problem can be overcome by numerical solutions. Nonetheless, simpler methods may be adequate in many cases [40, 68, 135, 151]. The approach introduced by Kuruoğlu, in particular, is simple and provides adequate estimates in general [121]. In Kuruoğlu's approach, the data of a general α-stable distributions is first transformed to data satisfying certain symmetric and skewness conditions. The parameters of the transformed data can then be estimated by the use of simple methods that use fractional lower-order statistics. Finally, the parameter estimates of the original data are obtained by using well-known relationships between these two sets of parameters. Kuruoğlu's approach is summarized next.

Let X_k be independent α-stable variates that are identically distributed with parameters α, δ, γ, and β. This stable law is denoted as

$$X_k \sim S_\alpha(\delta, \gamma, \beta). \tag{2.39}$$

The distribution of a weighted sum of these variables with weights a_k can be derived as [121]

$$Z = \sum_{k=1}^{n} a_k X_k \sim S_\alpha \left(\frac{\sum_{k=1}^{n} a_k^{<\alpha>}}{\sum_{k=1}^{n} |a_k|^\alpha} \delta, \left(\sum_{k=1}^{n} |a_k|^\alpha \right)^{\frac{1}{\alpha}} \gamma, \sum_{k=1}^{n} a_k \beta \right) \tag{2.40}$$

where $x^{<p>}$ denotes the signed pth power of a number x

$$x^{<p>} = \text{sign}(x)|x|^p. \tag{2.41}$$

This provides a convenient way to generate sequences of independent variables with zero β, zero δ, or with zero values for both β and δ (except when $\alpha = 1$). These are referred to as the centered, deskewed, and symmetrized sequences, respectively:

$$X_k^C = X_{3k} + X_{3k-1} - 2X_{3k-2} \sim S_\alpha \left(\left[\frac{2 - 2^\alpha}{2 + 2^\alpha} \right] \delta, [2 + 2^\alpha]^{\frac{1}{\alpha}} \gamma, 0 \right) \tag{2.42}$$

$$X_k^D = X_{3k} + X_{3k-1} - 2^{1/\alpha} X_{3k-2} \sim S_\alpha(0, 4^{\frac{1}{\alpha}} \gamma, [2 - 2^{1/\alpha}]\beta) \tag{2.43}$$

$$X_k^S = X_{2k} - X_{2k-1} \sim S_\alpha(0, 2^{\frac{1}{\alpha}} \gamma, 0). \tag{2.44}$$

Using such simpler sequences, moment methods for parameter estimation can be easily applied for variates with $\beta = 0$ or $\delta = 0$, or both, to the general variates at the cost of loss of some sample size. In turn, these estimates are used to calculate the estimates of the original α-stable distributions.

Moments of a distribution provide important statistical information about the distribution. Kuruoğlu's methods, in particular, exploit fractional lower-order or negative-order moments, which, for the skewed α-stable distributions, are finite for certain parameter values. First, the absolute and signed fractional-order moments of stable variates are calculated analytically as a generalization of Property 2.1 [121].

PROPERTY 2.2 *Let* $X \sim S_\alpha(\delta, \gamma, 0)$. *Then, for* $\alpha \neq 1$

$$
\mathrm{E}[|X|^p] = \frac{\Gamma\left(1 - \frac{p}{\alpha}\right)}{\Gamma(1 - p)} \left(\frac{\gamma}{|\cos\theta|^{\frac{1}{\alpha}}}\right)^p \frac{\cos\left(\frac{p\theta}{\alpha}\right)}{\cos\left(\frac{p\pi}{2}\right)}
\tag{2.45}
$$

for $p \in (-1, \alpha)$ *and where*

$$
\theta = \arctan\left(\delta \tan\frac{\alpha\pi}{2}\right).
\tag{2.46}
$$

As for the signed fractional moment of skewed α-stable distributions, the following holds [121].

PROPERTY 2.3 *Let* $X \sim S_\alpha(\delta, \gamma, 0)$. *Then*

$$
\mathrm{E}[X^{<p>}] = \frac{\Gamma(1 - \frac{p}{\alpha})}{\Gamma(1 - p)} \left(\frac{\gamma}{|\cos\theta|^{\frac{1}{\alpha}}}\right)^p \frac{\sin(\frac{p\theta}{\alpha})}{\sin(\frac{p\pi}{2})}.
\tag{2.47}
$$

Given n independent observations of a random variate X, the absolute and signed fractional moments can be estimated by the sample statistics:

$$
A_p = \frac{1}{n}\sum_{k=1}^n |X_k|^p, \quad S_p = \frac{1}{n}\sum_{k=1}^n X_k^{<p>}.
\tag{2.48}
$$

The presence of the gamma function in the formulae presented by the propositions hampers the direct solution of these expressions. However, by taking products and ratios of FLOMs and applying the following property of the gamma function:

$$
\Gamma(p)\Gamma(1 - p) = \frac{\pi}{\sin(p\pi)}
\tag{2.49}
$$

a number of simple closed-form estimators for α, δ, and γ can be obtained.

FLOM estimate for α**:** Noting that (2.48) is only the approximation of the absolute and signed fractional order moments, the analytic formulas (2.45), (2.47) are used. From (2.45), the product $A_p A_{-p}$ is given by

$$
A_p A_{-p} = \frac{\Gamma(1 - \frac{p}{\alpha})\Gamma(1 + \frac{p}{\alpha})}{\Gamma(1 - p)\Gamma(1 + p)} \frac{\cos^2\frac{p\theta}{\alpha}}{\cos^2\frac{p\pi}{2}}.
\tag{2.50}
$$

Using (2.49), the above reduces to

$$A_p A_{-p} = \frac{\sin^2(p\pi)\Gamma(p)\Gamma(-p)}{\sin^2(\frac{p\pi}{\alpha})\Gamma(\frac{p}{\alpha})\Gamma(-\frac{p}{\alpha})} \frac{\cos^2 \frac{p\theta}{\alpha}}{\cos^2 \frac{p\pi}{2}}.$$ (2.51)

The function $\Gamma(\cdot)$ has the property,

$$\Gamma(p+1) = p\Gamma(p)$$ (2.52)

thus, using equations (2.49) and (2.52), the following is obtained

$$\Gamma(p)\Gamma(-p) = -\frac{\pi}{p\sin(p\pi)}$$ (2.53)

and

$$\Gamma(\frac{p}{\alpha})\Gamma(-\frac{p}{\alpha}) = -\frac{\alpha\pi}{p\sin(p\pi)}.$$ (2.54)

Taking (2.53) and (2.54) into equation (2.51) results in

$$\frac{A_p A_{-p}}{\tan\frac{p\pi}{2}} = \frac{2\cos^2\frac{p\pi}{\alpha}}{\alpha\sin\frac{p\pi}{2}}.$$ (2.55)

In a similar fashion, the product $S_p S_{-p}$ can be shown to be equal to

$$S_p S_{-p} \tan\frac{p\pi}{2} = \frac{2\sin^2\frac{p\pi}{\alpha}}{\alpha\sin\frac{p\pi}{2}}.$$ (2.56)

Equations (2.55) and (2.56) combined lead to the following equality.

$$\text{sinc}\left(\frac{p\pi}{\alpha}\right) = \left[q\left(\frac{A_p A_{-p}}{\tan q} + S_p S_{-p}\tan q\right)\right]^{-1}$$ (2.57)

where $q = \frac{p\pi}{2}$.

Using the properties of Γ functions, and the first two propositions, other closed-form expressions for α, β, and γ can be derived assuming in all cases that $\delta = 0$. These FLOM estimation relations are summarized as follows.

Sinc Estimation for α: Estimate α as the solution to

$$\text{sinc}\left(\frac{p\pi}{\alpha}\right) = \left[q\left(\frac{A_p A_{-p}}{\tan q} + S_p S_{-p}\tan q\right)\right]^{-1}.$$ (2.58)

It is suggested in [121] that given a lower bound α_{LB} on α, a sensible range for p is $(0, \alpha_{LB}/2)$.

Ratio Estimate for δ: Given an estimate of α, estimate θ by solving

$$S_p/A_p = \tan\left(\frac{p\theta}{\alpha}\right) / \tan\left(\frac{p\pi}{2}\right). \tag{2.59}$$

Given this estimate of θ, obtain the following estimate of δ:

$$\delta = \frac{\tan(\theta)}{\tan\left(\frac{\alpha\pi}{2}\right)}. \tag{2.60}$$

FLOM Estimate for γ: Given an estimate of α, θ, solve

$$\gamma = |\cos\theta| \left(\frac{\Gamma(1-p)}{\Gamma\left(1-\frac{p}{\alpha}\right)} \frac{\cos(p\pi/2)}{\cos(p\theta/\alpha)} A_p\right)^{1/p}. \tag{2.61}$$

Note that the estimators above are all for zero-location cases, that is, $\beta = 0$. For the more general case where $\beta \neq 0$, the data must be transformed into a centered sequence by use of (2.42), then the FLOM estimation method should be applied on the parameters of the centered sequence, and finally the resulting δ and γ must be transformed by dividing by $(2 - 2^\alpha)/(2 + 2^\alpha)$ and $(2 + 2^\alpha)^{\frac{1}{\alpha}}$ respectively.

However, there are two possible problems with the FLOM method. First, since the value of a sinc function is in a finite range, when the value of the right size of (2.58) is out of this range, there is no solution for (2.58). Secondly, estimating α needs a proper value of p, which in turn depends on the value of α; in practice this can lead to errors in choosing p.

EXAMPLE 2.4
___

Consider the first-order modeling of the RTT time series in Figure 1.3 using the estimators of the α-stable parameters. The modeling results are shown in Table 2.1. Figure 2.10 shows histograms of the data and the pdfs associated with the parameters estimated.

Table 2.1   Estimated parameters of the distribution of the RTT time series measured between a host at the University of Delaware and hosts in Australia, Sydney, Japan, and the United Kingdom.

Parameter	Australia	Sydney	Japan	UK
α	1.0748	1.5026	1.0993	1.2180
δ	-0.3431	1	0.6733	1
γ	0.0010	7.6170×10^{-4}	0.0025	0.0014
β	0.2533	0.2359	0.2462	0.1091

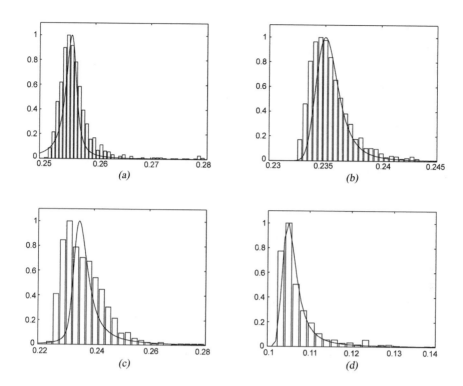

Figure 2.10 Histogram and estimated PDF of the RTT time series measured between a host at the University of Delaware and hosts in (*a*) Australia, (*b*) Sydney, (*c*) Japan, and (*d*) the United Kingdom.

■

Problems

2.1 Let $\hat{\beta}_p$ denote the L_p estimator defined by

$$\hat{\beta}_p = \arg\min_\beta \sum_{i=1}^{N} |x_i - \beta|^p.$$

(a) Show that when $0 < p \leq 1$, the estimator is selection-type (i.e., $\hat{\beta}_p$ is always equal to one of the input samples x_i).

(b) Define $\hat{\beta}_0 = \lim_{p \to 0} \hat{\beta}_p$. Prove that $\hat{\beta}_0$ is selection-type, and that it is always equal to one of the most repeated values in the sample set.

2.2 The set of well-behaved samples $\{-5, 5, -3, 3, -1, 1\}$ has been contaminated with an outlier sample of value 200.

(a) Plot the value of the L_p estimator $\hat{\beta}_p$ as a function of p, for $0 \leq p \leq 3$.

(b) Assuming that the ideal location of this distribution is $\beta = 0$, interpret the qualitative robustness of the L_p estimator as a function of p.

2.3   For $X \sim \text{Cauchy}(\gamma)$, find the mean and variance of X.

2.4   Let $X_1, \ldots, X_N$ denote a set of independent and identically distributed random variables with $X_i \sim \text{Cauchy}(1)$. Show that the sample mean

$$\bar{X} = \frac{1}{N} \sum_{i=1}^{N} X_i$$

posses the same distribution as any of the samples X_i. What does this tell about the efficiency of $\bar{X}$ in Cauchy noise? Can we say $\bar{X}$ is robust?

2.5   Show that Gaussian distributions are stable (i.e., show that $a^2 + b^2 = c^2$, so $\alpha = 2$).

2.6   Show that Cauchy distributions are stable (i.e., show that $a + b = c$, so $\alpha = 1$).

2.7   Find the asymptotic order of the tails of:

(a) A Gaussian distribution.

(b) A Laplacian distribution.

2.8   Show that the geometric power $S_0(X)$ satisfies:

(a) $S_0(X) \geq 0$.

(b) $S_0(cX) = |c| S_0(X)$.

2.9   Find the geometric power of $X \sim \text{Uniform}(-\sigma/2, \sigma/2)$.

2.10   Let W be exponentially distributed with mean 1. Show that $W = -\ln U$ with $U \sim \text{Uniform}(0, 1)$.

2.11   Show that the expressions in equations (2.42), (2.43), and (2.44) generate centered, deskewed, and symmetrized sequences with the parameters indicated.

3

Order Statistics

The subject of order statistics deals with the statistical properties and characteristics of a set of variables that have been ordered according to magnitude. Represent the elements of an observation vector $\mathbf{X} = [X(n), X(n-1), \ldots, X(n-N+1)]^T$, as $\mathbf{X} = [X_1, X_2, \ldots X_N]^T$. If the random variables $X_1, X_2, \ldots, X_N$ are arranged in ascending order of magnitude such that

$$X_{(1)} \leq X_{(2)} \leq \ldots \leq X_{(N)},$$

we denote $X_{(i)}$ as the *i*th-*order statistic* for $i = 1, \ldots, N$. The extremes $X_{(N)}$ and $X_{(1)}$, for instance, are useful tools in the detection of outliers. Similarly, the range $X_{(N)} - X_{(1)}$ is well known to be a quick estimator of the dispersion of a sample set.

An example to illustrate the applications of order statistics can be found in the ranking of athletes in Olympic sports. In this case, a set of N judges, generally from different nationalities, judge a particular athlete with a score bounded by a minimum assigned to a poor performance, and a maximum for a perfect score. In order to compute the overall score for a given athlete, the scores of the judges are not simply averaged. Instead, the maximum and the minimum scores given by the set of judges are discarded and the remaining scores are then averaged to provide the final score. This *trimming* of the data set is consistently done because of the possible bias of judges for a particular athlete. Since this is likely to occur in an international competition, the trimmed-average has evolved into the standard method of computing Olympic scores. This simple example shows the benefit of discarding, or discriminating against, a subset of samples from a larger data set based on the information provided by the sorted data.

Sorting the elements in the observation vector $\mathbf{X}$ constitutes a nonlinear permutation of the input vector. Consequently, even if the statistical characteristics of the input vector are exactly known, the statistical description of the sorted elements is often difficult to obtain. Simple mathematical expressions are only possible for samples which are mutually independent. Note that even in this simple case where the input samples $X_1, \ldots, X_N$ are statistically independent, the order statistics are necessarily dependent because of the ordering on the set.

The study of order statistics originated as a result of mathematical curiosity. The appearance of Sarhan and Greenberg's edited volume (1962) [171], and H. A. David's treatise on the subject (1970) [58] have changed this. Order statistics have since received considerable attention from numerous researchers. A classic and masterful survey is found in H. A. David (1981) [58]. Other important references include the work on extreme order statistics by Galambos (1978) [77], Harter's treatment in testing and estimation (1970) [96], Barnett and Lewis' (1984) [28] use of order statistics on data with outliers, and the introductory text of Arnold, Balakrishnan, and Nagaraja (1992) [16]. Parallel to the theoretical advances in the area, order statistics have also found important applications in diverse areas including life-testing and reliability, quality control, robustness studies, and signal processing. *The Handbook of Statistics Vol. 17*, edited by Balakrishnan and Rao (1998) [24], provides an encyclopedic survey of the field of order statistics and their applications.

3.1 DISTRIBUTIONS OF ORDER STATISTICS

When the variables are independent and identically distributed (i.i.d.), and when the parent distribution is continuous, the density of the rth order statistic is formed as follows. First, decompose the event that $x < X_{(r)} \le x + dx$ into three exclusive parts: that $r - 1$ of the samples X_i are less than or equal to x, that one is between x and $x + dx$, and that $N - r$ are greater than $x + dx$. Figure 3.1a depicts the configuration of such event. The probability that $N - r$ are greater than or equal to $x + dx$ is simply $[1 - F(x + dx)]^{N-r}$, the probability that one is between x and $x + dx$ is $f_x(x)\, dx$, and the probability that $r - 1$ are less than or equal to x is $F(x)^{r-1}$. The probability corresponding to the event of having more than one sample in the interval $(x, x + dx]$ is on the order of $(dx)^2$ and is negligible as dx approaches zero. The objective is to enumerate all possible outcomes of the $X_i's$ such that the ordering partition is satisfied. Counting all possible enumerations of N samples in the three respective groups and using the fact that $F(x + dx) \to F(x)$ as $dx \to 0$, we can write

$$
\begin{aligned}
f_{(r)}(x)\, dx &= Pr\left[\, x < X_{(r)} \le x + dx \,\right] \\
&= \frac{N!}{(r-1)!\,(N-r)!} F(x)^{r-1} \left[1 - F(x)\right]^{N-r} f_x(x)\, dx. \quad (3.1)
\end{aligned}
$$

The density function of the rth order statistic, $f_{(r)}(x)$, follows directly from the above. The coefficient in the right side of (3.1) is the trinomial coefficient whose structure follows from the general multinomial coefficient as described next. Given a set of N objects, k_1 labels of type 1, k_2 labels of type 2, ..., and k_m labels of type m and suppose that $k_1 + k_2 + \ldots + k_m = N$, the number of ways in which we may assign the labels to the N objects is given by the multinomial coefficient

$$\frac{N!}{k_1! \, k_2! \cdots k_m!} \, . \tag{3.2}$$

The trinomial coefficient in (3.1) is a special case of (3.2) with $k_1 = r - 1$, $k_2 = 1$, and $k_3 = N - r$.

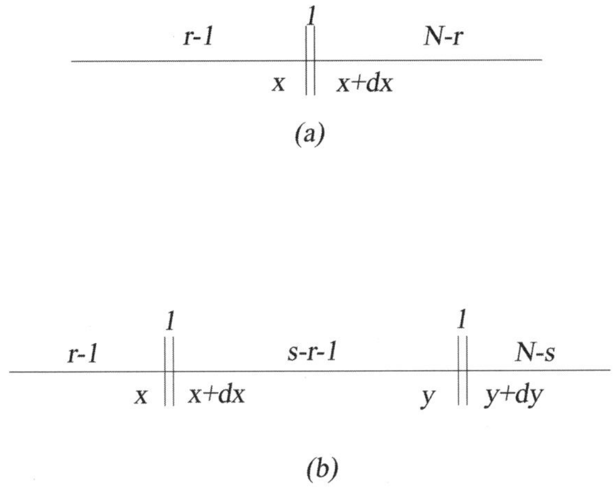

(a)

(b)

Figure 3.1   (*a*) The event $x < X_{(r)} \leq x + dx$ can be seen as $r - 1$ of the samples X_i are less than or equal to x, that one is between x and $x + dx$, and that $N - r$ are greater than or equal to x. (*b*) The event $x < X_{(r)} \leq x + dx$ and $y < X_{(s)} \leq y + dy$ can be seen as $r - 1$ of the samples X_i are less than x, that one of the samples is between x and $x + dx$, that $s - r - 1$ of the samples X_i are less than y but greater than x, that one of the samples is between y and $y + dy$, and finally that $N - s$ of the samples are greater than y.

The joint density function of the order statistics $X_{(r)}$ and $X_{(s)}$, for $1 \leq r < s \leq N$, can be found in a similar way. In this case, for $x \leq y$, the joint density is denoted as $f_{(r,s)}(x, y)$ and is obtained by decomposing the event

$$x < X_{(r)} \leq x + dx < y < X_{(s)} \leq y + dy \tag{3.3}$$

into five mutually exclusive parts: that $r - 1$ of the samples X_i are less than x, that one of the samples is between x and $x + dx$, that $s - r - 1$ of the samples X_i are less than y but greater than $x + dx$, that one of the samples is between y and $y + dy$, and finally that $N - s$ of the samples are greater than $y + dy$. The decomposition of the event in (3.3) is depicted in Figure 3.1b. The probability of occurrence for each of the five listed parts is $F(x)^{r-1}$, $f_x(x)\,dx$, $[F(y) - F(x+dx)]^{s-r-1}$, $f_x(y)\,dy$, and

$[1 - F(y + dy)]^{N-s}$. The probability corresponding to the events of having more than one sample in either of the intervals $(x, x + dx]$ and $(y, y + dy]$ is negligible as dx and dy approach zero. Using the multinomial counting principle to enumerate all possible occurrences in each part, and the fact that $F(x + dx) \sim F(x)$ and $F(y + dy) \sim F(y)$ as $dx,\ dy \to 0$ we obtain the joint density function

$$f_{(r,s)}(x, y) \quad = \quad \frac{N!}{(r - 1)!\,(s - r - 1)!\,(N - s)!} F(x)^{r-1}\, f_x(x) \qquad (3.4)$$

$$[F(y) - F(x)]^{s-r-1}\, f_x(y)\, [1 - F(y)]^{N-s}.$$

These density functions, however, are only valid for continuous random variables, and a different approach must be taken to find the distribution of order statistics with discontinuous parent distributions. The following approach is valid for both, continuous and discontinuous distributions: let the i.i.d. variables $X_1, X_2, \ldots, X_N$ have a parent distribution function $F(x)$, the distribution function of the largest order statistic $X_{(N)}$ is

$$\begin{aligned} F_{(N)}(x) \quad &= \quad Pr\{X_{(N)} \le x\} \\ &= \quad Pr\{\text{all } X_{(i)} \le x\} \\ &= \quad Pr\{\text{all } X_i \le x\} = [F(x)]^N, \end{aligned}$$

due to the independence property of the input samples. Similarly, the distribution function of the minimum sample $X_{(1)}$ is

$$\begin{aligned} F_{(1)}(x) \quad &= \quad Pr\{X_{(1)} \le x\} = 1 - Pr\{X_{(1)} > x\} \\ &= \quad 1 - Pr\{\text{all } X_i > x\} = 1 - [1 - F(x)]^N, \end{aligned}$$

since $X_{(1)}$ is less than, or equal to, all the samples in the set. The distribution function for the general case is

$$\begin{aligned} F_{(r)}(x) \quad &= \quad Pr\{X_{(r)} \le x\} \\ &= \quad Pr\{\text{at least } r \text{ of the } X_i \text{ are less than or equal to } x\} \\ &= \quad \sum_{i=r}^{N} Pr\{\text{exactly } i \text{ of the } X_i \text{ are less than or equal to } x\} \\ &= \quad \sum_{i=r}^{N} \binom{N}{i} [F(x)]^i [1 - F(x)]^{N-i}. \qquad (3.5) \end{aligned}$$

Letting the joint distribution function of $X_{(r)}$ and $X_{(s)}$, for $1 \le r < s \le N$, be denoted as $F_{(r,s)}(x, y)$ then for $x < y$ we have for discrete and continuous random variables

$$
\begin{aligned}
F_{(r,s)}(x,y) &= Pr\{\text{at least } r \text{ of the } X_i \leq x, \text{ at least } s \text{ of the } X_i \leq y\} \\
&= \sum_{j=s}^{N}\sum_{i=r}^{j} Pr\{\text{exactly } i \text{ of } X_1, X_2 \ldots, X_n \text{ are at most } x \text{ and} \\
&\qquad\qquad \text{exactly } j \text{ of } X_1, X_2 \ldots, X_n \text{ are at most } y\} \qquad (3.6) \\
&= \sum_{j=s}^{N}\sum_{i=r}^{j} \frac{N!}{i!(j-i)!(N-j)!}[F(x)]^i[F(y)-F(x)]^{j-i}[1-F(y)]^{N-j}.
\end{aligned}
$$

Notice that for $x \geq y$, the ordering $X_{(r)} < x$ with $X_{(s)} \leq y$, implies that $F_{(r,s)}(x,y) = F_{(s)}(y)$.

An alternate representation of the distribution function $F_{(r)}(x)$ is possible, which will prove helpful later on in the derivations of order statistics. Define the set of N samples from a uniform distribution in the closed interval $[0,1]$ as $U_1, U_2, \ldots, U_N$. The order statistics of these variates are then denoted as $U_{(1)}, U_{(2)}, \ldots, U_{(N)}$. For any distribution function $F(x)$, we define its corresponding inverse distribution function or quantile function F^{-1} as

$$
F^{-1}(y) = \text{supremum } [x : F(x) \leq y], \qquad (3.7)
$$

for $0 < y < 1$. It is simple to show that if $X_1, \ldots, X_N$ are i.i.d. with a parent distribution $F(x)$, then the transformation $F^{-1}(U_i)$ will lead to variables with the same distribution as X_i [157]. This is written as

$$
F^{-1}(U_i) \overset{d}{=} X_i \qquad (3.8)
$$

where the symbol $\overset{d}{=}$ represents equality in distribution. Since cumulative distribution functions are monotonic, the smallest U_i will result in the smallest X_i, the largest U_i will result in the largest X_i, and so on. It follows that

$$
F^{-1}(U_{(r)}) \overset{d}{=} X_{(r)}. \qquad (3.9)
$$

The density function of $U_{(r)}$ follows from (3.1) as

$$
f_{U_{(r)}}(u) = \frac{N!}{(r-1)!(N-r)!}u^{r-1}(1-u)^{N-r} \quad 0 < u < 1. \qquad (3.10)
$$

Integrating the above we can obtain the distribution function

$$
F_{U_{(r)}}(u) = \int_0^u \frac{N!}{(r-1)!(N-r)!}t^{r-1}(1-t)^{N-r}dt.
$$

Using the relationship $F^{-1}(U_{(r)}) \overset{d}{=} X_{(r)}$, we obtain from the above the general expression

$$
F_{(r)}(x) = \int_0^{F(x)} \frac{N!}{(r-1)!(N-r)!}t^{r-1}(1-t)^{N-r}dt,
$$

which is an incomplete Beta function valid for any parent distribution $F(x)$ of the i.i.d. samples X_i [16].

The statistical analysis of order statistics in this section has assumed that the input samples are i.i.d. As one can expect, if the i.i.d. condition is relaxed to the case of dependent variates, the distribution function of the ordered statistics are no longer straightforward to compute. Procedures to obtain these are found in [58].

Recursive Relations for Order Statistics Distributions   Distributions of order statistics can also be computed recursively, as in Boncelet (1987) [36]. No assumptions are made about the random variables. They can be discrete, continuous, mixed, i.i.d. or not.

Let $X_{(r):N}$ denote the rth order statistic out of N random variables. For first order distributions let $-\infty = t_0 < t_1 < t_2 = +\infty$ and, for second order distributions, let $-\infty = t_0 < t_1 < t_2 < t_3 = +\infty$ and let $r_1 \leq r_2$. Then, for events of order statistics:

$$\{X_{(r):N+1} \leq t_1\} = \{X_{(r):N} \leq t_1\}\{X_{N+1} > t_1\} \tag{3.11}$$
$$+ \{X_{(r-1):N} \leq t_1\}\{X_{N+1} \leq t_1\}$$
$$\{X_{(r_1):N+1} \leq t_1, X_{(r_2):N+1} \leq t_2\} = \tag{3.12}$$
$$\{X_{(r_1):N} \leq t_1, X_{(r_2):N} \leq t_2\}\{X_{N+1} > t_2\}$$
$$+ \{X_{(r_1):N} \leq t_1, X_{(r_2-1):N} \leq t_2\}\{t_1 < X_{N+1} \leq t_2\}$$
$$+ \{X_{(r_1-1):N} \leq t_1, X_{(r_2-1):N} \leq t_2\}\{X_{N+1} \leq t_1\}$$

In the first order case, (3.11) states that there are two ways the rth order statistic out of $N + 1$ random variables can be less or equal than t_1: one, that the $N + 1$st is larger than t_1 and the rth order statistic out of N is less or equal than t_1 and two, the $N + 1$st is less or equal than t_1 and the $r - 1$st order statistic out of N is less or equal than t_1. In the second order case, the event in question is similarly decomposed into three events.

Notice that the events on the right hand side are disjoint since the events on X_{N+1} partition the real line into nonoverlapping segments. A direct consequence of this is a recursive formula for calculating distributions for independent X_i:

$$P(X_{(r):N+1} \leq t_1) = P(X_{(r):N} \leq t_1)(1 - F_{N+1}(t_1)) \tag{3.13}$$
$$+ P(X_{(r-1):N} \leq t_1)F_{N+1}(t_1)$$
$$P(X_{(r_1):N+1} \leq t_1, X_{(r_2):N+1} \leq t_2) =$$
$$P(X_{(r_1):N} \leq t_1, X_{(r_2):N} \leq t_2)(1 - F_{N+1}(t_2)) \tag{3.14}$$
$$+ P(X_{(r_1):N} \leq t_1, X_{(r_2-1):N} \leq t_2)(F_{N+1}(t_2) - F_{N+1}(t_1))$$
$$+ P(X_{(r_1-1):N} \leq t_1, X_{(r_2-1):N} \leq t_2)F_{N+1}(t_1).$$

3.2 MOMENTS OF ORDER STATISTICS

The Nth order density function provides a complete characterization of a set of N ordered samples. These distributions, however, can be difficult to obtain. Moments of order statistics, on the other hand, can be easily estimated and are often sufficient to characterize the data. The moments of order statistics are defined in the same fashion as moments of arbitrary random variables. Here we always assume that the sample size is N. The mean or expected value of the rth order statistic is denoted as $\mu_{(r)}$ and is found as

$$
\begin{aligned}
\mu_{(r)} &= \int_{-\infty}^{\infty} x\, f_{(r)}(x)\, dx \qquad\qquad (3.15) \\
&= \frac{N!}{(r-1)!\,(N-r)!} \int_{-\infty}^{\infty} x\, F(x)^{r-1}[1 - F(x)]^{N-r}\, f_x(x)\, dx.
\end{aligned}
$$

The pth raw moment of the rth-order statistic can also be defined similarly from (3.9) and (3.10) as

$$
\begin{aligned}
\mu_{(r)}^{(p)} &= EX_{(r)}^p \\
&= \frac{N!}{(r-1)!\,(N-r)!} \int_{0}^{1} \left[F^{-1}(u)\right]^p u^{r-1}(1-u)^{N-r}\, du, \quad (3.16)
\end{aligned}
$$

for $1 \le r \le N$.

Expectation of order statistic products, or *order statistic correlation*, can also be defined, for $1 \le r \le s \le N$, as

$$
\begin{aligned}
\mu_{(r,s):N} &= E\left(X_{(r)}X_{(s)}\right) \qquad\qquad (3.17) \\
&= [B(r, s-r, N-s+1)]^{-1} \int_{0}^{1}\int_{0}^{1} \left[F_x^{-1}(u)F^{-1}(v)u^{r-1}\right. \\
&\qquad\qquad\qquad\qquad\qquad \left. (v-u)^{s-r-1}(1-v)\right] dv\, du
\end{aligned}
$$

where $B(a, b, c) = \frac{(a-1)!(b-1)!(c-1)!}{(a+b+c-1)!}$. Note that (3.17) does not allude to a time shift correlation, but to the correlation of two different order-statistic variates taken from the same sample set. The statistical characteristics of the order-statistics $X_{(1)}, X_{(2)}, \ldots, X_{(N)}$ are not homogeneous, since

$$
EX_{(r)} \ne EX_{(s)} \qquad\qquad (3.18)
$$

for $r \ne s$, as expected since the expected value of $X_{(r)}$ should be less than the expected value of $X_{(r+1)}$. In general, the expectation of products of order statistics are not symmetric

$$E(X_{(r)}X_{(r+s)}) \neq E(X_{(r)}X_{(r-s)}). \tag{3.19}$$

This symmetry only holds in very special cases. One such case is when the parent distribution is symmetric and where $r = (N + 1)/2$ such that $X_{(r)}$ is the median. The covariance of $X_{(r)}$ and $X_{(s)}$ is written as

$$\text{cov}\,[X_{(r)}X_{(s)}] = E\left\{(X_{(r)} - \mu_{(r)})\,(X_{(s)} - \mu_{(s)})\right\}. \tag{3.20}$$

Tukey (1958) [187], derived the nonnegative property for the covariance of order statistics: $cov[X_{(r)}X_{(s)}] \geq 0$.

3.2.1 Order Statistics From Uniform Distributions

In order to illustrate the concepts presented above, consider N samples of a standard uniform distribution with density function $f_u(u) = 1$ and distribution function $F_u(u) = u$ for $0 \leq u \leq 1$. Letting $U_{(r)}$ be the rth smallest sample, or order statistic, the density function of $U_{(r)}$ is obtained by substituting the corresponding values in (3.1) resulting in

$$f_{(r)}(u) = \frac{N!}{(r - 1)!\,(N - r)!}u^{r-1}(1 - u)^{N-r}, \tag{3.21}$$

also in the interval $0 \leq u \leq 1$. The distribution function follows immediately as

$$F_{(r)}(u) = \int_0^u \frac{N!}{(r - 1)!\,(N - r)!}t^{r-1}(1 - t)^{N-r}\,dt, \tag{3.22}$$

or alternatively using (3.5) as

$$F_{(r)}(u) = \sum_{i=r}^N \binom{N}{i} u^i[1 - u]^{N-i}. \tag{3.23}$$

The mode of the density function can be found at $(r - 1)/(N - 1)$. The kth moment of $U_{(r)}$ is found from the above as

$$\begin{aligned}
\mu_{(r)}^{(k)} &= \int_0^1 u^k f_{(r)}(u)\,du \\
&= \frac{N!}{(r - 1)!(N - r)!}\int_0^1 u^k u^{r-1}(1 - u)^{N-r}\,du \tag{3.24} \\
&= B(r + k, N - r + 1)/B(r, N - r + 1), \tag{3.25}
\end{aligned}$$

where we make use of the complete beta function

$$B(p, q) = \int_0^1 t^{p-1}(1 - t)^{q-1}\,dt \tag{3.26}$$

for $p, q > 0$. Simplifying (3.25) leads to the kth moment

$$\mu_{(r)}^{(k)} = \frac{N! \ (r + k - 1)!}{(N + k)! \ (r - 1)!}. \tag{3.27}$$

In particular, the first moment of the rth-order statistic can be found as

$$\mu_{(r)}^{(1)} = r/(N + 1).$$

To gain an intuitive understanding of the distribution of order statistics, it is helpful to plot $f_{(r)}(u)$ in (3.21) for various values of r. For $N = 11$, Figure 3.2 depicts the density functions of the 2nd-, 3rd-, 6th- (median), 9th-, and 10th-order statistics of the samples. With the exception of the median, all other order statistics exhibit asymmetric density functions. Other characteristics of these density functions, such as their mode and shape, can be readily observed and interpreted in an intuitive fashion.

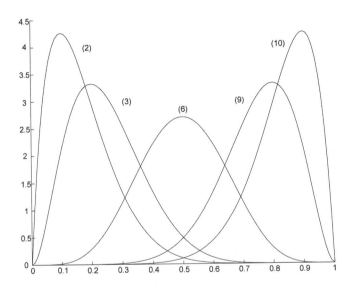

Figure 3.2 Density functions of $X_{(2)}, X_{(3)}, X_{(6)}$ (median), $X_{(9)}$, and $X_{(10)}$ for a set of eleven uniformly distributed samples.

Next consider the joint density function of $U_{(r)}$ and $U_{(s)}$ $(1 \leq r < s \leq N)$. From (3.4) we find

$$f_{(r,s)}(u, v) = \frac{N!}{(r - 1)!(s - r - 1)! \ (N - s)!} u^{r-1}(v - u)^{s-r-1}(1 - v)^{N-s} \tag{3.28}$$

Again there are two equivalent expressions for the joint cumulative distribution function, the first is obtained integrating (3.28) and the second from Eq. (3.6)

$$
\begin{aligned}
F_{(r,s)}(u, v) &= \int_0^u \int_{t_1}^v \frac{N!}{(r-1)!(s-r-1)!\,(N-s)!} \\
&\quad \times\, t_1^{r-1}(t_2 - t_1)^{s-r-1}(1 - t_2)^{N-s}\, dt_2\, dt_1 \\
&= \sum_{j=s}^N \sum_{i=r}^j \frac{N!}{i!\,(j-i)!\,(N-j)!} u^i (v-u)^{j-i}\,(1-v)^{N-j}
\end{aligned}
$$

for $0 \le u < v \le 1$. The joint density function allows the computation of the (k_r, k_s)th product moment of $(U_{(r)}, U_{(s)})$, which, after some simplifications, is found as

$$
\mu_{(r,s)}^{(k_r, k_s)} = \frac{N!\,(r + k_r - 1)!}{(N + k_r + k_s)!(r-1)!}\,\frac{(s + k_r + k_s - 1)!}{(s + k_r - 1)!}. \tag{3.29}
$$

In particular, for $k_r = k_s = 1$, the joint moment becomes

$$
\mu_{(r,s)} = \frac{r\,(s+1)}{(N+1)(N+2)}. \tag{3.30}
$$

As with their marginal densities, an intuitive understanding of bivariate density functions of order statistics can be gained by plotting $f_{(r,s)}(u, v)$. Figure 3.3 depicts the bivariate density function, described in (3.28) for the 2nd- and 6th- (median) order statistics of a set of eleven uniformly distributed samples. Note how the marginal densities are satisfied as the bivariate density is integrated over each variable. Several characteristics of the bivariate density, such as the constraint that only regions where $u < v$ will have mass, can be appreciated in the plot.

3.2.2 Recurrence Relations

The computation of order-statistic moments can be difficult to obtain for observations of general random variables. In such cases, these moments must be evaluated by numerical procedures. Moments of order statistics have been given considerable importance in the statistical literature and have been numerically tabulated extensively for several distributions [58, 96]. Order-statistic moments satisfy a number of recurrence relations and identities, which can reduce the number of direct computations. Many of these relations express higher-order moments in terms of lower-order moments, thus simplifying the evaluation of higher-order moments. Since the recurrence relations between moments often involve sample sets of lower orders, it is convenient to introduce the notation $X_{(i):N}$ to represent the ith-order statistic taken from a set of N samples. Similarly, $\mu_{(i):N}$ represents the expected value of $X_{(i):N}$.

Many recursive relations for moments of order-statistics are derived from the identities

$$
\sum_{i=1}^N X_{(i):N}^k = \sum_{i=1}^N X_i^k \tag{3.31}
$$

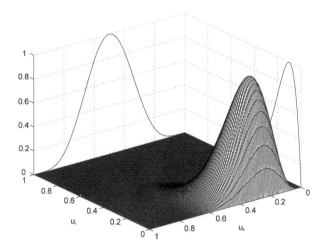

Figure 3.3 Bivariate density function of $X_{(6)}$ (median) and $X_{(2)}$ for a set of eleven uniformly distributed samples.

for $k \geq 1$, and

$$\sum_{i=1}^{N}\sum_{j=1}^{N} X_{(i):N}^{k_i} \, X_{(j):N}^{k_j} = \sum_{i=1}^{N}\sum_{j=1}^{N} X_i^{k_i} X_j^{k_j} \tag{3.32}$$

for k_i, $k_j \geq 1$, which follows from the principle that the sum of a set of samples raised to the kth power is unchanged by the order in which they are summed. Taking expectations of (3.31) leads to:

$$\sum_{i=1}^{N} \mu_{(i):N}^{(k)} = NE(X_i^k) = N\mu_{(1):1}^{(k)}$$

for $N \geq 2$ and $k \geq 1$. Similarly, from (3.32) the following is obtained:

$$
\begin{aligned}
\sum_{i=1}^{N}\sum_{j=1}^{N} \mu_{(i,j):N}^{(k_i,k_j)} &= NE(X^{k_i+k_j}) + N(N-1)E(X^{k_i})E(X^{k_j}) \\
&= N\mu_{(1):1}^{(k_i+k_j)} + N(N-1)\mu_{(1):1}^{(k_i)} \, \mu_{(1):1}^{(k_j)}
\end{aligned}
$$

for $k_i, k_j \geq 1$.

These identities are simple and can be used to check the accuracy of computation of moments of order statistics. Some other useful recurrence relations are presented in the following properties.

PROPERTY 3.1 *For* $1 \leq i \leq N-1$ *and* $k \geq 1$.

$$i\mu_{(i+1):N}^{(k)} + (N-i)\mu_{(i):N}^{(k)} = N\mu_{(i):N-1}^{(k)}.$$

This property can be obtained from equation (3.16) as follows:

$$
\begin{aligned}
N\mu_{(i):N-1}^{(k)} &= \frac{N!}{(i-1)!(N-i-1)!} \int_0^1 \left[F^{-1}(u)\right]^k u^{i-1}(1-u)^{N-i-1} du \\
&= \frac{N!}{(i-1)!(N-i-1)!} \int_0^1 \left[F^{-1}(u)\right]^k u^{i-1}(1-u)^{N-i-1} \\
&\qquad\qquad (u+1-u)du \\
&= \frac{N!}{(i-1)!(N-i-1)!} \left[\int_0^1 \left[F^{-1}(u)\right]^k u^i(1-u)^{N-i-1} du \right.\\
&\qquad\qquad \left. + \int_0^1 \left[F^{-1}(u)\right]^k u^{i-1}(1-u)^{N-i} du\right] \\
&= i\mu_{(i+1):N}^{(k)} + (N-i)\mu_{(i):N}^{(k)} .
\end{aligned}
$$

Property 3.1 describes a relation known as the triangle rule [16], which allows one to compute the kth moment of a single order statistic in a sample of size N, if these moments in samples of size less than N are already available. By repeated use of the same recurrence relation, the kth moment of the remaining $N-1$ order statistics can be subsequently obtained. Hence, one could start with $\mu_{(N):N}^{(k)}$ or $\mu_{(1):N}^{(k)}$ and recursively find the moments of the smaller-or larger-order statistics.

A different recursion, published by Srikantan [180], can also be used to recursively compute single moments of order statistics by expressing the kth moment of the ith-order statistic in a sample of size N in terms of the kth moments of the largest order statistics in samples of size N and less.

PROPERTY 3.2 *For* $1 \le i \le N-1$ *and* $k \ge 1$.

$$
\mu_{(i):N}^{(k)} = \sum_{j=i}^{N} (-1)^{j-i} \binom{N}{j} \binom{j-1}{i-1} \mu_{(j):j}^{(k)} .
$$

The proof of this property is left as an exercise.

3.3 ORDER STATISTICS CONTAINING OUTLIERS

Order statistics have the characteristic that they allow us to discriminate against outlier contamination. Hence, when properly designed, statistical estimates using ordered statistics can ignore clearly inappropriate samples. In the context of robustness, it is useful to obtain the distribution functions and moments of order-statistics arising from a sample containing outliers. Here, the case where the contamination consists of a single outlier is considered. These results can be easily generalized to higher

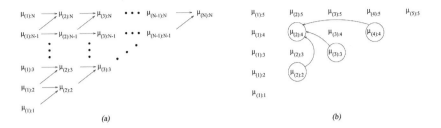

Figure 3.4 (*a*) Triangle recursion for single moments; (*b*) recurrence relation from moments of maxima of lower orders.

orders of contamination. The importance of a systematic study of order statistics from an outlier model has been demonstrated in several extensive studies [3, 59].

First, the distributions of order statistics obtained from a sample of size N when an unidentified single outlier contaminates the sample are derived. Let the N long sample set consist of $N-1$ i.i.d. variates X_i, $i = 1, \ldots, N-1$, and the contaminant variable Y, which is also independent from the other samples in the sample set. Let $F(x)$ and $G(x)$ be the continuous parent distributions of X_i and Y, respectively. Furthermore, let

$$Z_{(1):N} \leq Z_{(2):N} \leq \cdots \leq Z_{(N):N} \tag{3.33}$$

be the order statistics obtained by arranging the N independent observations in increasing order of magnitude. The distribution functions of these ordered statistics are now obtained. The distribution of the maxima denoted as $H_{(N):N}(x)$ is

$$
\begin{aligned}
H_{(N):N}(x) &= Pr\{\text{all of } X_1, \ldots, X_{N-1}, \text{ and } Y \leq x\} \\
&= F(x)^{N-1} G(x).
\end{aligned}
$$

The distribution of the ith-order statistic, for $1 < i \leq N-1$, can be obtained as follows:

$$
\begin{aligned}
H_{(i):N}(x) &= Pr\{\text{at least } i \text{ of } X_1, X_2, \ldots, X_{N-1}, Y \leq x\} \\
&= Pr\{\text{exactly } i-1 \text{ of } X_1, X_2, \ldots, X_{N-1} \leq x \text{ and } Y \leq x\} \\
&\quad + Pr\{\text{at least } i \text{ of } X_1, X_2, \ldots, X_{N-1} \leq x\} \\
&= \binom{N-1}{i-1}(F(x))^{i-1}(1 - F(x))^{N-i}G(x) + F_{(i):N-1}(x)
\end{aligned}
$$

where $F_{(i):N-1}(x)$ is the distribution of the ith-order statistic in a sample of size $N-1$ drawn from a parent distribution $F(x)$. The density function of $Z_{(i):N}$ can be obtained by differentiating the above or by direct derivation, which is left as an exercise:

$$
\begin{aligned}
h_{(i):N}(x) = \ & \frac{(N-1)!}{(i-2)!(N-i)!}(F(x))^{i-2}(1-F(x))^{N-i}G(x)f(x) \\
& + \frac{(N-1)!}{(i-1)!(N-i)!}(F(x))^{i-1}(1-F(x))^{N-i}g(x) \\
& + \frac{(N-1)!}{(i-1)!(N-i-1)!}(F(x))^{i-1}(1-F(x))^{N-i-1}(1-G(x))f(x)
\end{aligned}
$$

where the first term drops out if $i=1$, and the last term if $N=i$.

The effect of contamination on order statistics is illustrated in Figure 3.5 depicting the densities of $Z_{(2)}$, $Z_{(6)}$ (median), and $Z_{(10)}$ for a sample set of size 11, zero-mean, double-exponential random variables. The dotted curves are the densities where no contamination exists. In the contaminated case, one of the random variables is modified such that its mean is shifted to 20. The effect of the contamination on the second-order statistic is negligible, the density of the median is only slightly affected as expected, but the effect on the 10th-order statistic, on the other hand, is severe.

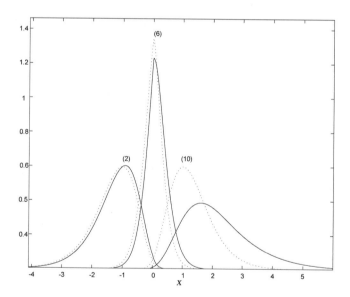

Figure 3.5 Density functions of $Z_{(2)}$, $Z_{(6)}$ (median), and $Z_{(10)}$ with (solid) and without contamination (dotted).

3.4   JOINT STATISTICS OF ORDERED AND NONORDERED SAMPLES

The discussion of order statistics would not be complete if the statistical relationships between the order statistics and the nonordered samples are not described. To begin, it is useful to describe the statistics of ranks. Sorting the elements $X_1, \ldots, X_N$ defines a set of N keys r_i, for $i = 1, \ldots, N$, where the rank key r_i identifies the location of X_i among the sorted set of samples $X_{(1)}, \ldots, X_{(N)}$. If the input elements to the sorter are i.i.d., each sample X_i is equally likely to be ranked first, second, or any arbitrary rank. Hence

$$
P_r\{r_i = r\} = \begin{cases} \frac{1}{N} & \text{for } r = 1, 2, \ldots, N \\ 0 & \text{else.} \end{cases}
\tag{3.34}
$$

The expected value of each rank key is then $E\{r_i\} = (N+1)/2$. Similarly, the bivariate distribution of the two keys r_i and r_j, is given by

$$
P_r\{r_i = r, \ r_j = s\} = \begin{cases} \frac{1}{N(N-1)} & \text{for } r \neq s = 1, 2, \ldots, N \\ 0 & \text{else.} \end{cases}
$$

The joint distribution function of the rth order statistic $X_{(r)}$ and the ith input sample is derived next. Again, let the sample set $X_1, X_2, \ldots, X_N$ be i.i.d. with a parent distribution $F(x)$. Since the observation samples are i.i.d., the joint distribution for $X_{(r)}$ and X_i is valid for any arbitrary value of i. The joint distribution function of X_i and $X_{(r)}$ is found for $X_i \leq X_{(r)}$ as

$$
\begin{aligned}
F_{X_i X_{(r)}}(x, z) &= F(x) Pr\{X_{(r)} \leq z | X_i \leq x\} \\
&= F(x) Pr\{\text{at least } r \text{ of the } X_i's \leq z | X_i \leq x\}.
\end{aligned}
$$

Since $x \leq z$, then given that $X_i < x$ we have that the second term in the right side of the above equation is simply the probability of at least $r - 1$ of the remaining $N - 1$ samples $X_i < z$; thus,

$$
F_{X_i X_{(r)}}(x, z) = F(x) \sum_{i=r-1}^{N-1} \binom{N-1}{i} F^i(z)(1 - F(z))^{N-1-i}.
\tag{3.35}
$$

For the case $X_i > X_{(r)}$, the following holds:

$$
\begin{aligned}
F_{X_i X_{(r)}}(x, z) &= Pr\{X_i \leq x, X_{(r)} \leq z\} \\
&= [F(x) - F(z)] Pr\{X_{(r)} \leq z | z \leq X_i < x\} \\
&\quad + F(z) Pr\{X_{(r)} \leq z | X_i \leq z\}.
\end{aligned}
$$

These probabilities can be shown to be

$$F_{X_i X_{(r)}}(x, z) = [F(x) - F(z)] \sum_{i=r}^{N-1} \binom{N-1}{i} F^i(z)(1 - F(z))^{N-1-i}$$

$$+ F(z) \sum_{i=r-1}^{N-1} \binom{N-1}{i} F^i(z)(1 - F(z))^{N-1-i}$$

for $z < x$.

The cross moments of X_i and $X_{(r)}$ can be found through the above equations, but an easier alternative method has been described in [154] as stated in the next property.

PROPERTY 3.3 *The cross moment of the rth order statistic and the nonordered sample X_i for an N i.i.d. sample set satisfies the relation*

$$E\{X_i X_{(r)}\} = \frac{1}{N} \sum_{s=1}^{N} E\{X_{(r)} X_{(s)}\}. \tag{3.36}$$

This property follows from the relation

$$\sum_{s=1}^{N} X_{(s)} = \sum_{s=1}^{N} X_s. \tag{3.37}$$

Substituting the above into the right hand side of (3.36) leads to

$$\sum_{s=1}^{N} E\{X_{(r)} X_{(s)}\} = E\left\{X_{(r)} \sum_{s=1}^{N} X_{(s)}\right\}$$

$$= E\left\{X_{(r)} \sum_{s=1}^{N} X_s\right\}$$

$$= \sum_{s=1}^{N} E\{X_{(r)} X_s\}.$$

Since all the input samples are i.i.d. then the property follows directly.

Problems

3.1 Let $X_1, \ldots, X_N$, be i.i.d. variates, X_i having a geometric density function

$$f(x) = q^x p \text{ with } q = 1 - p,$$

for $0 < p < 1$, and for $x \geq 0$. Show that $X_{(1)}$ is distributed geometrically.

3.2 For a random sample of size N from a continuous distribution whose density function is symmetrical about $x = \mu$.

(a) Show that $f_{(r)}(x)$ and $f_{(N-r+1)}(x)$ are mirror images of each other in $x = \mu$ as mirror. That is

$$f_{(r)}(\mu + x) = f_{(N-r+1)}(\mu - x).$$

(b) Generalize (a) to joint distributions of order statistics.

3.3   Let X_1, X_2, X_3 be independent and identically distributed observations taken from the density function $f(x) = 2x$ for $0 < x < 1$, and 0 elsewhere.

(a) Show that the median of the distribution is $\frac{\sqrt{2}}{2}$.

(b) What is the probability that the smallest sample in the set exceeds the median of the distribution.

3.4   Given the N marginal density functions $f_{(i)}(x)$, $1 \le i \le N$, of a set of i.i.d. variables, show that the average probability density function $\bar{f}(x)$ is identical to the parent density function $f(x)$. That is show

$$\bar{f}(x) = (1/N) \sum_{i=1}^{N} f_{(i)}(x) = f(x). \tag{3.38}$$

3.5   Let $X_1, X_2, \ldots, X_N$ be N i.i.d. samples with a Bernoulli parent density function such that $P_r\{X_i = 1\} = p$ and $P_r\{X_i = 0\} = 1 - p$ with $0 < p < 1$.

(a) Find $P_r\{X_{(i)} = 1\}$ and $P_r\{X_{(i)} = 0\}$.

(b) Derive the bivariate distribution function of $X_{(i)}$ and $X_{(j)}$.

(c) Find the moments $\mu_{(i)}$ and $\mu_{(i,j)}$.

3.6   Show that in odd-sized random samples from i.i.d continuous distributions, the expected value of the sample median equals the median of the parent distribution.

3.7   Show that the distribution function of the midrange $m = \frac{1}{2}(X_{(1)} + X_{(N)})$ of N i.i.d. continuous variates is

$$F(m) = N \int_{-\infty}^{m} [F_X(2m - x) - F_X(x)]^{N-1} f_X(x) dx.$$

3.8   For the geometric distribution with

$$Pr(X_i = x) = p\, q^x \quad \text{for } x \ge 0 \tag{3.39}$$

where $q = 1 - p$, show that for $1 \le i \le N$

$$\mu_{(i):N} = \sum_{j=N-i+1}^{N} (-1)^{j-N+i-1} \binom{j-1}{N-i} \binom{N}{j} \frac{1}{(1-q^j)} \tag{3.40}$$

and

$$\mu_{(i):N}^{(2)} = \sum_{j=N-i+1}^{N} (-1)^{j-N+i-1} \binom{j-1}{N-1} \binom{N}{j} \frac{(1+q)^j}{(1-q^j)}. \tag{3.41}$$

3.9   Consider a set of 3 samples $\{X_1, X_2, X_3\}$. While the sample X_3 is independent and uniformly distributed in the interval $[0,1]$, the other two samples are mutually dependent with a joint density function $f(X_1, X_2) = \frac{2}{3}\delta(X_1 - 1, X_2 - 1) + \frac{1}{3}\delta(X_1 - 1, X_2)$, where $\delta(\cdot, \cdot)$ is a 2-Dimensional Dirac delta function.

(a) Find the distribution function of $X_{(3)}$.

(b) Find the distribution function of the median.

(c) Is the distribution of $X_{(1)}$ symmetric to that of $X_{(3)}$, explain.

3.10   Prove the relation in Property 3.2.

$$\mu_{(i):N}^{(k)} = \sum_{j=i}^{N} (-1)^{j-i} \binom{N}{j} \binom{j-1}{i-1} \mu_{(j):j}^{(k)} \tag{3.42}$$

<u>Hint:</u> From the definition of $\mu_{(i):N}^{(k)}$ we get

$$\begin{aligned}
\mu_{(i):N}^{(k)} &= \frac{N!}{(i-1)!(N-i)!} \int_0^1 [F^{-1}(u)]^k u^{i-1}(1-u)^m (1-u)^{N-i-m} du \\
&= \frac{N!}{(i-1)!(N-i)!} \sum_{r=0}^{m} (-1)^r \binom{m}{r} \int_0^1 [F^{-1}(u)]^k u^{i+r-1}(1-u)^{N-i-m} du,
\end{aligned}$$

which can be simplified to

$$(N-i)\mu_{(i):N}^{(k)} = \sum_{r=0}^{m} (-i)^{(r)}(n)^{(m-r)} \binom{m}{r} \mu_{(i+r):N-m+r}^{(k)} \tag{3.43}$$

where $(n)^m$ denotes the terms $n(n-1)\ldots(n-m+1)$.

3.11   Consider a sequence $X_1, X_2, \ldots$ of independent and identically distributed random variables with a continuous parent distribution $F(x)$. A sample X_k is called *outstanding* if $X_k > max(X_1, X_2, \ldots, X_{k-1})$ (by definition X_1 is outstanding). Prove that $Pr\{X_k > max(X_1, X_2, \ldots, X_{k-1})\} = \frac{1}{k}$.

4

Statistical Foundations of Filtering

Filtering and parameter estimation are intimately related due to the fact that information is carried, or can be inserted, into one or more parameters of a signal at hand. In AM and FM signals, for example, the information resides in the envelope and instantaneous frequency of the modulated signals respectively. In general, information can be carried in a number of signal parameters including but not limited to the mean, variance, phase, and of course frequency. The problem then is to determine the value of the information parameter from a set of observations in some optimal fashion. If one could directly observe the value of the parameter, there would be no difficulty. In practice, however, the observation contains noise, and in this case, a statistical procedure to estimate the value of the parameter is needed.

Consider a simple example to illustrate the formulation and concepts behind parameter estimation. Suppose that a constant signal β is transmitted through a channel that adds Gaussian noise Z_i. For the sake of accuracy, several independent observations X_i are measured, from which the value of β can be inferred. A suitable model for this problem is of the form

$$X_i = \beta + Z_i \qquad i = 1, 2, \ldots, N.$$

Thus, given the sample set $X_1, X_2, \ldots, X_N$, the goal is to derive a rule for processing the observations samples that will yield a good estimate of β. It should be emphasized that the parameter β, in this formulation, is unknown but fixed — there is no randomness associated with the parameter itself. Moreover, since the samples in this example deviate about the parameter β, the estimate seeks to determine the value of the *location* parameter. Estimates of this kind are known as *location estimates*. As it will become clear later on, the location estimation problem is *key* in the formulation of the optimal filtering problem.

Several methods of estimating β are possible for the example at hand. The sample mean $\bar{X}$, given by

$$\bar{\beta}_N = \bar{X} = \frac{1}{N} \sum_{i=1}^{N} X_i$$

is a natural choice. An alternative would be the sample median $\tilde{\beta}_N = \tilde{X}$ in which we order the observation samples and then select the one in the middle. We might also use a *trimmed mean* where the largest and smallest samples are first discarded and the remaining $N - 2$ samples are averaged. All of these choices are valid estimates of location. Which of these estimators, if any, is best will depend on the criterion which is selected. In this Chapter, several types of location estimates are discussed. After a short introduction to the properties of estimators, the method of *maximum-likelihood estimation* is presented with criteria for the "goodness" of an estimate. The class of *M-estimators* is discussed next, generalizing the concepts behind maximum-likelihood estimation by introducing the concept of *robust* estimation. The application of location estimators to the smoothing of signals is introduced at the end of the Chapter.

4.1 PROPERTIES OF ESTIMATORS

For any application at hand, as in our example, there can be a number of possible estimators from which one can choose. Of course, one estimator may be adequate for some applications but not for others. Describing how good an estimator is, and under which circumstances, is important. Since estimators are in essence procedures that use observations that are random variables, then the estimators themselves are random variables. The estimates, as for any random variable, can be described by a probability density function. The probability density function of the estimate $\hat{\beta}$ is denoted as $f_{\hat{\beta}}(y|\beta)$, where y is a possible value for the estimate. Since this density function can change for different estimation rules, the densities alone provide a cumbersome description. Instead, we can recourse to the statistical properties of the estimates as a mean to quantify their characteristics. The statistical properties can, in turn, be used for purposes of comparison among various estimation alternatives.

Unbiased Estimators   A typical probability density $f_{\hat{\beta}}(y|\beta)$ associated with an estimate is given in Figure 4.1, where the actual value of the parameter β is shown. It would be desirable for the estimate $\hat{\beta}$ to be relatively close to the actual value of β. It follows that a good estimator will have its density function as clustered together as possible about β. If the density is not clustered or if it is clustered about some other point, it is a less good estimator. Since the mean and variance of the density are good measures of where and how clustered the density function is, a good estimator is one for which the mean of $\hat{\beta}$ is close to β and for which the variance of $\hat{\beta}$ is small.

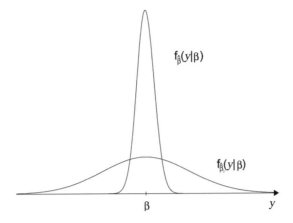

Figure 4.1 Probability density function associated with an unbiased location estimator.

In some cases, it is possible to design estimators for which the mean of $\hat{\beta}$ is always equal to the true value of β. When this desirable property is true for all values of β, the estimator is referred to as *unbiased*. Thus, the N-sample estimate of β, denoted as $\hat{\beta}_N$, is said to be *unbiased* if

$$E\{\hat{\beta}_N\} = \beta.$$

In addition, the variance of the estimate determines the precision of the estimate. If an unbiased estimate has low variance, then it will provide a more reliable estimate than other unbiased estimates with inherently larger variances. The sample mean in the previous example, is an unbiased estimate since $E\{\hat{\beta}_N\} = \beta$, with a variance that follows

$$\begin{aligned} \text{var}\left\{\hat{\beta}_N\right\} &= \text{var}\left\{\frac{1}{N}\sum_{i=1}^{N}X_i\right\} \\ &= \frac{\sigma^2}{N}, \end{aligned} \tag{4.1}$$

where σ^2 is the channel noise variance. Clearly, the precision of the estimate improves as the number of observations increases.

Efficient Estimators The mean and variance of an estimate are indicators of quality. If we restrict our attention to only those estimators that are unbiased, we are in effect reducing the measure of quality to one dimension where we can define the best estimator in this class as the one that attains the minimum variance. Although at first, this may seem partially useful since we would have to search among all unbiased estimators to determine which has the lowest variance, it turns out that a lower bound on the variance of *any* unbiased estimator exists. Thus, if a given estimator is found to have a variance equal to that of the bound, the *best* estimator has been identified.

The bound is credited to Cramér and Rao [56]. Let $f(\mathbf{X}; \beta)$ be the density function of the observations $\mathbf{X}$ given the value of β. For a scalar real parameter, if $\hat{\beta}$ is an unbiased estimate of β, its variance is bounded by

$$\text{var}\{\hat{\beta}\} \geq \left(E\left\{ \left[\frac{\partial}{\partial \beta} \ln f(\mathbf{X}; \beta) \right]^2 \right\} \right)^{-1} \tag{4.2}$$

provided that the partial derivative of the log likelihood function exists and is absolutely integrable. A second form of the Cramér-Rao bound can be written as

$$\text{var}\{\hat{\beta}\} \geq \left(-E\left\{ \frac{\partial^2}{\partial \beta^2} \ln f(\mathbf{X}; \beta) \right\} \right)^{-1}, \tag{4.3}$$

being valid if the second partial derivative of the log likelihood exists and is absolutely integrable. Proofs of these bounds can be found in [32, 126]. Although there is no guarantee that an unbiased estimate exists whose variance satisfies the Cramér-Rao bound with equality, if one is found, we are certain that it is the best estimator in the sense of minimum variance and it is referred to as an *efficient estimator*.

Efficiency can also be used as a relative measure between two estimators. An estimate is said to be efficient with respect to another estimate if it has a lower variance. If this *relative efficiency* is coupled with the order of an estimate the following concept emerges: If $\hat{\beta}_N$ is unbiased and efficient with respect to $\hat{\beta}_{N-1}$ for all N, then $\hat{\beta}_N$ is said to be *consistent*.

4.2   MAXIMUM LIKELIHOOD ESTIMATION

Having a set of observation samples, a number of approaches can be taken to derive an estimate. Among these, the method of *maximum likelihood (ML)* is the most popular approach since it allows the construction of estimators even for uncommonly challenging problems. ML estimation is based on a relatively simple concept: different distributions generate different data samples and any given data sample is more likely to have come from some population than from others [99]. Conceptually, a set of observations, $X_1, X_2, \ldots, X_N$, are postulated to be values taken on by random variables assumed to follow the joint distribution function $f(X_1, X_2, \ldots, X_N; \beta)$, where β is a parameter of the distributions. The parameter β is assumed unknown but fixed, and in parameter estimation one tries to specify the best procedure to estimate the value of the parameter β from a given set of measured data.

In the method of maximum likelihood the best estimate of β is the value $\hat{\beta}_{ML}$ for which the function $f(X_1, X_2, \ldots, X_N; \beta)$ is at its maximum

$$\hat{\beta}_{ML} = \arg\max_{\beta} f(X_1, X_2, \ldots, X_N; \beta) \tag{4.4}$$

where the parameter β is variable while the observation samples $X_1, X_2, \ldots, X_N$ are fixed. The density function when viewed as a function of β, for fixed values of the observations, is known as the *likelihood function*.

The philosophy of maximum likelihood estimation is elegant and simple. Maximum likelihood estimates are also very powerful due to the notable property they enjoy that relates them to the Cramér-Rao bound. It can be shown that if an efficient estimate exists, the maximum likelihood estimate is efficient [32]. Thanks to this property, maximum likelihood estimation has evolved into one of the most popular methods of estimation.

In maximum likelihood location estimates, the parameter of interest is the *location*. Assuming independence in this model, each of the samples in the set follows some distribution

$$P(X_i \leq x) = F(x - \beta), \qquad (4.5)$$

where $F(\cdot)$ corresponds to a distribution that is symmetric about 0.

Location Estimation in Gaussian Noise Assume that the observation samples $X_1, X_2, \ldots, X_N$, are i.i.d. Gaussian with a constant but unknown mean β. The maximum-likelihood estimate of location is the value $\hat{\beta}$ which maximizes the likelihood function

$$
\begin{aligned}
f(X_1, X_2, \ldots, X_N; \beta) &= \prod_{i=1}^{N} f(X_i - \beta) \\
&= \prod_{i=1}^{N} \frac{1}{\sqrt{2\pi}\sigma}\, e^{-(X_i - \beta)^2/2\sigma^2} \qquad (4.6)\\
&= \left(\frac{1}{2\pi\sigma^2}\right)^{N/2} e^{-\sum_{i=1}^{N}(X_i - \beta)^2/2\sigma^2}.
\end{aligned}
$$

The likelihood function in (4.6) can be maximized by minimizing the argument in the exponential. Thus, the maximum-likelihood estimate of location is the value $\hat{\beta}$ that minimizes the least squares sum

$$\hat{\beta}_{ML} = \arg\min_{\beta} \sum_{i=1}^{N}(X_i - \beta)^2. \qquad (4.7)$$

The value that minimizes the sum, found through differentiation, results in the sample mean

$$\hat{\beta}_{ML} = \frac{1}{N}\sum_{i=1}^{N} X_i. \qquad (4.8)$$

Note that the sample mean is unbiased in the assumed model since $E\{\hat{\beta}_{ML}\} = \frac{1}{N}\sum_{i=1}^{N} E\{X_i\} = \beta$. Furthermore, as a maximum-likelihood estimate, it is efficient having its variance, in (4.1), reach the Cramér-Rao bound.

Location Estimation in Generalized Gaussian Noise Now suppose that the observed data includes samples that clearly deviate from the central data cluster. The large deviations contradict a Gaussian model. The alternative is to model the deviations with a more appropriate distribution that is more flexible in capturing the characteristics of the data. One approach is to adopt the generalized Gaussian distribution. The function used to construct the maximum-likelihood estimate of location in this case is

$$f(X_1, X_2, \ldots, X_N; \beta) = \prod_{i=1}^{N} f_k(X_i - \beta) \tag{4.9}$$

$$= \prod_{i=1}^{N} C\, e^{-(\alpha|X_i - \beta|)^k} \tag{4.10}$$

$$= C^N e^{-\alpha^k \sum_{i=1}^{N}|X_i - \beta|^k}, \tag{4.11}$$

where C and α are normalizing constants and k is the fixed parameter that models the dispersion of the data. Maximizing the likelihood function is equivalent to minimizing the argument of the exponential, leading to the following estimate of location

$$\tilde{\beta}_{ML} = \arg\min_{\beta} \sum_{i=1}^{N}|X_i - \beta|^k. \tag{4.12}$$

Some intuition can be gained by plotting the cost function in (4.12) for various values of k. Figure 4.2 depicts the different cost function characteristics obtained for $k = 2$, 1, and 0.5.

When the dispersion parameter is given the value 2, the model reduces to the Gaussian assumption, the cost function is quadratic, and the estimator is, as expected, equal to the sample mean. For $k < 1$, it can be shown that the cost function exhibits several local minima. Furthermore, the estimate is of *selection* type as its value will be that of one of the samples $X_1, X_2, \ldots, X_N$. These characteristics of the cost function are shown in Figure 4.2.

When the dispersion parameter is given the value 1, the model is Laplacian, the cost function is piecewise linear and continuous, and the optimal estimator minimizes the sum of absolute deviations

$$\tilde{\beta}_{ML} = \arg\min_{\beta} \sum_{i=1}^{N}|X_i - \beta|. \tag{4.13}$$

Although not immediately seen, the solution to the above is the sample median as it is shown next.

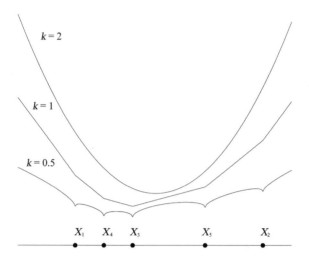

Figure 4.2 Cost functions for the observation samples $X_1 = -3, X_2 = 10, X_3 = 1, X_4 = 1, X_5 = 6$ for $k = 0.5, \ 1,$ and 2.

Define the cost function being minimized in (4.13) as $L_1(\beta)$. For values of β in the interval $-\infty < \beta \leq X_{(1)}$, $L_1(\beta)$ is simplified to

$$
\begin{aligned}
L_1(\beta) &= \sum_{i=1}^{N} \left(X_{(i)} - \beta \right) \\
&= \sum_{i=1}^{N} X_{(i)} - N\beta.
\end{aligned} \tag{4.14}
$$

This, as a direct consequence that in this interval, $X_{(1)} \geq \beta$. For values of β in the range $X_{(j)} < \beta \leq X_{(j+1)}$, $L_1(\beta)$ can be written as

$$
\begin{aligned}
L_1(\beta) &= \sum_{i=1}^{j} \left(\beta - X_{(i)} \right) + \sum_{i=j+1}^{N} \left(X_{(i)} - \beta \right) \\
&= \left(\sum_{i=j+1}^{N} X_{(i)} - \sum_{i=1}^{j} X_{(i)} \right) - (N - 2j)\beta,
\end{aligned} \tag{4.15}
$$

for $j = 1, 2, \ldots, N - 1$. Similarly, for $X_{(N)} < \beta < \infty$,

$$
L_1(\beta) = -\sum_{i=1}^{N} X_{(i)} + N\beta. \tag{4.16}
$$

Letting $X_{(0)} = -\infty$ and $X_{(N+1)} = \infty$, and defining $\sum_{i=m}^{n} X_{(i)} = 0$ if $m > n$, we can combine (4.14)–(4.16) into the following compactly written cost function

$$L_1(\beta) = \left(\sum_{i=j+1}^{N} X_{(i)} - \sum_{i=1}^{j} X_{(i)} \right) - (N - 2j)\,\beta, \qquad j = 0, 1, \dots, N \quad (4.17)$$

for $\beta \in (X_{(j)}, X_{(j+1)}]$. When expressed as in (4.17), $L_1(\beta)$ is clearly piecewise linear and continuous. It starts with slope $-N$ for $-\infty < \beta \leq X_{(1)}$, and as each $X_{(j)}$ is crossed, the slope is increased by 2. At the extreme right the slope ends at N for $X_{(N)} < \beta < \infty$.

For N odd, this implies that there is an integer m, such that the slopes over the intervals $(X_{(m-1)}, X_{(m)}]$ and $(X_{(m)}, X_{(m+1)}]$, are negative and positive, respectively. From (4.17), these two conditions are satisfied if both

$$m < \frac{N}{2} \quad \text{and} \quad m > \frac{N}{2} - 1$$

hold. Both constraints are met when $m = \frac{N+1}{2}$

For N even, (4.17) implies that there is an integer m, such that the slope over the interval $(X_{(m)}, X_{(m+1)}]$ is zero. This condition is satisfied in (4.17) if

$$-(N - 2m) = 0,$$

which is possible for $m = N/2$. Thus, the maximum-likelihood estimate of location under the Laplacian model is the sample median

$$
\begin{aligned}
\hat{\beta}_{ML} &= \arg\min_{\beta} \sum_{i=1}^{N} |X_i - \beta| \\
&= \begin{cases} X_{\left(\frac{N+1}{2}\right)} & N \text{ odd} \\ \left(X_{\left(\frac{N}{2}\right)}, X_{\left(\frac{N}{2}+1\right)} \right] & N \text{ even} \end{cases} \\
&= \text{MEDIAN}(X_1, X_2, \dots, X_N).
\end{aligned}
\qquad (4.18)
$$

In the case of N being even the output of the median can be any point in the interval shown above, the convention is to take the mean of the extremes $\hat{\beta}_{ML} = \dfrac{X_{\left(\frac{N}{2}\right)} + X_{\left(\frac{N}{2}+1\right)}}{2}$.

Location Estimation in Stable Noise   The formulation of maximum likelihood estimation requires the knowledge of the model's closed-form density function. Among the class of symmetric stable densities, only the Gaussian ($\alpha = 2$) and Cauchy ($\alpha = 1$) distributions enjoy closed-form expressions. Thus, to formulate the non-Gaussian maximum likelihood estimation problem in a stable distribution framework, it is logical to start with the only non-Gaussian distribution for which we have a closed form expression, namely the Cauchy distribution. Although at first, this approach may seem too narrow to be effective over the broad class of stable processes, maximum-likelihood estimates under the Cauchy model can be made tunable, acquiring remarkably efficiency over the entire spectrum of stable distributions.

Given a set of i.i.d. samples $X_1, X_2, \ldots, X_N$ obeying the Cauchy distribution with scaling factor K,

$$f(x - \beta) = \frac{K}{\pi} \frac{1}{K^2 + (x - \beta)^2}, \tag{4.19}$$

the location parameter β is to be estimated from the data samples as the value $\hat{\beta}_K$, which maximizes the likelihood function

$$\hat{\beta}_K = \arg\max_{\beta} \prod_{i=1}^{N} f(X_i - \beta) = \arg\max_{\beta} \left(\frac{K}{\pi}\right)^N \prod_{i=1}^{N} \frac{1}{K^2 + (X_i - \beta)^2}. \tag{4.20}$$

This is equivalent to minimizing

$$G_K(\beta) = \prod_{i=1}^{N} [K^2 + (X_i - \beta)^2]. \tag{4.21}$$

Thus given $K > 0$, the ML location estimate is known as the sample *myriad* and is given by [82]

$$\begin{aligned} \hat{\beta}_K &= \arg\min_{\beta} \prod_{i=1}^{N} \left(K^2 + (X_i - \beta)^2\right) \tag{4.22} \\ &= \text{MYRIAD}\{K; X_1, X_2, \ldots, X_N\}. \end{aligned}$$

Note that, unlike the sample mean or median, the definition of the sample myriad involves the free parameter K. For reasons that will become apparent shortly, we will refer to K as the *linearity parameter* of the myriad. The behavior of the myriad estimator is markedly dependent on the value of its linearity parameter K. Some intuition can be gained by plotting the cost function in (4.23) for various values of K. Figure 4.3 depicts the different cost function characteristics obtained for $K = 20, 2, 0.2$ for a sample set of size 5.

Although the definition of the sample myriad in (4.23) is straightforward, it is not intuitive at first. The following interpretations provide additional insight.

LEAST LOGARITHMIC DEVIATION

The sample myriad minimizes $G_K(\beta)$ in (4.21), which consists of a set of products. Since the logarithm is a strictly monotonic function, the sample myriad will also minimize the expression $\log G_K(\beta)$. The sample myriad can thus be equivalently written as

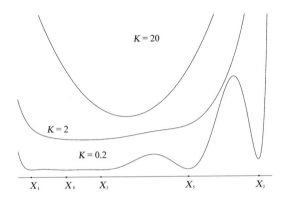

Figure 4.3   Myriad cost functions for the observation samples $X_1 = -3, X_2 = 10, X_3 = 1, X_4 - 1, X_5 = 6$ for $K = 20, 2, 0.2$.

$$\text{MYRIAD}\{K; X_1, X_2, \ldots, X_N\} = \arg\min_{\beta} \sum_{i=1}^{N} \log\left[K^2 + (X_i - \beta)^2\right]. \quad (4.23)$$

Upon observation of the above, if an observation in the set of input samples has a large magnitude such that $|X_i - \beta| \gg K$, the cost associated with this sample is approximately $\log(X_i - \beta)^2$ —the log of the square deviation. Thus, much as the sample mean and sample median respectively minimize the sum of square and absolute deviations, the sample myriad (approximately) minimizes the sum of logarithmic square deviations, referred to as the LLS criterion, in analogy to the Least Squares (LS) and Least Absolute Deviation (LAD) criteria.

Figure 4.4 illustrates the cost incurred by each sample as it deviates from the location parameter β. The cost of the sample mean (LS) is quadratic, severely penalizing large deviations. The sample median (LAD) assigns a cost that is linearly proportional to the deviation. The family of cost functions for the sample myriad assigns a penalty proportional to the logarithm of the deviation, which leads to a much milder penalization of large deviations than that imposed by the LAD and LS cost functions. The myriad cost function structure, thus, rests importance on clearly inappropriate samples.

GEOMETRICAL INTERPRETATION

A second interpretation of the sample myriad that adds additional insight lies in its geometrical properties. First, the observations samples $X_1, X_2, \ldots, X_N$ are placed along the real line. Next, a vertical bar that runs horizontally through the real line is added as depicted in Figure 4.5. The length of the vertical bar is equal to the linearity parameter K. In this arrangement, each of the terms

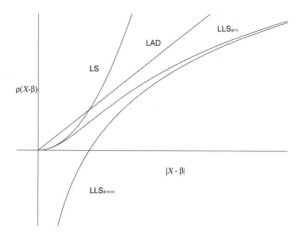

Figure 4.4   Cost functions of the mean (LS), the median (LAD), and the myriad (LLS)

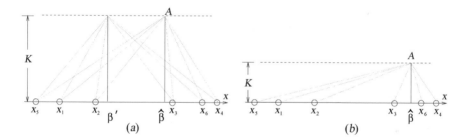

Figure 4.5   (a) The sample myriad, $\hat{\beta}$, minimizes the product of distances from point A to all samples. Any other value, such as $x = \beta'$, produces a higher product of distances; (b) the myriad as K is reduced.

$$\left(K^2 + (X_i - \beta)^2\right) \tag{4.24}$$

in (4.23), represents the distance from point A, at the end of the vertical bar, to the sample point X_i. The sample myriad, $\hat{\beta}_K$, indicates the position of the bar for which the product of distances from point A to the samples $X_1, X_2, \ldots, X_N$ is minimum. Any other value, such as $x = \beta'$, produces a higher product of distances.

If the value of K is reduced as shown in Figure 4.5b, the sample myriad will favor samples that are clustered together. The sample myriad has a mode-like behavior for small values of K. The term "myriad" was coined as a result of this characteristic of the estimator.

4.3   ROBUST ESTIMATION

The maximum-likelihood estimates derived so far have assumed that the form of the distribution is known. In practice, we can seldom be certain of such distributional assumptions and two types of questions arise:

(1) How sensitive are optimal estimators to the precise nature of the assumed probability model?

(2) Is it possible to construct *robust* estimators that perform well under deviations from the assumed model?

Sensitivity of Estimators   To answer the first question, consider an observed data set $Z_1, Z_2, \ldots, Z_N$, and let us consider the various location estimators previously derived, namely, the mean, median, and myriad. In addition, we also consider two simple M-estimators, namely the *trimmed-mean* defined as

$$T_N(\alpha) = \frac{1}{N - 2\alpha} \sum_{i=\alpha+1}^{N-\alpha} Z_{(i):N} \tag{4.25}$$

for $\alpha = 0, 1, \ldots, \lfloor N/2 \rfloor$, and the *Windsorized mean* defined as:

$$W_N(r) = \frac{1}{N} \left[\sum_{i=r+2}^{N-r-1} Z_{(i):N} + (r+1) \left[Z_{(r+1):N} + Z_{(N-r):N} \right] \right] \tag{4.26}$$

The median, is a special case of trimmed mean where $\alpha = \lfloor N/2 \rfloor$.

The effects of data contamination on these estimators is then tested. In the first set of experiments, a sample set of size 10 including one outlier is considered. The nine i.i.d. samples are distributed as $N(\mu, 1)$ and the outlier is distributed as $N(\mu + \lambda, 1)$. Table 4.1, adapted from David [58], depicts the bias of the estimation where eight different values of λ were selected.

This table clearly indicates that the mean is highly affected by the outlier. The trimming improves the robustness of the estimate. Clearly the median performs best, although it is still biased.

The expected value of the biases shown in Table 4.1 are not sufficient to compare the various estimates. The variances of the different estimators of μ are needed. These have also been tabulated in [58] and are shown on Table 4.2.

This table shows that the Windsorized mean performs better than the trimmed mean when λ is small. It also shows that, although the bias of the median is smaller, the variance is larger than the trimmed and Windsorized means. The mean is also shown to perform poorly in the MSE, except when there is no contamination.

Another useful test is to consider the contamination sample having the same mean as the other $N - 1$ samples, but in this case the variance of the outlier is much larger. Hence, Table 4.3 tabulates the variance of the various estimates of μ for $N = 10$.

Table 4.1 Bias of estimators of μ for $N = 10$ when a single observation is from $N(\mu+\lambda, 1)$ and the others from $N(\mu, 1)$.

				λ				
Estimator	0.0	0.5	1.0	1.5	2.0	3.0	4.0	∞
$\bar{X}_{10}$	0.0	0.05000	0.10000	0.15000	0.20000	0.30000	0.40000	∞
$T_{10}(1)$	0.0	0.04912	0.09325	0.12870	0.15400	0.17871	0.18470	0.18563
$T_{10}(2)$	0.0	0.04869	0.09023	0.12041	0.13904	0.15311	0.15521	0.15538
Med_{10}	0.0	0.04932	0.08768	0.11381	0.12795	0.13642	0.13723	0.13726
$W_{10}(1)$	0.0	0.04938	0.09506	0.13368	0.16298	0.19407	0.20239	0.20377
$W_{10}(2)$	0.0	0.04889	0.09156	0.12389	0.14497	0.16217	0.16504	0.16530

Table 4.2 Mean squared error of various estimators of μ for $N = 10$, when a single observation is from $N(\mu + \lambda, 1)$ and the others from $N(\mu, 1)$.

				λ				
Estimator	0.0	0.5	1.0	1.5	2.0	3.0	4.0	∞
$\bar{X}_{10}$	0.10000	0.10250	0.11000	0.12250	0.14000	0.19000	0.26000	∞
$T_{10}(1)$	0.10534	0.10791	0.11471	0.12387	0.13285	0.14475	0.14865	0.14942
$T_{10}(2)$	0.11331	0.11603	0.12297	0.13132	0.13848	0.14580	0.14730	0.14745
Med_{10}	0.13833	0.14161	0.14964	0.15852	0.16524	0.17072	0.17146	0.17150
$W_{10}(1)$	0.10437	0.10693	0.11403	0.12405	0.13469	0.15039	0.15627	0.15755
$W_{10}(2)$	0.11133	0.11402	0.12106	0.12995	0.13805	0.14713	0.14926	0.14950

Table 4.3 shows that the mean is a better estimator than the median as long as the variance of the outlier is not large. The trimmed mean, however, outperforms the median regardless of the variance of the outlier. The Windsorized mean performs comparably to the trimmed mean.

These tables illustrate that by trimming the observation sample set, we can effectively increase the robustness of estimation.

M-Estimation *M-estimation* aims at answering the second question raised at the beginning of this section: Is it possible to construct estimates of location which perform adequately under deviations from distributional assumptions? According to the theory of M-estimation this is not only possible, but a well defined set of design guidelines can be followed. A brief summary of M-estimation is provided below. The interested reader can further explore the theory and applications of M-estimation in [91, 105].

Table 4.3   Variance of various estimators of μ for $N = 10$, where a single observation is from $N(\mu, \sigma^2)$ and the others from $N(\mu, 1)$.

			σ			
Estimator	0.5	1.0	2.0	3.0	4.0	∞
X_{10}	0.09250	0.10000	0.13000	0.18000	0.25000	∞
$T_{10}(1)$	0.09491	0.10534	0.12133	0.12955	0.13417	0.14942
$T_{10}(2)$	0.09953	0.11331	0.12773	0.13389	0.13717	0.14745
Med_{10}	0.11728	0.13833	0.15373	0.15953	0.16249	0.17150
$W_{10}(1)$	0.09571	0.10437	0.12215	0.13221	0.13801	0.15754
$W_{10}(2)$	0.09972	0.11133	0.12664	0.13365	0.13745	0.14950

Given a set of samples $X_1, X_2, \ldots, X_N$, an M-estimator of location is defined as the parameter $\hat{\beta}$ that minimizes a sum of the form

$$\sum_{i=1}^{N} \rho(X_i - \beta) \tag{4.27}$$

where ρ is referred to as a cost function. The behavior of the M-estimate is determined by the shape of ρ. When $\rho(x) = x^2$, for example, the associated M-estimator minimizes the sum of square deviations, which corresponds to the sample mean. For $\rho(x) = |x|$, on the other hand, the M-estimator is equivalent to the sample median. In general, if $\rho(x) = -\log f(x)$, where f is a density function, the M-estimate $\hat{\beta}$ corresponds to the maximum likelihood estimator associated with f. Accordingly, the cost function associated with the sample myriad is proportional to

$$\rho(X) = \log[k^2 + X^2]. \tag{4.28}$$

The flexibility associated with shaping $\rho(x)$ has been the key for the success of M-estimates.

Some insight into the operation of M-estimates is gained through the definition of the *influence function*. The influence function roughly measures the effect of contaminated samples on the estimates and is defined as

$$\psi(X_i - \beta) = \frac{\partial}{\partial \beta} \rho(X_i - \beta), \tag{4.29}$$

provided the derivative exists. Denoting the sample deviation $X_i - \beta$ as U_i, the influence functions for the sample mean and median are proportional to $\psi_{MEAN}(U_i) = (U_i)$ and $\psi_{MEDIAN}(U_i) = \text{sign}(U_i)$, respectively. Since the influence function of the mean is unbounded, a gross error in the observations can lead to severe distortion in the estimate. On the other hand, a similar gross error has a limited effect on the median estimate. The influence function of the sample myriad is

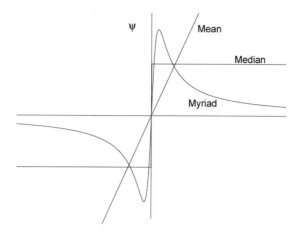

Figure 4.6   Influence functions of the mean, median and myriad

$$\psi_{MYRIAD}(U_i) = \frac{U_i}{K^2 + U_i^2}. \qquad (4.30)$$

As shown in Figure 4.6, the myriad's influence function is *re-descending* reaching its maxima (minima) at $|U_i| = K$. Thus, the further away an observation sample is from the value K, the less it is considered in the estimate. Intuitively, the myriad must be more resistant to outliers than the median, and the mean is linearly sensitive to these.

Problems

4.1   Given N independent and identically distributed samples obeying the Poisson distribution:

$$f(x) = \frac{e^{-\lambda}\lambda^x}{x!} \qquad (4.31)$$

where x can take on positive integer values, and where λ is a positive parameter to be estimated:

(a) Find the mean and variance of the random variables X_i.

(b) Derive the maximum-likelihood estimate (MLE) of λ based on a set of N observations.

(c) Is the ML estimate unbiased?

(d) Find the Cramér-Rao bound for the variance of an unbiased estimate.

(e) Find the variance of the ML estimate. Is the ML estimate efficient?

4.2   Consider N independent and identically distributed samples from a Gaussian distribution with zero mean and variance σ^2. Find the maximum likelihood estimate of σ^2 (unknown deterministic parameter). Is the estimate unbiased? Is the estimate consistent? What can you say about the ML estimate in relation to the Cramer-Rao bound.

4.3   Let X be a uniform random variable on $[\theta, \theta + 1]$, where the real-valued parameter θ is constant but unknown, and let $T(X) = \lfloor X \rfloor =$ greatest integer less than or equal to X. Is $T(X)$ an unbiased estimate of θ. Hint: consider two cases: θ is an integer and θ is not an integer.

4.4   A random variable X has the uniform density

$$f(x) = 1/a \quad \text{for } 0 \le x \le a \tag{4.32}$$

and zero elsewhere.

(a) For independent samples of the above random variable, determine the likelihood function $f(X_1, X_2, \ldots, X_N : a)$ for $N = 1$ and $N = 2$ and sketch it. Find the maximum-likelihood estimate of the parameter a for these two cases. Find the ML estimate of the parameter a for an arbitrary number of observations N.

(b) Are the ML estimates in (a) unbiased.

(c) Is the estimate unbiased as $N \to \infty$?

4.5   Let the zero-mean random variables X and Y obey the Gaussian distribution,

$$f(x, y) = \frac{1}{2\pi\sigma_1\sigma_2\sqrt{1-\rho^2}} \exp\left(-\frac{1}{2(1-\rho^2)}\left[\frac{x^2}{\sigma_1^2} - 2\rho\frac{xy}{\sigma_1\sigma_2} + \frac{y^2}{\sigma_2^2}\right]\right) \tag{4.33}$$

where $\rho = \frac{E[XY]}{\sigma_1\sigma_2}$ is the correlation coefficient and where $E[XY]$ is the correlation parameter. Given a set of observation pairs $(X_1, Y_1), (X_2, Y_2), \ldots, (X_n, Y_n)$, drawn from the joint random variables X and Y. Find the maximum likelihood estimate of the correlation parameter $E[XY]$ or of the correlation coefficient ρ.

4.6   Consider a set of N independent and identically distributed observations X_i obeying the Rayleigh density function

$$f(x) = \frac{x}{\sigma^2}e^{-x^2/2\sigma^2}. \tag{4.34}$$

(a) Find the mean and variance of the X_i variables. Note

$$\int x^m e^{ax} dx = \frac{x^m e^{ax}}{a} - \frac{m}{a}\int x^{m-1}e^{ax} dx. \tag{4.35}$$

(b) If we assume that the parameter σ^2 is unknown but constant, derive the maximum-likelihood estimate of σ^2 obtained from the N observation samples. Is the estimate unbiased?

4.7   Find the maximum-likelihood estimate of θ (unknown constant parameter) from a single observation of the variable X where

$$X = \ln \theta + N \tag{4.36}$$

where N is a noise term whose density function is unimodal with $f_N(0) > f_N(\alpha)$ for all $\alpha \neq 0$.

4.8   Consider the data set

$$X(n) = AS(n) + W(n), \quad \text{for } n = 0, 1, \cdots, N-1, \tag{4.37}$$

where $S(n)$ is known, $W(n)$ is white Gaussian noise with known variance σ^2, and A is an unknown constant parameter.

(a) Find the maximum-likelihood estimate of A.

(b) Is the MLE unbiased?

(c) Find the variance of the MLE.

4.9   Consider N i.i.d. observations $\mathbf{X} = \{X_1, \ldots, X_N\}$ drawn from a parent distribution $F(x) = P_r(X \leq x)$. Let $\hat{F}(\mathbf{X})$ be the estimate of $F(X)$, where

$$\hat{F}(\mathbf{X}) = \frac{\text{number of } X_i's \leq x}{N} \tag{4.38}$$

$$\hat{F}(\mathbf{X}) = \frac{\sum_{i=1}^{N} U(x - X_i)}{N} \tag{4.39}$$

where $U(x) = 1$ if $x > 0$, and zero otherwise.

(a) Is this estimate unbiased.

(b) Prove that this estimate is the maximum-likelihood estimate. That is, let $z = \sum_{i=1}^{N} U(x - X_i)$, $\theta = F(x)$ and find $P(z/\theta)$.

Part II

Signal Processing with Order Statistics

5

Median and Weighted Median Smoothers

5.1 RUNNING MEDIAN SMOOTHERS

The running median was first suggested as a nonlinear smoother for time-series data by Tukey in 1974 [189], and it was largely popularized in signal processing by Gallagher and Wise's article in 1981 [78]. To define the running median smoother, let $\{X(\cdot)\}$ be a discrete time sequence. The running median passes a window over the sequence $\{X(\cdot)\}$ that selects, at each instant n, an odd number of consecutive samples to comprise the observation vector $\mathbf{X}(n)$. The observation window is centered at n, resulting in

$$\mathbf{X}(n) = [X(n - N_L), \ldots, X(n), \ldots, X(n + N_R)]^T, \tag{5.1}$$

where N_L and N_R may range in value over the nonnegative integers and $N = N_L + N_R + 1$ is the window size. In most cases, the window is symmetric about $X(n)$ and $N_L = N_R = N_1$. The median smoother operating on the input sequence $\{X(\cdot)\}$ produces the output sequence $\{Y\}$, defined at time index n as:

$$
\begin{aligned}
Y(n) &= \text{MEDIAN}\,[X(n - N_1), \ldots, X(n), \ldots, X(n + N_1)] \\
&= \text{MEDIAN}\,[X_1(n), \ldots, X_N(n)]
\end{aligned} \tag{5.2}
$$

where $X_i(n) = X(n - N_1 - 1 + i)$ for $i = 1, 2, \ldots, N$. That is, the samples in the observation window are sorted and the middle, or median, value is taken as the output. If $X_{(1)}, X_{(2)}, \ldots, X_{(N)}$ are the sorted samples in the observation window, the median smoother outputs

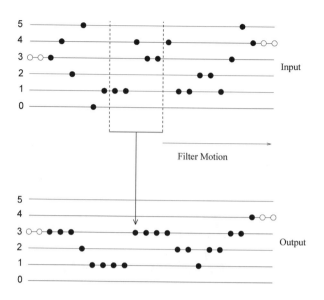

Figure 5.1 The operation of the window width 5 median smoother. o: appended points.

$$
Y(n) = \begin{cases} X_{\left(\frac{N+1}{2}\right)} & \text{if } N \text{ is odd} \\ \dfrac{X_{\left(\frac{N}{2}\right)} + X_{\left(\frac{N}{2}+1\right)}}{2} & \text{otherwise.} \end{cases}
\tag{5.3}
$$

The input sequence $\{X(\cdot)\}$ may be either finite or infinite in extent. For the finite case, the samples of $\{X(\cdot)\}$ can be indexed as $X(1),\ X(2),\ldots,\ X(L)$, where L is the length of the sequence. Because of the symmetric nature of the observation window, the window extends beyond the finite extent of the input sequence at both the beginning and end. When the window is centered at the first and last point in the signal, half of the window is empty. These end effects are generally accounted for by appending N_L samples at the beginning and N_R samples at the end of $\{X(\cdot)\}$. Although the appended samples can be arbitrarily chosen, typically these are selected so that the points appended at the beginning of the sequence have the same value as the first signal point, and the points appended at the end of the sequence all have the value of the last signal point.

To illustrate the appending of input sequences and the median smoother operation, consider the input signal $\{X(\cdot)\}$ of Figure 5.1. In this example, $\{X(\cdot)\}$ consists of 20 observations from a 6–level process, $\{X\ :\ X(n)\ \in\ \{0,\ 1,\ldots,\ 5\}, n = 1,\ 2,\ldots,\ 20\}$. The figure shows the input sequence and the resulting output sequence for a median smoother of window size 5. Note that to account for edge effects, two samples have been appended to both the beginning and end of the sequence. The median smoother output at the window location shown in the figure is

$$
Y(9) \;=\; \text{MEDIAN}[X(7),\ X(8),\ X(9),\ X(10),\ X(11)]
$$

$$= \text{MEDIAN}[\,1,\ 1,\ 4,\ 3,\ 3\,] = 3.$$

Running medians can be extended to a recursive mode by replacing the "causal" input samples in the median smoother by previously derived output samples. The output of the recursive median smoother is given by

$$
\begin{aligned}
Y(n) \quad &= \quad \text{MEDIAN}[Y(n - N_L),\ Y(n - N_L + 1), \dots, \\
&\qquad\qquad Y(n - 1),\ X(n), \dots,\ X(n + N_R)].
\end{aligned}
\tag{5.4}
$$

In recursive median smoothing, the center sample in the observation window is modified before the window is moved to the next position. In this manner, the output at each window location replaces the old input value at the center of the window. With the same amount of operations, recursive median smoothers have better noise attenuation capabilities than their nonrecursive counterparts [5, 8]. Alternatively, recursive median smoothers require smaller window lengths in order to attain a desired level of noise attenuation. Consequently, for the same level of noise attenuation, recursive median smoothers often yield less signal distortion.

The median operation is nonlinear. As such, the running median does not possess the superposition property and traditional impulse response analysis is not strictly applicable. The impulse response of a median smoother is, in fact, zero for all time. Consequently, alternative methods for analyzing and characterizing running medians must be employed. Broadly speaking, two types of analysis have been applied to the characterization of median smoothers: *statistical* and *deterministic*. Statistical properties examine the performance of the median smoother, through such measures as optimality and output variance, for the case of white noise time sequences. Conversely, deterministic properties examine the smoother output characteristics for specific types of commonly occurring deterministic time sequences.

5.1.1 Statistical Properties

The statistical properties of the running median can be examined through the derivation of output distributions and statistical conditions on the optimality of median estimates. This analysis generally assumes that the input to the running median is a constant signal with additive white noise. The assumption that the noise is additive and white is quite natural, and made similarly in the analysis of linear filters. The assumption that the underlying signal is a constant is certainly convenient, but more importantly, often valid. This is especially true for the types of signals median filters are most frequently applied to, such as images. Signals such as images are characterized by regions of constant value separated by sharp transitions, or edges. Thus, the statistical analysis of a constant region is valid for large portions of these commonly used signals. By calculating the output distribution of the median filter over a constant region, the noise smoothing capabilities of the median can be measured through statistics such as the filter output variance.

The calculation of statistics such as the output mean and variance from the expressions in (3.15) and (3.16) is often quite difficult. Insight into the smoothing

Table 5.1 Asymptotic output variances for the window size N mean and running median for white input samples with uniform, Gaussian, and Laplacian distributions.

Input Sample Probability Density Function		Filter Type	
		Mean	Median
Uniform $f_x(t) = \begin{cases} \frac{1}{\sqrt{12\sigma^2}} & \text{for } -\sqrt{3\sigma^2} \le t \le \sqrt{3\sigma^2} \\ 0 & \text{otherwise} \end{cases}$		$\frac{\sigma^2}{N}$	$\frac{3\sigma^2}{N+2}$
Gaussian $f_x(t) = \frac{1}{\sqrt{2\pi\sigma^2}} e^{-\frac{1}{2\sigma^2}(t-\mu)^2}$		$\frac{\sigma^2}{N}$	$\frac{\pi\sigma^2}{2N}$
Laplacian $f_x(t) = \frac{1}{\sqrt{2\sigma^2}} e^{-\frac{\sqrt{2}}{\sigma}|t-\mu|}$		$\frac{\sigma^2}{N}$	$\frac{\sigma^2}{2N}$

characteristics of the median filter can, however, be gained by examining the asymptotic behavior ($N \to \infty$) of these statistics, where, under some general assumptions, results can be derived. For the case of white noise input samples, the asymptotic mean, μ_{med}, and variance, σ^2_{med}, of the running median output are [126]

$$\mu_{med} = t_{0.5}, \tag{5.5}$$

and

$$\sigma^2_{med} = \frac{1}{4N(f_x(t_{0.5}))^2}, \tag{5.6}$$

where $t_{0.5}$ is the median parameter of the input samples.

Thus, the median smoother produces a consistent ($\lim_{N \to \infty} \sigma^2_{med} = 0$) and unbiased estimate of the median of the input distribution. Note that the output variance is not proportional to the input variance, but rather $1/f_x^2(t_{0.5})$. For heavy tailed noises, $1/f_x^2(t_{0.5})$ is not related to the input variance. Therefore, the variance is proportional to the impulse magnitude, not $1/f_x^2(t_{0.5})$. Thus, the output variance of the median in this case is not proportional to the input variance. This is not true for the sample mean, and further explains the more robust behavior of the median.

The variances for the sample mean and running median output are given in Table 5.1 for the uniform, Gaussian, and Laplacian input distribution cases [58]. The results hold for all N in the uniform case and are asymptotic for the Gaussian and Laplacian cases. Note that the median performs about 3 dB better than the sample mean for the Laplacian case and 2 dB worse in the Gaussian case.

Recursive median smoothers, as expected, are more efficient than their nonrecursive counterparts in attenuating noise due to the fact that half of the data points in the window of the recursive median have already been "cleaned." Consider the simplest scenario where the recursive median smoother is applied to an i.i.d. time

Table 5.2   Relative efficiency of recursive and non-recursive medians

N	$\frac{\sigma_s^2}{\sigma_r^2}$
3	1.09
5	1.39
7	1.83
9	2.40
11	3.04
13	3.73
15	4.43

series $\{X(n)\}$ described by the cumulative distribution function $F(x)$. It has been shown that the cumulative distribution function of the output of the recursive median filter $Y(n)$ with window size N, is [5, 8]

$$F_N(y) = \Pr(Y \le y) = \frac{F(y)^{N_1} + N(1 - F(y))^{N_1} F(y)^{N_1}}{F(y)^{N_1} + 2N(1 - F(y))^{N_1} F(y)^{N_1} + (1 - F(y))^{N_1}},$$
(5.7)

where $N_1 = (N + 1)/2$. The output distribution in (5.7) can be used to measure the relative efficiency between the recursive and non-recursive (standard) medians. For a window of size N and for uniformly distributed noise, the ratio σ_s^2/σ_r^2 of the nonrecursive variance estimate to the recursive variance estimate is given in Table 5.2, where the higher efficiency of the recursive median smoother is readily seen.

To further illustrate the improved noise attenuation capability of recursive medians, consider an i.i.d. input sequence, $\{X(n)\}$ consisting of a constant signal, C, embedded in additive white noise $Z(n)$. Without loss of generality, assume $C = 0$, and that the noise is symmetrically distributed. Figure 5.2a shows 1000 samples of the sequence $\{X(n)\}$, where the underlying distribution is double exponential (heavy tailed). Figures 5.2b,c show the noisy sequence after the application of a nonrecursive and a recursive median smoothers, respectively, both of window size 7. The improved noise attenuation provided by recursion is apparent in Figures 5.2b,c.

A phenomenon that occurs with median smoothers in impulsive noise environment is that if several impulsive noise samples are clustered together within the window, the impulses may not be removed from the signal. This phenomenon can be observed in Figures 5.2b,c. To quantify such events, Mallows (1980) [137] introduced the concept of *breakdown probability* as the probability of an impulse occurring at the output of the estimator, when the probability of impulses at the input is given. In essence, the breakdown probability is a measure that indicates the robustness of a particular estimator. To derive the breakdown probability of median smoothers, let us first arbitrarily select a threshold t, such that if a noise sample exceeds such level, the sample is regarded as an impulse. Let the symmetric distribution function of the noise be $F(\cdot)$, then the probability of a noise sample being an impulse (positive or negative) is $2F(-t)$. For the recursive median filter, half of the breakdown probability is given in (5.7) with $y = -t$. The breakdown probability of nonrecursive median smoothers is found through order statistics as

$$
2 \sum_{\ell=N_1}^{N} \binom{N}{\ell} F(-t)^{\ell}[1 - F(-t)]^{N-\ell},
\tag{5.8}
$$

where $N_1 = (N + 1)/2$. In Figure 5.2, the threshold $|t|$ is set to 1; thus, the probability of an impulse occurring at the input is 0.24. The breakdown probability, for the non-recursive median filter, in Figure 5.2*b* is 0.011. For the recursive median filter, this probability is 0.002. Thus, on the average, for every impulse occurring at the output of the recursive median smoother in this example, there will be 5.5 impulses at the output of the nonrecursive median smoother output.

Tables 5.3 and 5.4 show the breakdown probabilities for recursive and nonrecursive median smoothers for different values of input impulse probability, $2F(-t)$, and for different window sizes. The better noise suppression characteristics of the recursive median smoothers can be seen in Figure 5.2, and in a more quantitative way in Tables 5.3 and 5.4.

Table 5.3   Breakdown probabilities for the Non-Recursive Median Smoother

Probability p	$N = 3$	$N = 5$	$N = 7$	$N = 9$	$N = 11$	$N = 13$
0.1	0.0145	0.0023	0.0003	0.00006	0.00001	0.000003
0.2	0.0560	0.0171	0.0054	0.0017	0.00059	0.0001
0.3	0.1215	0.0532	0.0242	0.0112	0.0053	0.0025
0.4	0.2080	0.1158	0.0667	0.0391	0.0233	0.0140
0.5	0.3125	0.2070	0.1411	0.0978	0.0686	0.0048
0.6	0.4320	0.3261	0.2520	0.1976	0.1564	0.1247
0.7	0.5635	0.4703	0.3997	0.3434	0.2974	0.2589

Median smoothers are primarily used to remove undesired disturbances in data, thus their statistical characterization, in terms of output distributions, would provide the required information about the median smoothers' noise attenuation power. Unfortunately, the general output distribution can seldom be put in manageable form. Unlike linear smoothers, median smoothers have well defined deterministic properties that effectively complement their set of statistical properties. In particular, root signals (also referred to invariant and fixed points) play an important role revealing the deterministic behavior of median smoothers, and in this respect the set of root signals resemble the pass band characteristics of linear frequency-selective filters.

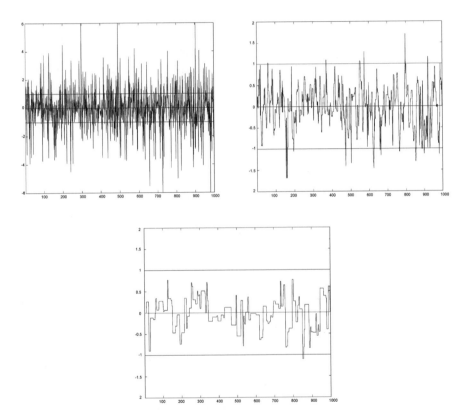

Figure 5.2 Impulse threshold $|t| = 1$: (a) Laplacian noisy sequence, (b) median smoothed sequence, and (c) recursive median smoothed sequence.

Table 5.4 Breakdown probabilities for the Recursive Median Smoothers

Probability p	$N = 3$	$N = 5$	$N = 7$	$N = 9$	$N = 11$	$N = 13$
0.1	0.0102	0.0007	0.00005	0.000003	0.00000002	0.00000001
0.2	0.0417	0.0066	0.0009	0.0001	0.00001	0.000001
0.3	0.0954	0.0239	0.0052	0.0010	0.0002	0.00004
0.4	0.1714	0.0604	0.0184	0.0052	0.0014	0.0003
0.5	0.2692	0.1253	0.0501	0.0187	0.0067	0.0022
0.6	0.3873	0.2285	0.1162	0.0552	0.0253	0.0113
0.7	0.5233	0.3782	0.2387	0.1417	0.0812	0.0455

5.1.2 Root Signals (Fixed Points)

Statistical properties give considerable insight into the performance of running medians. Running medians cannot, however, be sufficiently characterized through statistical properties alone. For instance, an important question not answered by the statistical properties is what type of signal, if any, is passed through a running median unaltered. Linear smoothers, when applied repeatedly to a signal, for instance, will increasingly smooth a signal. With the exception of some contrived examples, fixed points of linear smoothers are only those belonging to constant-valued sequences. On the other hand, Gallagher and Wise (1981) [78] showed that running medians have nontrivial fixed-point sequences referred to as *root signals* for reasons that will become clear shortly. The concept of root signals is important to the understanding of running medians and their effect on general signal structures. In noise smoothing, for instance, the goal is to attain maximum noise attenuation while preserving the desired signal features. An ideal situation would arise if the smoother could be tailored so that the desired signal features were invariant to the smoothing operation and only the noise would be affected. Since the median operation is nonlinear and lacks the superposition property, this idealized case is of course not possible. Nonetheless, when a signal consists of constant areas and step changes between these areas, a similar effect is achieved. Noise will be attenuated, but the signal features will remain intact. This concept is used extensively in image smoothing, where the median smoother is designed such that certain image patterns, such as lines and edges, are root signals and thus not affected by the smoothing operation [7, 147].

The definition of a root signal is quite simple: a signal is a running median root if the signal is invariant under the median smoothing operation. For simplicity assume that the window is symmetric about $X(n)$, with $N_L = N_R$ taking on the value N_1. Thus, a signal $\{X(\cdot)\}$ is a root of the window size $N = 2N_1 + 1$ median smoother if

$$X(n) = \text{MEDIAN}[X(n - N_1), \ldots, X(n), \ldots, X(n + N_1)] \qquad (5.9)$$

for all n. As an example, consider the signal shown in Figure 5.3. This signal is smoothed by three different window size running medians ($N_1 = 1, 2$, and 3). Note that for the window size three case ($N_1 = 1$), the output is a root. That is, further smoothing of this signal with the window size three running median does not alter the signal. Notice, however, that if this same signal is smoothed with a larger window running median, the signal will be modified. Thus, the second signal (from the top) in Figure 5.3 is in the pass band, or a root, of a $N_1 = 1$ running median but outside the pass band, or not a root, of the $N_1 = 2$ and $N_1 = 3$ smoothers.

The goal of root analysis is to relate the smoothing of desired signals corrupted by noise to root and nonroot signals. If it can be shown that certain types of desired signals are in the running median root set, while noise is outside the root set, then median smoothing of a time series will preserve desired structures while altering the noise. Such a result does in fact hold and will be made clear through the following definitions and properties. First note that, as the example above illustrates, whether or not a signal is a running median root depends on the window size of the smoother in question. Clearly, all signals are roots of the window size one running median (identity). To investigate this dependence on window size, running median root signals can be characterized in terms of local signal structures, where the local signal structures are related to the window size. Such a local structure based analysis serves two purposes. First, it defines signal structures that, when properly combined, form the running median root set. Second, by relating the local structures to the window size, the effect of window size on roots is made clear. The local structure analysis of running median roots relies on the following definitions [78].

Constant Neighborhood: A region of at least $N_1 + 1$ consecutive identically valued points.

An Edge: A monotonic region between two constant neighborhoods of different value. The connecting monotonic region cannot contain any constant neighborhoods.

An Impulse: A constant neighborhood followed by at least one, but no more than N_1 points, that are then followed by another constant neighborhood having the same value as the first constant neighborhood. The two boundary points of these at most N_1 points do not have the same value as the two constant neighborhoods.

An Oscillation: A sequence of points that is not part of a constant neighborhood, an edge, or an impulse.

These definitions may now be used to develop a description of those signals that do and those that do not pass through a running median without being perturbed. In particular, Gallagher and Wise [78] developed a number of properties which characterize these signal sets for the case of finite length sequences. First, any impulse will be eliminated upon median smoothing. Secondly, a finite length signal is a running median root if it consists of constant neighborhoods and edges only. Thus,

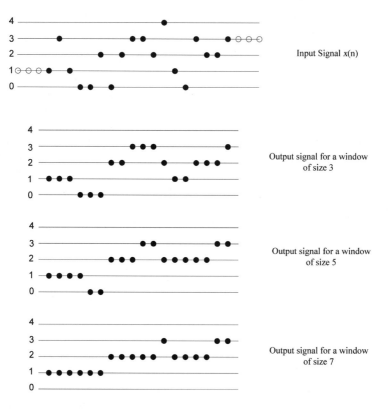

Figure 5.3 Effects of window size on a median smoothed signal. ○: appended points.

if a desired signal is constructed solely of constant neighborhoods and edges, then it will not be altered by the median smoothing operation. Conversely, if observation noise consists of impulses (as defined above), it will be removed by the median smoothing operation. These running median root properties are made exact by the following.

LOMO Sequence: A sequence $\{X(\cdot)\}$ is said to be locally monotonic of length m, denoted LOMO(m), if the subsequence $X(n),\ X(n+1), \cdots,\ X(n+m-1)$ is monotonic for all $n \geq 1$.

Root Signals: Given a length L sequence to be median smoothed with a length $N = 2N_1 + 1$ window, a necessary and sufficient condition for the signal to be invariant (a root) under median smoothing is that the extended (beginning and end appended) signal be LOMO($N_1 + 2$).

Thus, the set of root signals (invariant to smoothing) of a size N running median consists solely of those signals that are formed of constant neighborhoods and edges. Note that by the definition of LOMO(m), a change of trend implies that the sequence must stay constant for at least $m - 1$ points. It follows that for a running median root signal to contain both increasing and decreasing regions, these regions must be separated by a constant neighborhood of least $N_1 + 1$ identically valued samples. It is also clear from the definition of LOMO($\cdot$) that a LOMO(m_1) sequence is also LOMO(m_2) for any two positive integers $m_1 \geq m_2$. This implies that the roots for decreasing window size running medians are nested, that is, every root of a window size M smoother is also a root of a window sized N median smoother for all $N < M$. This is formalized by:

Root Signal Set: Let S denote a set of finite length sequences and R_{N_1} be the root set of the window size $N = 2N_1 + 1$ running median operating on S. Then the root sets are nested such that $\ldots R_{N_1+1} \subseteq R_{N_1} \subseteq R_{N_1-1} \subseteq \ldots \subseteq R_1 \subseteq R_0 = S$.

In addition to the above description of the root signal set for running medians, it can be shown that any signal of finite length is mapped to a root signal by repeated median smoothing. This property of median filters is very significant and is called the *root convergence property*. It can be shown that the first and last points to change value on a median smoothing operation remain invariant upon additional running median passes, where repeated smoother passes consist of using the output of the prior smoothing pass for the input of an identical smoother on the current pass. This fact, in turn, indicates that any L long nonroot signal (oscillations and impulses) will become a root structure after a maximum of $(L - 2)/2$ successive smoothings. This simple bound was improved in [194] where it was shown that at most

$$3 \left\lfloor \frac{L - 2}{2(N_1 + 2)} \right\rfloor \tag{5.10}$$

passes of the median smoother are required to reach a root. This bound is conservative in practice since in most cases root signals are obtained with much fewer smoothing passes.

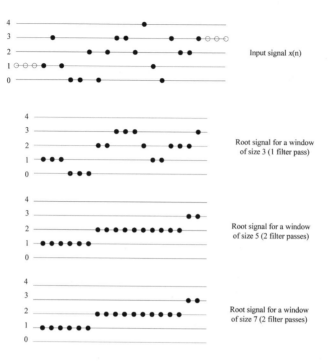

Figure 5.4 Root signals obtained by running medians of size 3, 5, and 7. ∘: appended points.

The running median root properties are illustrated through an example in Figure 5.4. This figure shows an original signal and the resultant root signals after multiple passes of window size $3, 5,$ and 7 running medians. Note that while it takes only a single pass of the window size 3 running median to obtain a root, it takes two passes for the window sizes 5 and 7 median smoothers. Clearly, the locally monotonic structure requirements of the root signals are satisfied in Figure 5.4. For the window size 3 case, the input sequence becomes LOMO(3) after a single pass of the smoother. Thus, this sequence is in the root set of the window size 3 running median, but not a root of the window size $N > 3$ running median, since it is not LOMO($N_1 + 2$) for $N_1 > 1$ $(N > 3)$.

Recursive median smoothers also possess the root convergence property [5, 150]. In fact, they produce root signals after a single filter pass. For a given window size, recursive and nonrecursive median filters have the same set of root signals. A given input signal, however, may be mapped to distinct root signals by the two filters [5, 150]. Figure 5.5 illustrates this concept where a signal is mapped to different root signals by the recursive and nonrecursive median smoothers. In this case, both roots are attained in a single smoother pass.

The deterministic and statistical properties form a powerful set of tools for describing the median smoothing operation and performance. Together, they show that

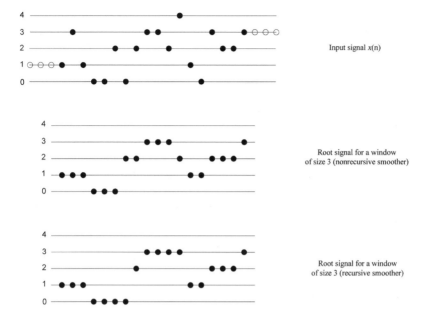

Figure 5.5 A signal and its recursive and non-recursive running median roots. ○: appended points.

the median is an optimal estimator of location for Laplacian noise and that common signal structures, for example, constant neighborhoods and edges in images, are in its pass-band (root set). Moreover, impulses are removed by the smoothing operation and repeated passes of the running median always result in the signal converging to a root, where a root consists of a well defined set of structures related to the smoother's window size. Further properties of root signals can be found in Arce and Gallagher (1982) [9], Bovik (1987) [37], Wendt et al. (1986) [194], Wendt (1990) [193]. Multiscale root signal analysis was developed by Bangham (1993) [25].

MAX-MIN Representation of Medians MAX-MIN representation of medians
The median has an interesting and useful representation where only minima and maxima operations are used. See Fitch (1987) [71]. This representation is useful in the software of hardware implementation of medians, but more important, it is also useful in the analysis of median operations. In addition, the max-min representation of medians provides a link between rank-order and *morphological* operators as shown in Maragos and Schafer (1987) [140]. Given the N samples $X_1, X_2, \ldots, X_N$, and defining $m = \frac{N+1}{2}$, the median of the sample set is given by

$$X_{\left(\frac{N+1}{2}\right)} = \min\left[\max(X_1, \ldots, X_m), \ldots, \max(X_{j_1}, X_{j_2}, \ldots, X_{j_m}),\right.$$
$$\left. \ldots, \max(X_{N-m+1}, \ldots, X_N)\right] \quad (5.11)$$

where $j_1, j_2, \ldots, j_m$ index all $C_N^m \equiv \frac{N!}{(N-m)!m!}$ combinations of N samples taken m at a time. The median of 3 samples, for instance, has the following min-max representation

$$\text{MEDIAN}(X_1, X_2, X_3) = \min\left[\max(X_1, X_2), \ \max(X_1, X_3), \ \max(X_2, X_3)\right].$$
(5.12)

The max-min representation follows by reordering the input samples into the corresponding order-statistics $X_{(1)}, X_{(2)}, \ldots, X_{(N)}$ and indexing the resultant samples in all the possible group combinations of size m. The maximum of the first subgroup $X_{(1)}, X_{(2)}, \ldots, X_{(m)}$ is clearly $X_{(m)}$. The maximum of the other subgroups will be greater than $X_{(m)}$ since these subgroups will include one of the elements in $X_{(m+1)}, X_{(m+2)}, \ldots, X_{(N)}$. Hence, the minimum of all these maxima will be the mth-order statistic $X_{(m)}$, that is, the median.

EXAMPLE 5.1

Consider the vector $X = [1, \ 3, \ 2, \ 5, \ 5]$, to calculate the median using the max-min representation we have:

$$
\begin{aligned}
\text{MEDIAN}(1, 3, 2, 5, 5) \ &= \ \min\left[\max(1, 3, 2), \ \max(1, 3, 5), \ \max(1, 3, 5),\right. \\
&\qquad \max(1, 2, 5), \ \max(1, 2, 5), \ \max(1, 5, 5), \\
&\qquad \left.\max(3, 2, 5), \ \max(3, 2, 5), \ \max(2, 5, 5)\right] \\
&= \ \min(3, 5, 5, 5, 5, 5, 5, 5, 5) \\
&= \ 3.
\end{aligned}
$$

∎

5.2 WEIGHTED MEDIAN SMOOTHERS

Although the median is a robust estimator that possesses many optimality properties, the performance of running medians is limited by the fact that it is temporally blind. That is, all observation samples are treated equally regardless of their location within the observation window. This limitation is a direct result of the i.i.d. assumption made in the development of the median. A much richer class of smoothers is obtained if this assumption is relaxed to the case of independent, but not identically distributed, samples.

Statistical Foundations   Although time-series samples, in general, exhibit temporal correlation, the independent but not identically distributed model can be used to synthesize the mutual correlation. This is possible by observing that the estimate

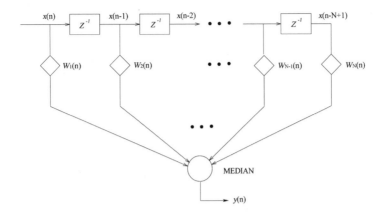

Figure 5.6   The weighted median smoothing operation.

$Y(n)$ can rely more on the sample $X(n)$ than on the other samples of the series that are further away in time. In this case, $X(n)$ is more reliable than $X(n-1)$ or $X(n+1)$, which in turn are more reliable than $X(n-2)$ or $X(n+2)$, and so on. By assigning different variances (reliabilities) to the independent but not identically distributed location estimation model, the temporal correlation used in time-series smoothing is captured. Thus, weighted median smoothers incorporate the reliability of the samples and temporal order information by weighting samples prior to rank smoothing.   The WM smoothing operation can be schematically described as in Figure 5.6.

Consider again the generalized Gaussian distribution where the observation samples have a common location parameter β, but where each X_i has a (possibly) unique scale parameter σ_i. Incorporating the unique scale parameters into the ML criteria for the generalized distribution, equation (4.9), shows that, in this case, the ML estimate of location is given by the value of β minimizing

$$G_p(\beta) = \sum_{i=1}^{N} \frac{1}{\sigma_i^p} |X_i - \beta|^p. \tag{5.13}$$

In the special case of the standard Gaussian distribution ($p = 2$), the ML estimate reduces to the normalized weighted average

$$\hat{\beta} = \arg\min_{\beta} \sum_{i=1}^{N} \frac{1}{\sigma_i^2} (X_i - \beta)^2 = \frac{\sum_{i=1}^{N} W_i \cdot X_i}{\sum_{i=1}^{N} W_i} \tag{5.14}$$

where $W_i = 1/\sigma_i^2 > 0$. In the case of a heavier–tailed Laplacian distribution ($p = 1$), the ML estimate is realized by minimizing the sum of weighted absolute deviations

$$G_1(\beta) = \sum_{i=1}^{N} \frac{1}{\sigma_i} |X_i - \beta|. \tag{5.15}$$

where again $1/\sigma_i > 0$. Note that $G_1(\beta)$ is piecewise linear and convex for $W_i \geq 0$. The value β minimizing (5.15) is thus guaranteed to be one of the samples $X_1, X_2, \ldots, X_N$. This is the weighted median (WM), originally introduced over a hundred years ago by Edgeworth [66]. The running weighted median output is defined as

$$Y(n) = \text{MEDIAN}[W_1 \diamond X_1(n), \ W_2 \diamond X_2(n), \cdots, \ W_N \diamond X_N(n)], \qquad (5.16)$$

where $W_i > 0$ and $\diamond$ is the replication operator defined as $W_i \diamond X_i = \overbrace{X_i, \ldots, \ X_i}^{W_i \text{ times}}$. Weighted median smoothers were introduced in the signal processing literature by Brownigg (1984) [41] and have since received considerable attention. Note that the formulation in (5.16) requires that the weights take on nonnegative values which is consistent with the statistical interpretation of the weighted median where the weights have an inverse relationship to the variances of the respective observation samples. A simplified representation of a weighted median smoother, specified by the set of N weights, is the list of the weights separated by commas within angle brackets [202]; thus the median smoother defined in (5.16) has the representation $\langle W_1, W_2, \ldots, W_N \rangle$.

Weighted Median Computation   As an example, consider the window size 5 WM smoother defined by the symmetric weight vector $\mathbf{W} = \langle 1, 2, 3, 2, 1 \rangle$. For the observation $\mathbf{X}(n) = [12, 6, 4, 1, 9]$, the weighted median smoother output is found as

$$
\begin{aligned}
Y(n) &= \text{MEDIAN}\,[\,1\diamond 12, \ 2\diamond 6, \ 3\diamond 4, \ 2\diamond 1, \ 1\diamond 9\,] \\[2mm]
&= \text{MEDIAN}\,[\,12, \ 6, \ 6, \ 4, \ 4, \ 4, \ 1, \ 1, \ 9\,] \\[2mm]
&= \text{MEDIAN}\,[\,1, \ 1, \ 4, \ 4, \ \underline{4}, \ 6, \ 6, \ 9, \ 12\,] \\[2mm]
&= 4
\end{aligned}
\qquad (5.17)
$$

where the median value is underlined in equation (5.17). The large weighting on the center input sample results in this sample being taken as the output. As a comparison, the standard median output for the given input is $Y(n) = 6$.

In general, the WM can be computed without replicating the sample data according to the corresponding weights, as this increases the computational complexity. A more efficient method to find the WM is shown next, which not only is attractive from a computational perspective but it also admits positive real–valued weights:

(1) Calculate the threshold $W_0 = \frac{1}{2}\sum_{i=1}^{N} W_i$;

(2) Sort the samples in the observation vector $\mathbf{X}(n)$;

(3) Sum the concomitant weights[1] of the sorted samples beginning with the maximum sample and continuing down in order;

(4) The output is the sample whose weight causes the sum to become $\geq W_0$.

The validity of this method can be supported as follows. By definition, the output of the WM smoother is the value of β minimizing (5.15). Suppose initially that $\beta \geq X_{(N)}$. (5.15) can be rewritten as:

$$G_1(\beta) = \sum_{i=1}^{N} W_{[i]}\left(X_{(i)} - \beta\right)$$

$$= \left(\sum_{i=1}^{N} W_{[i]}\right)\beta - \sum_{i=1}^{N} W_{[i]}X_{(i)}, \qquad (5.18)$$

which is the equation of a straight line with slope $m_N = \sum_{i=1}^{N} W_{[i]} \geq 0$. Now suppose that $X_{(N-1)} \leq \beta < X_{(N)}$. (5.15) is now equal to:

$$G_1(\beta) = \sum_{i=1}^{N-1} W_{[i]}\left(X_{(i)} - \beta\right) + W_{[N]}\left(\beta - X_{(N)}\right)$$

$$= \left(\sum_{i=1}^{N-1} W_{[i]} - W_{[N]}\right)\beta - \sum_{i=1}^{N-1} W_{[i]}X_{(i)} + W_{[N]}X_{(N)}. \quad (5.19)$$

This time the slope of the line is $m_{N-1} = \sum_{i=1}^{N-1} W_{[i]} - W_{[N]} \leq m_N$, since all the weights are positive. If this procedure is repeated for values of β in intervals lying between the order statistics, the slope of the lines in each interval decreases and so will the value of the cost function (5.15), until the slope reaches a negative value. The value of the cost function at this point will increase. The minimum is then reached when this change of sign in the slope occurs. Suppose the minimum (i.e., the weighted median) is the Mth-order statistic. The slopes of the cost function in the intervals before and after $X_{(M)}$ are given by:

$$m_M = \sum_{i=1}^{M} W_{[i]} - \sum_{i=M+1}^{N} W_{[i]} > 0 \qquad (5.20)$$

$$m_{M-1} = \sum_{i=1}^{M-1} W_{[i]} - \sum_{i=M}^{N} W_{[i]} \leq 0. \qquad (5.21)$$

---

[1]Represent the input samples and their corresponding weights as pairs of the form (X_i, W_i). If the pairs are ordered by their X variates, then the value of W associated with $X_{(m)}$, denoted by $W_{[m]}$, is referred to as the *concomitant of the mth order statistic*   [58].

From (5.20), we have

$$\sum_{i=1}^{M} W_{[i]} > \sum_{i=M+1}^{N} W_{[i]}$$

$$\sum_{i=1}^{N} W_{[i]} > 2 \sum_{i=M+1}^{N} W_{[i]}$$

$$W_0 = \frac{1}{2} \sum_{i=1}^{N} W_{[i]} > \sum_{i=M+1}^{N} W_{[i]}. \tag{5.22}$$

Similarly, form (5.21):

$$\sum_{i=1}^{M-1} W_{[i]} \leq \sum_{i=M}^{N} W_{[i]}$$

$$\sum_{i=1}^{N} W_{[i]} \leq 2 \sum_{i=M}^{N} W_{[i]}$$

$$W_0 = \frac{1}{2} \sum_{i=1}^{N} W_{[i]} \leq \sum_{i=M}^{N} W_{[i]}. \tag{5.23}$$

That is, if the concomitant weights of the order statistics are added one by one beginning with the last, the concomitant weight associated with the weighted median, $W_{[M]}$, will be the first to make the sum greater or equal than the threshold W_0.

EXAMPLE 5.2 (COMPUTATION OF THE WEIGHTED MEDIAN)

To illustrate the WM smoother operation for positive real–valued weights, consider the WM smoother defined by $\mathbf{W} = \langle 0.1,\ 0.1,\ 0.2,\ 0.2,\ 0.1 \rangle$. The output for this smoother operating on $\mathbf{X}(n) = [12,\ 6,\ 4,\ 1,\ 9]$ is found as follows. Summing the weights gives the threshold $W_0 = \frac{1}{2} \sum_{i=1}^{5} W_i = 0.35$. The observation samples, sorted observation samples, their corresponding weight, and the partial sum of weights (from each ordered sample to the maximum) are:

observation samples	12,	6,	4,	1,	9	
corresponding weights	0.1,	0.1,	0.2,	0.2,	0.1	
sorted observation samples	1,	4,	6,	9,	12	(5.24)
corresponding weights	0.2,	0.2,	0.1,	0.1,	0.1	
partial weight sums	0.7,	<u>0.5</u>,	0.3,	0.2,	0.1	

Thus, the output is 4 since when starting from the right (maximum sample) and summing the weights, the threshold $W_0 = 0.35$ is not reached until the weight associated with 4 is added. The underlined sum value above indicates that this is the first sum that meets or exceeds the threshold.   ■

An interesting characteristic of WM smoothers is that the nature of a WM smoother is not modified if its weights are multiplied by a positive constant. Thus, the same filter characteristics can be synthesized by different sets of weights. Although the WM smoother admits real–valued weights, it turns out that any WM smoother based on real–valued weights has an equivalent integer–valued weight representation [202]. Consequently, there are only a finite number of WM smoothers for a given window size. The number of WM smoothers, however, grows rapidly with window size [201].

Weighted median smoothers can also operate in a recursive mode. The output of a recursive WM smoother is given by

$$Y(n) = \text{MEDIAN} \left[W_{-N_1} \diamond Y(n - N_1), \ldots, W_{-1} \diamond Y(n - 1), \right.$$
$$\left. W_0 \diamond X(n), \ldots, W_{N_1} \diamond X(n + N_1) \right] \qquad (5.25)$$

where the weights W_i are as before constrained to be positive-valued. Recursive WM smoothers offer advantages over WM smoothers in the same way that recursive medians have advantages over their nonrecursive counterparts. In fact, recursive WM smoothers can synthesize nonrecursive WM smoothers of much longer window sizes. As with nonrecursive weighted medians, when convenient we use a simplified representation of recursive weighted median smoothers where the weights are listed separated by commas within a double set of angle brackets [202]; thus the recursive median smoother defined in (5.25) has the representation $\langle\langle W_{-N_1}, W_{-N_1+1}, \ldots, W_{N_1} \rangle\rangle$.

Using repeated substitution, it is possible to express recursive WM smoothers in terms of a nonrecursive WM series expansion [202]. For instance, the recursive three-point smoother can be represented as

$$Y(n) = \text{MEDIAN} \left[Y(n - 1), X(n), X(n + 1) \right] \qquad (5.26)$$
$$= \text{MEDIAN} \left[\text{MEDIAN} \left[Y(n - 2), X(n - 1), X(n) \right], X(n), X(n + 1) \right].$$

An approximation is found by truncating the recursion above by using $X(n - 2)$ instead of $Y(n - 2)$. This leads to

$$Y(n) = \text{MEDIAN} \left[\text{MEDIAN} \left[X(n - 2), X(n - 1), X(n) \right], X(n), X(n + 1) \right]. \qquad (5.27)$$

Using the max-min representation of the median above it can be shown after some simplifications that the resultant max-min representation is that of a 4-point median. Representing a recursive median smoother, its Pth order series expansion approximation, and a nonrecursive median smoother of size $N_L + N_R + 1$ by

$$\langle\langle W_{-N_R}, \ldots, \underline{W_0}, \ldots, W_{N_R}\rangle\rangle \qquad (5.28)$$

$$\langle\langle W_{-N_R}, \ldots, \underline{W_0}, \ldots, W_{N_R}\rangle\rangle_P \qquad (5.29)$$

$$\langle W_{-N_L}, \ldots, \underline{W_0}, \ldots, W_{N_R}\rangle, \qquad (5.30)$$

respectively, the second-order series expansion approximation of the 3-point recursive median is [202]

$$\langle\langle 1, \underline{1}, 1\rangle\rangle_2 = \langle 1, 1, \underline{1}, 1\rangle. \qquad (5.31)$$

The order of the series expansion approximation refers to truncation of the series after P substitutions. With this notation,

$$\langle\langle 1, \underline{1}, 1\rangle\rangle_0 = \langle\underline{1}\rangle, \qquad (5.32)$$

and

$$\langle\langle 1, \underline{1}, 1\rangle\rangle_1 = \langle 1, \underline{1}, 1\rangle. \qquad (5.33)$$

The fourth-order approximation is

$$\langle\langle 1, \underline{1}, 1\rangle\rangle_4 = \langle 1, 1, 2, 3, \underline{5}, 3\rangle \qquad (5.34)$$

illustrating that recursive WM smoothers can synthesize nonrecursive WM smoothers with more than twice their window size.

EXAMPLE 5.3 (IMAGE ZOOMING)
_____

Zooming is an important task used in many imaging applications. When zooming, pixels are inserted into the image in order to expand the size of the image, and the major task is the interpolation of the new pixels from the surrounding original pixels. Consider the zooming of an image by a factor of powers of two. General zooming with noninteger factors are also possible with simple modifications of the method described next.

To double the size of an image in both dimensions, first an empty array is constructed with twice the number of rows and columns as the original (Figure 5.7a), and the original pixels are placed into alternating rows and columns (the "00" pixels in Figure 5.7a). To interpolate the remaining pixels, the method known as polyphase interpolation is used. In the method, each new pixel with four original pixels at its four corners (the "11" pixels in Figure 5.7b) is interpolated first by using the weighted median of the four nearest original pixels as the value for that pixel. Since all original pixels are equally trustworthy and the same distance from the pixel being interpolated, a weight of 1 is used for the four nearest original pixels. The resulting array is shown in Figure 5.7c. The remaining pixels are determined by taking a weighted median of the four closest pixels. Thus each of the "01" pixels in Figure

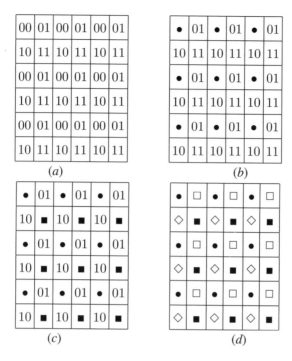

Figure 5.7 The steps of polyphase interpolation.

5.7c is interpolated using two original pixels to the left and right and two previously interpolated pixels above and below. Similarly, the "10" pixels are interpolated with original pixels above and below and interpolated pixels ("11" pixels) to the right and left.

Since the "11" pixels were interpolated, they are less reliable than the original pixels and should be given lower weights in determining the "01" and "10" pixels. Therefore the "11" pixels are given weights of 0.5 in the median to determine the "01" and "10" pixels, while the "00" original pixels have weights of 1 associated with them. The weight of 0.5 is used because it implies that when both "11" pixels have values that are not between the two "00" pixel values, one of the "00" pixels or their average will be used. Thus "11" pixels differing from the "00" pixels do not greatly affect the result of the weighted median. Only when the "11" pixels lie between the two "00" pixels will they have a direct effect on the interpolation. The choice of 0.5 for the weight is arbitrary, since any weight greater than 0 and less than 1 will produce the same result. When implementing the polyphase method, the "01" and "10" pixels must be treated differently due to the fact that the orientation of the two closest original pixels is different for the two types of pixels. Figure 5.7d shows the final result of doubling the size of the original array.

To illustrate the process, consider an expansion of the grayscale image represented by an array of pixels, the pixel in the ith row and jth column having brightness $a_{i,j}$.

The array $a_{i,j}$ is interpolated into the array $x_{i,j}^{pq}$, with p and q taking values 0 or 1 indicating in the same way as above the type of interpolation required:

$$
\begin{bmatrix} a_{1,1} & a_{1,2} & a_{1,3} \\ a_{2,1} & a_{2,2} & a_{2,3} \\ a_{3,1} & a_{3,2} & a_{3,3} \end{bmatrix} \Longrightarrow
\begin{bmatrix}
x_{1,1}^{00} & x_{1,1}^{01} & x_{1,2}^{00} & x_{1,2}^{01} & x_{1,3}^{00} & x_{1,3}^{01} \\
x_{1,1}^{10} & x_{1,1}^{11} & x_{1,2}^{10} & x_{1,2}^{11} & x_{1,3}^{10} & x_{1,3}^{11} \\
x_{2,1}^{00} & x_{2,1}^{01} & x_{2,2}^{00} & x_{2,2}^{01} & x_{2,3}^{00} & x_{2,3}^{01} \\
x_{2,1}^{10} & x_{2,1}^{11} & x_{2,2}^{10} & x_{2,2}^{11} & x_{2,3}^{10} & x_{2,3}^{11} \\
x_{3,1}^{00} & x_{3,1}^{01} & x_{3,2}^{00} & x_{3,2}^{01} & x_{3,3}^{00} & x_{3,3}^{01} \\
x_{3,1}^{10} & x_{3,1}^{11} & x_{3,2}^{10} & x_{3,2}^{11} & x_{3,3}^{10} & x_{3,3}^{11}
\end{bmatrix}
$$

The pixels are interpolated as follows:

$$
\begin{aligned}
x_{i,j}^{00} &= a_{i,j} \\
x_{i,j}^{11} &= \text{MEDIAN}[a_{i,j},\ a_{i+1,j},\ a_{i,j+1},\ a_{i+1,j+1}] \\
x_{i,j}^{01} &= \text{MEDIAN}[a_{i,j},\ a_{i,j+1},\ 0.5 \Diamond x_{i-1,j}^{11},\ 0.5 \Diamond x_{i,j}^{11}] \\
x_{i,j}^{10} &= \text{MEDIAN}[a_{i,j},\ a_{i+1,j},\ 0.5 \Diamond x_{i,j-1}^{11},\ 0.5 \Diamond x_{i,j}^{11}]
\end{aligned}
$$

An example of median interpolation compared with bilinear interpolation is given in Figure 5.8. The zooming factor is 4 obtained by two consecutive interpolations, each doubling the size of the input. Bilinear interpolation uses the average of the nearest two original pixels to interpolate the "01" and "10" pixels in Fig. 5.7*b* and the average of the nearest four original pixels for the "11" pixels. The edge-preserving advantage of the weighted median interpolation is readily seen in this figure. ∎

5.2.1 The Center-Weighted Median Smoother

The weighting mechanism of WM smoothers allows for great flexibility in emphasizing or deemphasizing specific input samples. In most applications, not all samples are equally important. Due to the symmetric nature of the observation window, the sample most correlated with the desired estimate is, in general, the center observation sample. This observation lead Ko and Lee (1991) [115] to define the center-weighted median (CWM) smoother, a relatively simple subset of WM smoothers that has proven useful in many applications.

The CWM smoother is realized by allowing only the center observation sample to be weighted. Thus, the output of the CWM smoother is given by

Figure 5.8 Zooming by 4: original is at the top with the area of interest outlined in white. On the lower left is the bilinear interpolation of the area, and on the lower right the weighted median interpolation.

$$Y(n) = \text{MEDIAN}[X_1, \ldots, X_{c-1}, W_c \diamond X_c, X_{c+1}, \ldots, X_N], \qquad (5.35)$$

where W_c is an odd positive integer and $c = (N+1)/2 = N_1 + 1$ is the index of the center sample. When $W_c = 1$, the operator is a median smoother, and for $W_c \geq N$, the CWM reduces to an identity operation.

The effect of varying the center sample weight is perhaps best seen by means of an example. Consider a segment of recorded speech. The voice waveform "a" is shown at the top of Figure 5.9. This speech signal is taken as the input of a CWM smoother of size 9. The outputs of the CWM, as the weight parameter W_c is progressively increased as $1, 3, 5$, and 7, are shown in the figure. Clearly, as W_c is increased less smoothing occurs. This response of the CWM smoother is explained by relating the weight W_c and the CWM smoother output to select order statistics (OS).

The CWM smoother has an intuitive interpretation. It turns out that the output of a CWM smoother is equivalent to computing

$$Y(n) = \text{MEDIAN}\left[X_{(k)}, X_c, X_{(N+1-k)}\right], \qquad (5.36)$$

where $k = (N + 2 - W_c)/2$ for $1 \leq W_c \leq N$, and $k = 1$ for $W_c > N$. Since $X(n)$ is the center sample in the observation window, that is, $X_c = X(n)$, the output of the smoother is identical to the input as long as the $X(n)$ lies in the interval $\left[X_{(k)}, X_{(N+1-k)}\right]$. If the center input sample is greater than $X_{(N+1-k)}$ the smoother outputs $X_{(N+1-k)}$, guarding against a high rank order (large) aberrant data point being taken as the output. Similarly, the smoother's output is $X_{(k)}$ if the sample $X(n)$ is smaller than this order statistic. This implementation of the CWM filter is also known as the LUM filter as described by Hardie and Boncelet (1993) [94]. This CWM smoother performance characteristic is illustrated in Figures 5.10 and 5.11. Figure 5.10 shows how the input sample is left unaltered if it is between the trimming statistics $X_{(k)}$ and $X_{(N+1-k)}$ and mapped to one of these statistics if it is outside this range. Figure 5.11 shows an example of the CWM smoother operating on a Laplacian sequence. Along with the input and output, the trimming statistics are shown as an upper and lower bound on the filtered signal. It is easily seen how increasing k will tighten the range in which the input is passed directly to the output.

Application of CWM Smoother To Image Cleaning   Median smoothers are widely used in image processing to clean images corrupted by noise. Median filters are particularly effective at removing outliers. Often referred to as "salt and pepper" noise, outliers are often present due to bit errors in transmission, or introduced during the signal acquisition stage. Impulsive noise in images can also occur as a result to damage to analog film. Although a weighted median smoother can be designed to "best" remove the noise, CWM smoothers often provide similar results at a much lower complexity. See Ko and Lee (1991) [115] and Sun et al. (1994) [184]. By simply tuning the center weight a user can obtain the desired level of smoothing. Of course, as the center weight is decreased to attain the desired level of impulse suppression, the output image will suffer increased distortion particularly around

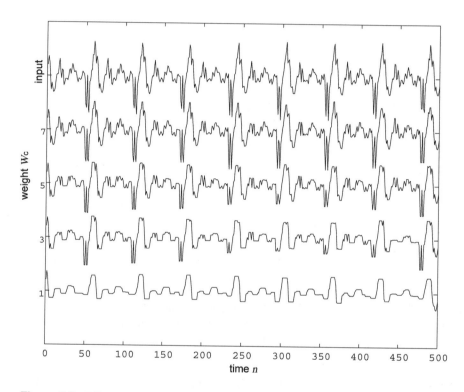

Figure 5.9 Effects of increasing the center weight of a CWM smoother of size $N = 9$ operating on the voiced speech "a". The CWM smoother output is shown for $W_c = 1,3,5$, and 7. Note that for $W_c = 1$ the CWM reduces to median smoothing.

$X_{(1)}$ $\qquad$ $X_{(k)}$ $\qquad$ $X_{(N+1-k)}$ $\qquad$ $X_{(N)}$

Figure 5.10 The center weighted median smoothing operation. The output is mapped to the order statistic $X_{(k)}$ ($X_{(N+1-k)}$) if the center sample is less (greater) than $X_{(k)}$ ($X_{(N+1-k)}$), and to the center sample otherwise.

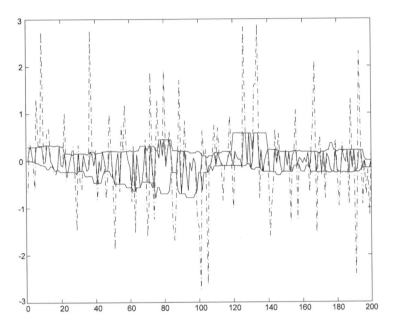

Figure 5.11 An example of the CWM smoother operating on a Laplacian distributed sequence with unit variance. Shown are the input and output sequences as well as the trimming statistics $X_{(k)}$ and $X_{(N+1-k)}$. The window size is 25 and $k = 7$.

the image's fine details. Nonetheless, CWM smoothers can be highly effective in removing "salt and pepper" noise while preserving the fine image details. Figures 5.12*a* and *b* depicts a noise free image and the corresponding image with "salt and pepper" noise. Each pixel in the image has a 10 percent probability of being contaminated with an impulse. The impulses occur randomly and were generated by MATLAB's imnoise function. Figures 5.12*c* and *d* depict the noisy image processed with a 5×5 window CWM smoother with center weights 15 and 5, respectively. The impulse-rejection and detail-preservation tradeoff in CWM smoothing is illustrated in these figures. Another commonly used measure of the quality of an image is the Peak Signal to Noise Ratio (PSNR) defined as:

$$\text{PSNR} = 10 \log_{10} \left(\frac{\text{max}^2}{\text{MSE}} \right) \tag{5.37}$$

where MSE is the mean squared error of the image and max is the maximum pixel value (255 for 8-bit images). The value of the PSNR of the pictures shown is included in the captions for illustrative purposes.

At the extreme, for $W_c = 1$, the CWM smoother reduces to the median smoother, which is effective at removing impulsive noise and preserving edges. It is, however, unable to preserve the image's fine details. Figure 5.13 shows enlarged sections of the noise-free image (left-top), the noisy image (right-top), and of the noisy image after the median smoother has been applied (left-bottom). Severe blurring is introduced by the median smoother and it is readily apparent in Figure 5.13. As a reference, the output of a running mean of the same size is also shown in Figure 5.13 (right-bottom). The image is severely degraded as each impulse is smeared to neighboring pixels by the averaging operation.

Figures 5.12 and 5.13 show that CWM smoothers are effective at removing impulsive noise. If increased detail-preservation is sought and the center weight is increased, CWM smoothers begin to break down and impulses appear on the output. One simple way to ameliorate this limitation is to employ a recursive mode of operation. In essence, past inputs are replaced by previous outputs as described in (5.25) with the only difference that only the center sample is weighted. All the other samples in the window are weighted by one. Figure 5.14 shows enlarged sections of the nonrecursive CWM filter (right-top) and of the corresponding recursive CWM smoother (left-bottom), both with the same center weight ($W_c = 15$). This figure illustrates the increased noise attenuation provided by recursion without the loss of image resolution.

Both recursive and nonrecursive CWM smoothers can produce outputs with disturbing artifacts, particularly when the center weights are increased to improve the detail-preservation characteristics of the smoothers. The artifacts are most apparent around the image's edges and details. Edges at the output appear jagged and impulsive noise can break through next to the image detail features. The distinct response of CWM smoother in different regions of the image is because images are nonstationary in nature. Abrupt changes in the image's local mean and texture carry most of the visual information content. CWM smoothers process the entire image with fixed weights and are inherently limited in this sense by their static nature. Although some improvement is attained by introducing recursion or by using more weights in a properly designed WM smoother structure, these approaches are also static and do not properly address the nonstationarity nature of images. A simple generalization of WM smoothers that overcomes these limitations is presented next.

5.2.2 Permutation-Weighted Median Smoothers

The principle behind the CWM smoother lies in the ability to emphasize, or deemphasize, the center sample of the window by tuning the center weight while keeping the weight values of all other samples at unity. In essence, the value given to the center weight indicates the reliability of the center sample. This concept can be further developed by adapting the value of the center weight in response to the rank of the center sample among the samples in the window. If the center sample does not

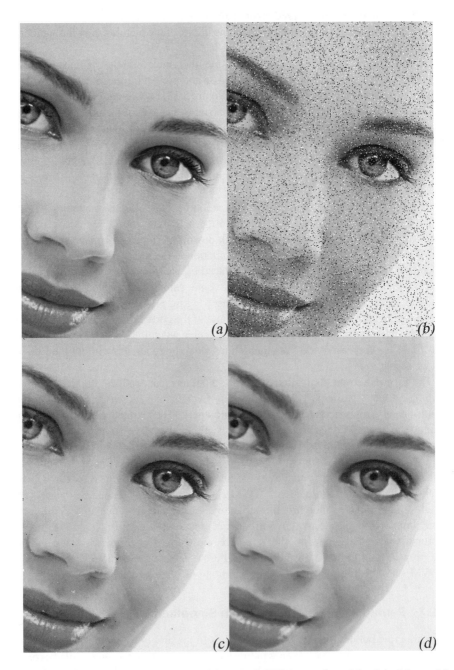

Figure 5.12 Impulse noise cleaning with a 5×5 CWM smoother: (*a*) original "portrait" image, (*b*) image with salt and pepper noise (PSNR = 14.66dB), (*c*) CWM smoother with $W_c = 15$ (PSNR = 32.90dB), (*d*) CWM smoother with $W_c = 5$ (PSNR = 35.26dB).

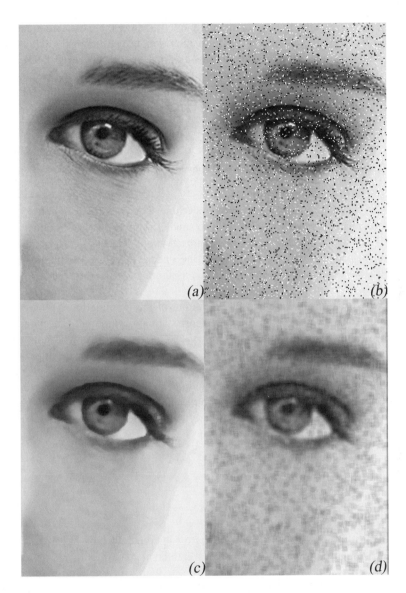

Figure 5.13 (Enlarged) Noise-free image, image with salt and pepper noise (PSNR=14.66dB), 5×5 median smoother output (PSNR=33.34dB), and 5×5 mean smoother (PSNR=23.52dB).

contain an impulse (high reliability), it would be desirable to make the center weight large such that no smoothing takes place (identity filter). On the other hand, if an impulse was present in the center of the window (low reliability), no emphasis should be give to the center sample (impulse), and the center weight should be given the smallest possible value, $W_c = 1$, reducing the CWM smoother structure to a simple median. Notably, this adaptation of the center weight can be easily achieved by considering the center sample's rank among all pixels in the window [10, 93]. More precisely, denoting the rank of the center sample of the window at a given location as $R_c(n)$, then the simplest *permutation* WM smoother is defined by the following modification of the CWM smoothing operation

$$W_c(n) = \begin{cases} N & \text{if } T_L \leq R_c(n) \leq T_U \\ 1 & \text{otherwise} \end{cases} \qquad (5.38)$$

where N is the window size and $1 \leq T_L \leq T_U \leq N$ are two adjustable threshold parameters that determine the degree of smoothing. Note that the weight in (5.38) is data adaptive and may change with n. The smaller (larger) the threshold parameter T_L (T_U) is set to, the better the detail-preservation. Generally, T_L and T_U are set symmetrically around the median. If the underlying noise distribution is not symmetric about the origin, a nonsymmetric assignment of the thresholds would be appropriate.

Figure 5.14 (right-bottom) shows the output of the permutation CWM filter in (5.38) when the "salt and pepper" degraded portrait image is inputted. The parameters were given the values $T_L = 6$ and $T_U = 20$. The improvement achieved by switching W_c between just two different values is significant. The impulses are deleted without exception, the details are preserved, and the jagged artifacts typical of CWM smoothers are not present in the output.

The data-adaptive structure of the smoother in (5.38) can be extended so that the center weight is not only switched between two possible values, but can take on N different values:

$$W_c(n) = \begin{cases} W_{c(j)}(n) & \text{if } R_c(n) = j, \quad j \in \{1, 2, \ldots, N\} \\ 1 & \text{otherwise} \end{cases} \qquad (5.39)$$

Thus, the weight assigned to X_c is drawn from the center weight set $\{W_{c(1)}, W_{c(2)}, \ldots, W_{c(N)}\}$. Thus the permutation center weighted median smoother is represented by $\langle W_1, \ldots, W_{c-1}, W_{c(j)}, W_{c+1}, \ldots, W_N \rangle$, with $W_{c(j)}$ taking on one of N possible values. With an increased number of weights, the smoother in (5.39) can perform better. However, the design of the weights is no longer trivial and optimization algorithms are needed [10, 93]. A further generalization of (5.39) is feasible, where weights are given to all samples in the window, but where the value of each weight is data-dependent and determined by the rank of the corresponding sample. In this case, the output of the permutation WM smoother is found as

$$Y(n) = \text{MEDIAN}[W_{1(R_1)} \diamond X_1(n), \; W_{2(R_2)} \diamond X_2(n), \ldots, \; W_{N(R_N)} \diamond X_N(n)], \qquad (5.40)$$

where $W_{i(R_i)}$ is the weight assigned to $X_i(n)$ and selected according to the sample's rank R_i. The weight assigned to X_i is drawn from the weight set $\{W_{i(1)}, W_{i(2)}, \ldots, W_{i(N)}\}$. Having N weights per sample, a total of N^2 weights need to be stored for the computation of (5.40). In general, optimization algorithms are needed to design the set of weights although in some cases the design is simple, as with the smoother in (5.38). Permutation WM smoothers can provide significant improvement in performance at the higher cost of memory cells [10].

5.3 THRESHOLD DECOMPOSITION REPRESENTATION

Threshold decomposition (TD) is a powerful theoretical tool used in the analysis of weighted median filters and smoothers. The use of this signal decomposition method allows the study of weighted median smoothers by merely studying their behavior on binary signals. Introduced by Fitch et al. (1984) [72], threshold decomposition was originally formulated to admit signals having only a finite number of positive-valued quantization levels. Threshold decomposition admitting integer-valued signals taking on positive or negative values is formulated as follows. Consider an integer-valued set of samples $X_1, X_2, \ldots, X_N$ forming the vector $\mathbf{X} = [X_1, X_2, \ldots, X_N]^T$, where $X_i \in \{-M, \ldots, 0, \ldots, M\}$. The threshold decomposition of $\mathbf{X}$ amounts to decomposing this vector into $2M$ binary vectors $\mathbf{x}^{-M+1}, \ldots, \mathbf{x}^0, \ldots, \mathbf{x}^M$, where the ith element of $\mathbf{x}^m$ is defined by

$$x_i^m = T^m(X_i) = \begin{cases} 1 & \text{if } X_i \geq m; \\ -1 & \text{if } X_i < m, \end{cases} \tag{5.41}$$

where $T^m(\cdot)$ is referred to as the thresholding operator. Using the sign function, the above can be written as $x_i^m = \text{sgn}(X_i - m^-)$ where m^- represents a real number approaching the integer m from the left. Although defined for integer-valued signals, the thresholding operation in (5.41) can be extended to noninteger signals with a finite number of quantization levels. The threshold decomposition of the vector

$$\mathbf{X} = [0, 0, 2, -2, 1, 1, 0, -1, -1]^T \tag{5.42}$$

with $M = 2$, for instance, leads to the 4 binary vectors

$$\begin{aligned} \mathbf{x}^2 &= [-1, -1, 1, -1, -1, -1, -1, -1, -1]^T \\ \mathbf{x}^1 &= [-1, -1, 1, -1, 1, 1, -1, -1, -1]^T \\ \mathbf{x}^0 &= [1, 1, 1, -1, 1, 1, 1, -1, -1]^T \\ \mathbf{x}^{-1} &= [1, 1, 1, -1, 1, 1, 1, 1, 1]^T. \end{aligned} \tag{5.43}$$

Threshold decomposition has several important properties. First, threshold decomposition is reversible. Given a set of thresholded signals, each of the samples in $\mathbf{X}$ can be exactly reconstructed as

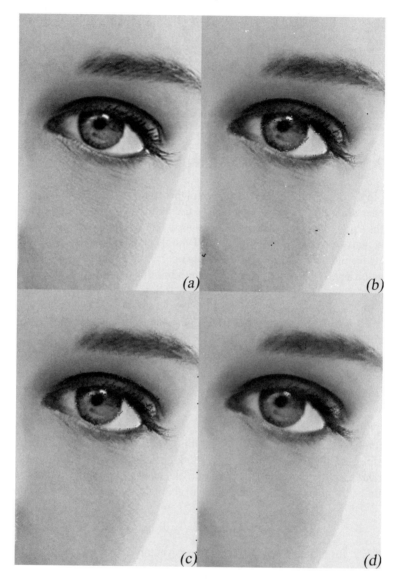

Figure 5.14 (Enlarged) Original image, CWM smoother output (PSNR=32.90dB), Recursive CWM smoother output (PSNR=32.11dB), and Permutation CWM smoother output (PSNR=35.01dB). Window size is 5×5.

$$X_i = \frac{1}{2} \sum_{m=-M+1}^{M} x_i^m. \qquad (5.44)$$

Thus, an integer-valued discrete-time signal has a unique threshold signal representation, and vice versa:

$$X_i \overset{T.D.}{\longleftrightarrow} \{x_i^m\},$$

where $\overset{T.D.}{\longleftrightarrow}$ denotes the one-to-one mapping provided by the threshold decomposition operation.

A second property of importance is the partial ordering obeyed by the threshold decomposed variables. For all thresholding levels $m > \ell$, it can be shown that $x_i^m \le x_i^\ell$. In particular, if $x_i^m = 1$ then $x_i^\ell = 1$ for all $\ell < m$. Similarly, if $x_i^\ell = -1$ then $x_i^m = -1$, for all $m > \ell$. The partial order relationships among samples across the various thresholded levels emerge naturally in thresholding and are referred to as the *stacking constraints* [195].

Threshold decomposition is of particular importance in median smoothing since they are commutable operations. That is, applying a median smoother to a $2M + 1$ valued signal is equivalent to decomposing the signal to $2M$ binary thresholded signals, processing each binary signal separately with the corresponding median smoother, and then adding the binary outputs together to obtain the integer-valued output. Thus, the median of a set of samples $X_1, X_2, \ldots, X_N$ is related to the set of medians of the thresholded signals as [72]

$$\text{MEDIAN}(X_1, \ldots, X_N) = \frac{1}{2} \sum_{m=-M+1}^{M} \text{MEDIAN}(x_1^m, \ldots, x_N^m). \quad (5.45)$$

The threshold decomposition property of median smoothers follows from the following. Let $\tilde{X}$ be the sample median of the set of samples $X_1, \ldots, X_N$, with N an odd integer. By definition of the sample median, out of N samples in the set there are at least $\frac{N+1}{2}$ samples having values that are equal or smaller than the value of $\tilde{X}$. From (5.41), for a given threshold value m in $m = -M + 1, \ldots, \tilde{X}$, at least $\frac{N+1}{2}$ of the thresholded binary samples x_i^m have value 1. Similarly, for a given threshold value m in $m = \tilde{X}, \ldots, M$, at least $\frac{N+1}{2}$ of the thresholded samples x_i^m have value -1. Hence.

$$\text{MEDIAN}(x_1^m, \ldots, x_N^m) = \begin{cases} 1 & \text{for } -M+1 \le m \le \tilde{X}; \\ -1 & \text{for } \tilde{X} < m \le M. \end{cases} \quad (5.46)$$

Consequently,

$$\sum_{m=-M+1}^{M} \text{MEDIAN}(x_1^m, \ldots, x_N^m) = 1\left(\tilde{X} + M\right) - 1\left(M - \tilde{X}\right) = 2\tilde{X}. \quad (5.47)$$

The reverse is also true. That is, half the sum of a set of filtered binary samples, which satisfy the stacking constraints, produce the median value of the samples synthesized by the unfiltered set of thresholded binary samples. The threshold decomposition property in (5.45) is thus verified.

To illustrate the concepts behind threshold decomposition consider the sample set $\mathbf{X} = [2, -1, 2, 0, 1]^T$. With $M = 2$, the set of $2M$ binary vectors leads to the array of binary vectors

$$
\begin{pmatrix} \mathbf{x}^2 \\ \mathbf{x}^1 \\ \mathbf{x}^0 \\ \mathbf{x}^{-1} \end{pmatrix} = \begin{pmatrix} 1 & -1 & 1 & -1 & -1 \\ 1 & -1 & 1 & -1 & 1 \\ 1 & -1 & 1 & 1 & 1 \\ 1 & 1 & 1 & 1 & 1 \end{pmatrix}. \tag{5.48}
$$

Applying the binary median operation to each binary vector

$$
\begin{pmatrix} \mathrm{MEDIAN} & [1, & -1, & 1, & -1, & -1] \\ \mathrm{MEDIAN} & [1, & -1, & 1, & -1, & 1] \\ \mathrm{MEDIAN} & [1, & -1, & 1, & 1, & 1] \\ \mathrm{MEDIAN} & [1, & 1, & 1, & 1, & 1] \end{pmatrix} = \begin{pmatrix} -1 \\ 1 \\ 1 \\ 1 \end{pmatrix}. \tag{5.49}
$$

Summing the outputs of the binary medians and scaling by 2 leads to the desired multivalued median.

Since $X_i \overset{T.D.}{\longleftrightarrow} \{x_i^m\}$ and $\mathrm{MEDIAN}(X_i|_{i=1}^N) \overset{T.D.}{\longleftrightarrow} \{\mathrm{MEDIAN}(x_i^m|_{i=1}^N)\}$, the relationship in (5.45) establishes a *weak* superposition property satisfied by the nonlinear median operator, which is important from the fact that the effects of median smoothing on binary signals are much easier to analyze than that on multilevel signals. In fact, the median operation on binary samples reduces to a simple Boolean operation. The median of three binary samples x_1, x_2, and x_3, for example, is equivalent to: $x_1 x_2 + x_2 x_3 + x_1 x_3$, where the $+$ (OR) and $x_i x_j$ (AND) "Boolean" operators in the $\{-1, 1\}$ domain are defined as

$$
x_i + x_j = \max(x_i, x_j)
$$
$$
x_i x_j = \min(x_i, x_j). \tag{5.50}
$$

Note that the operations in (5.50) are also valid for the standard Boolean operations in the $\{0, 1\}$ domain.

5.3.1 Stack Smoothers

The framework of threshold decomposition and Boolean operations has led to the general class of nonlinear smoothers referred to as stack smoothers, introduced by Wendt et al. (1986) [195]. The output of a stack smoother is the result of a sum of a stack of binary operations acting on thresholded versions of the samples spanned by the smoother's running window. The stack smoother output is defined by

$$S_f(X_1, \ldots, X_N) = \frac{1}{2} \sum_{m=-M+1}^{M} f(x_1^m, \ldots, x_N^m) \tag{5.51}$$

where x_i^m, $i = 1, \ldots, N$, are the thresholded input samples $x_i^m = T^m(X_i)$ defined in (5.41), and where $f(\cdot)$ is a Positive Boolean Function (PBF) which, by definition, contains only uncomplemented input variables. More precisely, if two binary vectors $\mathbf{u} \in \{-1, 1\}^N$ and $\mathbf{v} \in \{-1, 1\}^N$ stack, that is, $u_i \geq v_i$ for all $i \in \{1, \ldots, N\}$, then it is said that $\mathbf{u} \geq \mathbf{v}$. For a PBF $f(\cdot)$ it holds that

$$f(\mathbf{u}) \geq f(\mathbf{v}) \quad \text{if } \mathbf{u} \geq \mathbf{v}. \tag{5.52}$$

The property in (5.52) is called the stacking property. A complete proof of (5.52) can be found in [145]. From (5.51), it can be seen that stack smoothers are completely characterized by their independent operation on a set of binary vectors. The importance of this property lies in the fact that the analysis and manipulation of the nonlinear operations behind stack smoothers is simpler with binary vectors than with real-valued vectors. The computation of a binary median of size N (odd), for instance, reduces to the sign of the sum of the N binary inputs.

Given an input vector $\mathbf{X}$ and its set of thresholded binary vectors $\mathbf{x}^{-M+1}, \ldots, \mathbf{x}^0,$ $\ldots, \mathbf{x}^M$, it follows from the definition of threshold decomposition that the set of thresholded binary vectors satisfy the partial ordering

$$\mathbf{x}^i \leq \mathbf{x}^j \quad \text{if } i \geq j. \tag{5.53}$$

Consequently, the stack smoothing of the thresholded binary vectors by the PBF $f(\cdot)$ also satisfy the partial ordering

$$f(\mathbf{x}^i) \leq f(\mathbf{x}^j) \quad \text{if } i \geq j. \tag{5.54}$$

The relation above leads to an interesting interpretation on the stack smoother operation. Since stack smoothers operate independently on each of the thresholded input vectors. The stack smoother decides whether the desired signal is less than a given level j or not based on the noisy observations provided by the binary vectors at the various thresholds. The stacking property in (5.54) ensures that the decisions on different levels are consistent. Thus, if the smoother at a given time location decides that the signal is less than j, then the smoother outputs at levels $j + 1$ and greater must draw the same conclusion. The estimation consistency is illustrated in Figure 5.15 where a 5-level sequence is median smoothed via threshold decomposition.

As defined in (5.51), stack smoothers input signals are assumed to be quantized to a finite number of signal levels. For practical and analytical reasons it is desirable to extend their definition to a class of smoothers admitting real-valued input signals.

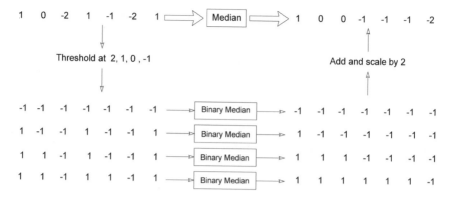

Figure 5.15   Threshold decomposition for integer-valued signals.

DEFINITION 5.1 (CONTINUOUS STACK SMOOTHERS) *Given a set of N real-valued samples* $\mathbf{X} = (X_1, X_2, \ldots, X_N)$, *the output of a stack smoother defined by a PBF* $f(\cdot)$ *is given by*

$$S_f(\mathbf{X}) = \max\{\ell \in \mathbf{R} : f(T^\ell(X_1), \ldots, f(T^\ell(X_N)) = 1\}, \qquad (5.55)$$

where the thresholding function $T^\ell(\cdot)$ *is defined in (5.41).*

Although very general, Definition 5.1 is somewhat cumbersome. The following property yields an important link between the continuous stack smoother $S_f(\cdot)$ and the corresponding PBF $f(\cdot)$.

PROPERTY 5.1 (MAX-MIN REPRESENTATION OF STACK SMOOTHERS) *Let* $\mathbf{X} = (X_1, X_2, \ldots, X_N)$ *be a real-valued input vector to a stack smoother* $S_f(\cdot)$ *defined by the positive Boolean function* $f(x_1, x_2, \ldots, x_N)$. *Then, the PBF with the sum of products expression*

$$f(x_1, x_2, \ldots, x_N) = \sum_{i=1}^{K} \prod_{j \in P_i} x_j, \qquad (5.56)$$

where P_i *are subsets of* $\{1, 2, \ldots, N\}$, *has the stack filter representation*

$$S_f(\mathbf{X}) = \max\left\{\min\{X_j : j \in P_1\}, \min\{X_j : j \in P_2\}, \ldots, \min\{X_j : j \in P_K\}\right\}. \qquad (5.57)$$

Thus, given a positive Boolean function $f(x_1, \ldots, x_N)$ which characterizes a stack smoother, it is possible to find the equivalent smoother in the integer domain by replacing the binary AND and OR Boolean functions acting on the x_i's with *max* and *min* operations acting on the real-valued X_i samples. A more intuitive class of smoothers is obtained, however, if the positive Boolean functions are further restricted, as described in the study of Yli-Harja et al. (1991) [202]. When self-duality[2] and separability are imposed, for instance, the equivalent integer domain stack smoothers reduce to the well known class of weighted median smoothers with positive weights. For example, if the Boolean function in the stack smoother representation is selected as $f(x_1, x_2, x_3, x_4) = x_1 x_3 x_4 + x_2 x_4 + x_2 x_3 + x_1 x_2$, the equivalent WM smoother takes on the positive weights $\langle W_1, W_2, W_3, W_4 \rangle = \langle 1, 2, 1, 1 \rangle$. The procedure of how to obtain the weights W_i from the PBF is described below.

Statistical Properties of Stack Smoothers Stack smoothers are primarily used for noise attenuation; thus, it is natural to resort to their output statistics for their characterization. Consider a set of N real-valued input samples $\mathbf{X} = [X_1, X_2, \ldots, X_N]^T$ applied to a stack smoother $S_f(\cdot)$ defined by a positive Boolean function $f(x_1, x_2, \ldots, x_N)$. From Definition 5.1, if we select a real-valued threshold ℓ such that

$$f(x_1^\ell, x_2^\ell, \ldots, x_N^\ell) = 1, \tag{5.58}$$

where $x_i^\ell \ \ i = 1, \ldots, N$ are the thresholded input variables, then it follows from the stacking property that the output of the stack smoother must satisfy

$$S_f(\mathbf{X}) \geq \ell. \tag{5.59}$$

Likewise, the event

$$f(x_1^\ell, x_2^\ell, \ldots, x_N^\ell) = -1, \tag{5.60}$$

indicates that the stack smoother output satisfies

$$S_f(\mathbf{X}) < \ell. \tag{5.61}$$

The deterministic partial ordering described in (5.58)–(5.61) was used by Arce (1986) [5] and Wendt et al. (1986) [194] to determine the output distribution for the output of a stack smoother as follows

$$
\begin{aligned}
F_S(\ell) &= Pr\{S_f(\mathbf{X}) \leq \ell\} & (5.62) \\
&= Pr\{f(x_1^\ell, x_2^\ell, \ldots, x_N^\ell) = -1\}, & (5.63)
\end{aligned}
$$

where it is assumed that the distribution functions of the input random variables are continuous. Likewise, the relation

$$Pr\{S_f(\mathbf{X}) > \ell\} = Pr\{f(x_1^\ell, x_2^\ell, \ldots, x_N^\ell) = 1\}, \tag{5.64}$$

---

[2]Self-duality refers to the Boolean function structure that when complementing the input variables leads to an output that is the complement of the original output.

follows from the stacking property of stack smoothers. The probabilistic relations in (5.63) and (5.64) are referred to as the *statistical threshold decomposition* property of stack smoothers, which provides the key to determining their output distribution [8, 5, 195]. Note that (5.63) provides a general statistical description of the output where no statistical assumptions have been made about the input random variables.

The computation of $F_S(\ell)$, however, can be difficult to obtain in closed form expression except for cases where some assumptions are placed on the statistics of the input samples. $F_S(\ell)$ for general input random variables can still be computed in the general case, but numerical methods may be required. If the observation samples $X_1, X_2, \ldots, X_N$, are assumed independent for example, with X_i obeying the distribution function $F_i(x)$, for $i = 1, \ldots, N$, the binary thresholded samples x_i^ℓ are distributed as

$$Pr\{x_i^\ell = -1\} = F_i(\ell) \tag{5.65}$$
$$Pr\{x_i^\ell = 1\} = 1 - F_i(\ell). \tag{5.66}$$

The distribution function for the stack smoother output $S_f(\ell)$ is then found as

$$
\begin{aligned}
F_S(\ell) &= Pr\{f(x_1^\ell, x_2^\ell, \ldots, x_N^\ell) = -1\} \\
&= \sum_{f(\boldsymbol{x})=-1} \prod_{i=1}^{N} F_i(\ell)^{(1-x_i)/2} (1 - F_i(\ell))^{(1+x_i)/2} \\
&= \sum_{\boldsymbol{x} \in f^{-1}(-1)} \prod_{i=1}^{N} F_i(\ell)^{(1-x_i)/2} (1 - F_i(\ell))^{(1+x_i)/2}
\end{aligned}
\tag{5.67}
$$

where $f^{-1}(-1)$ is the pre-image of -1, such that, $f^{-1} = \{\boldsymbol{x} : f(\boldsymbol{x}) = -1\}$, and the binary valued variable in the exponents are to be understood as real -1s and 1s. In essence, the expression in (5.67) adds the probabilities of all possible binary inputs (events) that lead the PBF $f(\cdot)$ to output a -1. Since these events are disjoint, the expression in (5.67) simply adds the probability of each event.

If the input samples are not only independent from each other but identically distributed as well, the output distribution in (5.67) can be further simplified. In this case, denote the common distribution function of the input variables as $F(x)$, and let $w_H(\boldsymbol{x})$ be the Hamming weight of the vector $[x_1 + 1, x_2 + 1, \ldots, x_N + 1]/2$, that is, the number of 1s in $\boldsymbol{x}$. The distribution function $F_S(\ell)$ of the output of a stack filter is then

$$F_S(\ell) = \sum_{i=0}^{N} A_i (1 - F(\ell))^i F(\ell)^{N-i}, \tag{5.68}$$

where the numbers A_i are defined by

$$A_i = |\{\mathbf{x} : f(\mathbf{x}) = -1, w_H(\mathbf{x}) = i\}|_C, \tag{5.69}$$

where $|\Omega|_C$ denotes the cardinality of the set Ω. The expression in (5.68) is simply a reordering of the terms in (5.67), where the product operation in (5.67) is absorbed in the coefficients A_i. This is only possible if all the samples share a common distribution function $F(\cdot)$. In order to guarantee that the stack smoother is not defined by the trivial PBF $f(x) = -1$ for all x, or $f(x) = 1$ for all x, it is required that

$$A_0 = 1 \quad \text{and} \quad A_N = 0. \tag{5.70}$$

Since A_i is the cardinality of a set of binary vectors of length N with hamming weight i, it follows that

$$0 \le A_i \le \binom{N}{i}. \tag{5.71}$$

The coefficients A_i embody important deterministic information about the underlying PBF defining a particular stack smoother. A number of properties for the coefficients A_i have been derived, some of which we list here without proof [119, 198]. Because of the stacking constraints of PBFs, the coefficients A_i satisfy the symmetric ordering

$$A_{i+1} \le \frac{N-i}{i+1} A_i$$

for $i = 0, 1, \ldots, N-1$. Since $(N-i)/(i+1) \le 1$ when $i \ge (N-1)/2$, the above implies that

$$A_{i+1} \le A_i \quad \text{for} \quad i \ge \frac{N-1}{2}.$$

For self-dual PBFs the A_is satisfy

$$A_i = \binom{N}{i} - A_{N-i},$$

for $i = 0, 1, \ldots, N$.

Moments of Stack Smoothers In principle, the statistical properties of the stack smoother output are embodied in the probability distribution function. In some cases, it may be simpler or more advantageous to characterize the stack smoother output through expectations or moments. The second-order central moment, for instance, is often used to measure the noise attenuation capability of a smoother. It measures the spread of the output variable about its expected value. A criterion for stack smoother design may take advantage of this fact and would try to minimize the second central moment of the output [119].

The pth-order moment of the output of a stack smoother can be expressed as

$$E\{Y^2\} = \sum_{i=0}^{N-1} A_i M(F, p, N, i), \tag{5.72}$$

where

$$M(F, p, N, i) = \int_{-\infty}^{\infty} x^p \frac{d}{dx} \left((1 - F(\ell))^i F(\ell)^{N-i} \right) dx \tag{5.73}$$

for $i = 1, \ldots, N - 1$. $M(\cdot)$ is thus a function of the input distribution $F(\cdot)$, the window size N and the index i. The central moment about the mean of the output is then

$$\sigma_S^2 = \sum_{i=0}^{N-1} A_i M(F, 2, N, i) - \left(\sum_{i=0}^{N-1} A_i M(F, 1, N, i) \right)^2. \tag{5.74}$$

As can be seen in the above equations, the function $M(F, p, N, i)$ plays a key role in determining the output moments. Notably, these functions have a number of properties that can facilitate their computation. The following property provides a recursive tool for computation.

PROPERTY 5.2 *The numbers $M(F, p, N, i)$ satisfy*

$$M(F, p, N, i) = \sum_{j=0}^{i} \binom{i}{j} (-1)^{i-j} M(F, p, N - j, 0) \tag{5.75}$$

for $0 \le i \le N$.

Property 5.2 follows from the identity

$$(1 - F(x))^i = \sum_{j=0}^{i} \binom{i}{j} (-1)^{i-j} F(x)^{i-j}. \tag{5.76}$$

Using (5.76) in the definition of $M(F, p, N, i)$ leads to

$$M(F, p, N, i) = \sum_{j=0}^{i} \binom{i}{j} (-1)^{i-j} \int_{-\infty}^{\infty} x^p \frac{d}{dx} \left(F(x)^{N-j} \right) dx. \tag{5.77}$$

The integral in (5.77) is simply $M(F, p, N - j, 0)$ which proves Property 5.2.

The recursion in Property 5.2 allows the computation of all $M(F, p, N, i)$ numbers provided that the numbers $M(F, p, j, 0)$ are known. Notice, however, that $M(F, p, j, 0)$ are by definition the moments of the jth order-statistic of N i.i.d. variates with a parent distribution $F(\cdot)$. These have been extensively studied in statistics and have been tabulated for various distributions including the Gaussian and uniform distributions.

If the input distribution is symmetric with respect to its mean, a number of additional properties for $M(F, p, N, i)$ emerge. Using these set of properties for $M(F, p, N, i)$, it was shown in [119, 198] that among all stack smoothers of a fixed window size, the standard median smoother attains the smallest output variance.

Not surprisingly, the standard median also minimizes the breakdown probability [119]. Thus, the median smoother is optimal in the class of stack smoothers for noise cleaning. In more demanding tasks, where the underlying desired signals are not constant, the standard median is no longer optimal and more elaborated stack smoother design procedures are needed.

Optimization of Stack Smoothers Having the framework of stack smoothers, it is necessary to develop tools for their design and optimization under some error criteria. This problem has been extensively studied and consequently a number of stack smoother optimization algorithms exist today. A number of error criteria have been proposed including the traditional mean square error (MSE) and mean absolute error (MAE). See Coyle and Lin (1988) [55], Coyle (1988) [54], Lin and Kim (1994) [131], Lin et al. [132], Yin et al. (1993) [199], and T ăbus et al. (1996) [186]. Other less common approaches have also been considered. These include optimization under the associate memory criterion, a shape preservation criterion, and optimization under a set of constraints. They can be found in Gabbouj and Coyle (1990) [76], Kuosmanen et al. (1995) [120], Yu and Coyle (1992) [203], and Yang et al. (1995) [198]. Optimization under a set of constraints take on many forms. Structural, rank-selection, breakdown point constraints can be imposed leading to a number of optimization algorithms as in Kuosmanen and Astola [119], Gabbouj and Coyle (1990) [76], and Yin et al. (1993) [199]. A review on these approaches is given in Astola and Kuosmanen (1999) [22]. The performance of the various algorithms depends on the optimality criterion, which should be carefully chosen. Here we present two approaches for the optimization of stack smoothers: (a) design under structural constraints, and (b) the adaptive optimization approach to be covered in Chapter 6.

Stack Smoothers Under Structural Constraints Zeng (1994) [205] introduced a simple stack smoother design procedure that can preserve certain structural features of an image, such as straight lines or corners, while taking advantage of the optimal noise reduction properties of the median smoother. The structural constraints consist of a list of different structures to be preserved, deleted or modified. Since stack smoothers obey threshold decomposition, the structural constraints only need to be considered in the context of binary signals. That is, they can be specified by a set of binary vectors and the corresponding stack smoother outputs. Zeng's algorithm leads to stack smoothers that compare favorably with other stack smoothers such as the center weighted median, but, with their closed-form Boolean function representations, they are generally easier to derive.

To examine the structural constraints that one may desire in the design of a stack smoother, consider the following 2-dimensional example. Assume that a binary image of interest includes a horizontal line of width one. An unweighted median filter of size 3 or greater applied to the image will annihilate the line.

Let $\mathbf{x}$ be the 3×3 observation vector:

$$\mathbf{x} = \begin{bmatrix} x_1 & x_2 & x_3 \\ x_4 & x_5 & x_6 \\ x_7 & x_8 & x_9 \end{bmatrix}$$

where x_5 is the center point of the window. To preserve the desired feature (horizontal lines) define the Boolean function

$$g_H^{(3\times3)}(\mathbf{x}) = x_4 x_5 x_6 + (x_4 + x_5 + x_6)m^{(3\times3)}(\mathbf{x})$$

where $m^{(3\times3)}(\cdot)$ is the Boolean function for the unweighted median smoother of size 3×3. We find that if x_4, x_5, x_6 are all 1 then $g_H^{(3\times3)}$ will be 1 no matter the values of the other points[3]. Also, if there was a line of -1s instead of 1s then the term $(x_4 + x_5 + x_6)$, as well as the product $x_4 x_5 x_6$, would be -1, and thus $g_H^{(3\times3)}(\cdot) = -1$. Hence, $g_H^{(3\times3)}(\cdot)$ preserves straight horizontal lines of either -1s or 1s through the center point of the binary image. If there was no such line, then $x_4 x_5 x_6 = -1$ and $x_4 + x_5 + x_6 = 1$, so that $g_H^{(3\times3)}(\cdot)$ will return the value of the median $m^{(3\times3)}(\cdot)$, thus employing the noise reduction property of the latter smoother.

The above concept can be extended in a straightforward manner to create smoothers that will preserve horizontal, vertical, or diagonal lines of any length, as well as corners. In general, let the observation window be $\mathbf{x} = x_1, x_2, \ldots, x_{N^2}$ and denote subsets

$$X_i = x_{i_1}, x_{i_2}, \ldots, x_{i_k}, \quad i = 1, \ldots M$$

that include the center point of the smoother. Further, let P_i be the product of the points in X_i and Z_i be the sum of those points. Then, to create a smoother that will return a 1 (or -1) when any subset X_i is made up entirely of 1s (or -1s), and return the result of the median otherwise, we let

$$g(\mathbf{x}) = \sum_{i=1}^{M} P_i + (\prod_{i=1}^{M} Z_i)m^{(N\times N)}(\mathbf{x}).$$

For example, given the 5×5 mask

$$\mathbf{x} = \begin{bmatrix} x_1 & x_2 & x_3 & x_4 & x_5 \\ x_6 & x_7 & x_8 & x_9 & x_{10} \\ x_{11} & x_{12} & x_{13} & x_{14} & x_{15} \\ x_{16} & x_{17} & x_{18} & x_{19} & x_{20} \\ x_{21} & x_{22} & x_{23} & x_{24} & x_{25} \end{bmatrix}$$

---

[3]Recall that $x_i x_j = \min(x_i x_j)$ and $x_i + x_j = \max(x_i, x_j)$.

we can preserve horizontal, vertical, and diagonal lines of length 5 by using

$$g_{HVD}^{(5\times5)}(\mathbf{x}) = x_{11}x_{12}x_{13}x_{14}x_{15} + x_3x_8x_{13}x_{18}x_{23}$$

$$+ x_1x_7x_{13}x_{19}x_{25} + x_5x_9x_{13}x_{17}x_{21}$$

$$+ (x_{11} + x_{12} + x_{13} + x_{14} + x_{15})(x_3 + x_8 + x_{13} + x_{18} + x_{23})$$

$$(x_1 + x_7 + x_{13} + x_{19} + x_{25})(x_5 + x_9 + x_{13} + x_{17} + x_{21})m^{(5\times5)}(\mathbf{x}).$$

Certain center weighted smoothers will also preserve straight lines. For instance, the 5×5 center-weighted median smoother with center weight 17 will preserve the same features as $g_{HVD}^{(5\times5)}$. However, there will be some loss in noise attenuation. In contrast to Zeng's design method, the design of center-weighted medians, and in general of weighted median smoothers, call for more elaborated optimization procedures. The design advantages of Zeng's approach are further illustrated in the following example. Suppose one wishes to increase the window mask to improve noise attenuation while fixing the length of the feature to be preserved. For instance, the design for a 5×5 mask that preserves lines of length 3 is

$$g_{3HVD}^{(5\times5)}(\mathbf{x}) = x_{11}x_{12}x_{13} + x_{12}x_{13}x_{14} + x_{13}x_{14}x_{15}+$$

$$x_3x_8x_{13} + x_8x_{13}x_{18} + x_{13}x_{18}x_{23}+$$

$$x_1x_7x_{13} + x_7x_{13}x_{19} + x_{13}x_{19}x_{25}+$$

$$x_5x_9x_{13} + x_9x_{13}x_{17} + x_{13}x_{17}x_{21}+$$

$$(x_{11} + x_{12} + x_{13})(x_{12} + x_{13} + x_{14})(x_{13} + x_{14} + x_{15})$$

$$(x_3 + x_8 + x_{13})(x_8 + x_{13} + x_{18})(x_{13} + x_{18} + x_{23})$$

$$(x_1 + x_7 + x_{13})(x_7 + x_{13} + x_{19})(x_{13} + x_{19} + x_{25})$$

$$(x_5 + x_9 + x_{13})(x_9 + x_{13} + x_{17})(x_{13} + x_{17} + x_{21})m^{(5\times5)}(\mathbf{x}).$$

This derivation is much easier than that of an equivalent weighted median smoother. Figure 5.16 illustrates the principles behind Zeng's algorithm. Figure 5.16a shows a 460×480 test image consisting of several horizontal, vertical, and diagonal lines. Figure 5.16b shows the image with additive 10% salt and pepper noise. An unweighted and a center-weighted median smoother of size 3×3 were applied to the noisy image. The median smoother annihilated all image features as expected. The CWM smoother output with center weight 5 is shown in Fig. 5.16c. The output of Zeng's stack smoother ($g_{HVD}^{(3x3)}$) is shown in Fig. 5.16d. We see that Zeng's stack smoother performs as expected, preserving structural features while possessing high noise attenuation. The center-weighted median smoother preserved details, but at

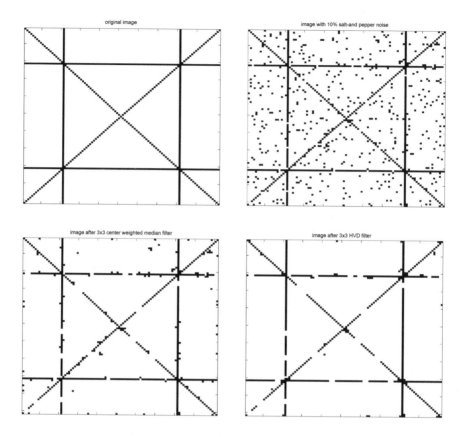

Figure 5.16 (*a*) Test pattern with line features, (*b*) noisy pattern, (*c*) output of 3×3 CWM, (*d*) Zeng's stack smoother preserving line features.

the cost of some reduction in noise attenuation. The noise attenuation of the CWM can be improved by lowering the center weight, although the feature preservation characteristics of the CWM becomes inadequate.

5.4 WEIGHTED MEDIANS IN LEAST ABSOLUTE DEVIATION (LAD) REGRESSION

Linear regression is widely used today in science and engineering applications. It attempts to regulate more than one set of data which are presumed to be linear related, that is, the sets statistically differ by their scale and a possible shift. In microarray data normalization, for example, the goal is to recognize certain contributing gene set by assaying the expression levels of thousands of genes from different arrays

[51]. The underlying assumption is that one or more reference genes have constant expression levels across batches. Thus, variation between arrays must be taken into account before any meaningful biological comparison can be made. Microarray data normalization is often accomplished through a linear regression between two arrays of gene expressions. When the reference gene set is recognized and the linear regression parameters are obtained, normalization of two arrays can then be carried out simply by scaling and shifting one of the arrays using the obtained regression parameters. Comparison of two arrays can then be executed easily.

Historically, linear regression has long been dominated by Least Squares (LS) techniques, mostly because of their elegant theoretical foundation and ease of implementation. The assumption in this method is that the model has normally distributed errors. In many applications, such as in microarray data normalization, heavier-than-Gaussian tailed distributions may be encountered, where outliers in the measurements may easily ruin the estimates [35]. To address this problem, robust regression methods have been developed so as to mitigate the influence of outliers. Among all the approaches to robust regression, the Least Absolute Deviations (LAD) method, or norm, L_1, is considered conceptually the simplest, since it does not require a tuning mechanism like most other robust regression procedures. As a result, LAD regression has drawn significant attention in statistics, finance, engineering, and other applied sciences as detailed in a series of studies on L_1-norm methods [60, 61, 62, 63]. LAD regression is based on the assumption that the model has Laplacian distributed errors. Unlike the LS approach though, LAD regression has no closed-form solution, hence numerical and iterative algorithms must be resorted to.

Surprisingly to many, the LAD regression method first suggested by Boscovich (1757) and studied by Laplace (1793) predated the LS technique originally developed by Legendre (1805) and Gauss (1823) [35, 60]. It was not until nearly a century later that Edgeworth (1887) [66] proposed a general numerical method to solve the unconstrained LAD problem, where the *weighted median* was introduced as the basic operation in each iteration. Edgeworth's method, however, suffers from cycling when data has degeneracies [97]. A breakthrough came in the 1950's when Harris (1950) [95] brought in the notion that linear programming techniques could be used to solve the LAD regression, and Charnes et al. (1955) [44] actually utilized the simplex method to minimize the LAD objective function. Many simplex-like methods blossomed thereafter, among which Barrodale and Roberts (1973) [29] and Armstrong, Frome, and Kung (1979) [15] are the most representative. Other efficient approaches include the active set method by Bloomfield and Steiger (1980) [34], the direct decent algorithm by Wesolowsky (1981) [196], and the interior point method proposed by Zhang (1993) [206]. More historical background on LAD estimates can be found in [60].

5.4.1   Foundation and Cost Functions

The simple LAD regression problem is formulated as follows. Consider N observation pairs (X_i, Y_i) modeled in a linear fashion

$$Y_i = aX_i + b + U_i, \qquad i = 1, 2, \ldots, N \tag{5.78}$$

where a is the unknown slope of the fitting line, b the intercept, and U_i are unobservable errors drawn from a random variable U obeying a zero-mean Laplacian distribution $f(U) = \frac{1}{2\lambda} e^{\frac{-|U|}{\lambda}}$ with variance $\sigma^2 = 2\lambda^2$. The Least Absolute Deviation regression is found by choosing a pair of parameters a and b that minimizes the objective function

$$F_1(a, b) = \sum_{i=1}^{N} |Y_i - aX_i - b|, \tag{5.79}$$

which has long been known to be continuous and convex [35]. Moreover, the cost surface is of a polyhedron shape, and its edge lines are characterized by the sample pairs (X_i, Y_i). As a comparison, the objective function of the Least Squares regression is well known as

$$F_2(a, b) = \sum_{i=1}^{N} (Y_i - aX_i - b)^2, \tag{5.80}$$

which has a closed-form solution

$$\begin{cases} a^* = \dfrac{\sum (X_i - \bar{X})(Y_i - \bar{Y})}{\sum (X_i - \bar{X})^2} \\[2mm] b^* = \bar{Y} - a^* \bar{X}. \end{cases} \tag{5.81}$$

Since the LAD objective function $F_1(a, b)$ is the only one being concerned throughout the section, we will drop the subscript 1 hereafter. Figure 5.17 depicts the different line-fitting characteristics between LAD and LS. The LS fitting line is greatly offset by the single outlier in the samples, the sign of the slope can be even reverted given a big enough outlier, as shown in Figure 5.17c.

If the value of a is fixed at first, say $a = a_0$, the objective function (5.79) now becomes a one-parameter function of b

$$F(b) = \sum_{i=1}^{N} |Y_i - a_0 X_i - b|. \tag{5.82}$$

Assuming a Laplace distribution for the errors U_i, the above cost function reduces to a Maximum Likelihood estimator of location for b. That is, we observe the sequence of random samples $\{Y_i - a_0 X_i\}$, and the goal is to estimate the fixed but unknown location parameter b. Thus the parameter b^* in this case can be obtained by

$$b^* = \text{MED}(Y_i - a_0 X_i \,|\, {}_{i=1}^{N}). \tag{5.83}$$

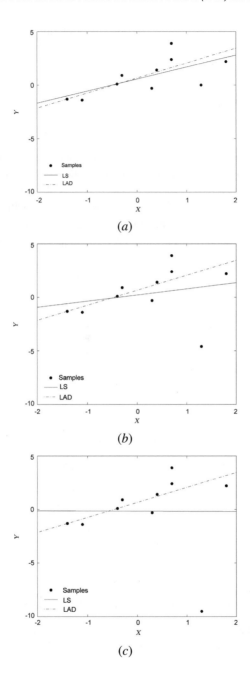

Figure 5.17 LAD and LS linear regression results as the value of an outlier sample is increased.

If, on the other hand, we fix $b = b_0$, the objective function reduces to

$$F(a) = \sum_{i=1}^{N} |Y_i - b_0 - aX_i|$$

$$= \sum_{i=1}^{N} |X_i| \left| \frac{Y_i - b_0}{X_i} - a \right|. \tag{5.84}$$

Again, if the error random variable U_i obeys a Laplacian distribution, the observed samples $\{\frac{Y_i - b_0}{X_i}\}$ are also Laplacian distributed, but with the difference that each sample in this set has different variance. The reason is obvious since for each known X_i and zero mean U_i, $\frac{U_i}{X_i}$ remains a zero mean Laplacian with variance scaled by $\frac{1}{X_i^2}$. Thus the parameter a^* minimizing the cost function (5.84) can still be seen as the ML estimator of location for a, and can be calculated out as the *weighted median*,

$$a^* = \text{MED}\left(|X_i| \diamond \left. \frac{Y_i - b_0}{X_i} \right|_{i=1}^{N} \right), \tag{5.85}$$

where $\diamond$ is the replication operator. For a positive integer $|X_i|$, $|X_i| \diamond Y_i$ means Y_i is replicated $|X_i|$ times.

A simple and intuitive way of solving the LAD regression problem is through the following iterative algorithm:

(1) Set $k = 0$. Find an initial value a_0 for a, such as the Least Squares (LS) solution.

(2) Set $k = k + 1$ and obtain a new estimate of b for a fixed a_{k-1} using

$$b_k = \text{MED}(Y_i - a_{k-1}X_i \,|\, _{i=1}^{N}).$$

(3) Obtain a new estimate of a for a fixed b_k using

$$a_k = \text{MED}\left(|X_i| \diamond \left. \frac{Y_i - b_k}{X_i} \right|_{i=1}^{N} \right).$$

(4) Once a_k and b_k do not deviate from a_{k-1} and b_{k-1} within a tolerance range, end the iteration. Otherwise, go back to step 2).

Since the median and weighted median operations are both ML location estimators under the least absolute criterion, the cost functions will be nonincreasing throughout the iterative procedure, that is

$$F(a_{k-1}, b_{k-1}) \geq F(a_{k-1}, b_k) \geq F(a_k, b_k).$$

The algorithm then, converges iteratively. Since the objective function $F(a, b)$ is continuous and convex, one may expect that the algorithm converges to the global

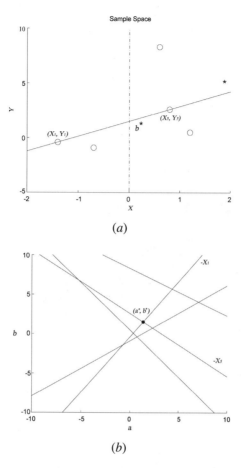

Figure 5.18 Illustration of the sample space and the parameter space in the simple linear regression problem. The circles in the upper plot represent the samples, the dot in the lower plot represents the global minimum.

minimum. However, careful inspection reveals that there are cases where the algorithm does not reach the global minimum. To see this, it is important to describe the relationship between the sample space and the parameter space.

As shown in Figure 5.18, the two spaces are dual to each other. In the sample space, each sample pair (X_i, Y_i) represents a point on the plane. The solution to the problem (5.78), namely (a^*, b^*), is represented as a line with slope a^* and intercept b^*. If this line goes through some sample pair (X_i, Y_i), then the equation $Y_i = a^* X_i + b^*$ is satisfied. On the other hand, in the parameter space, (a^*, b^*) is a point on the plane, and $(-X_i, Y_i)$ represents a line with slope $(-X_i)$ and intercept Y_i. When $b^* = (-X_i)a^* + Y_i$ holds, it can be inferred that the point (a^*, b^*) is on the line defined by $(-X_i, Y_i)$. As can be seen in Figure 5.18, the line going through

(X_1, Y_1) and (X_5, Y_5) in the sample space has a slope a^* and an intercept b^*, but in the parameter space it is represented as a point which is the intersection of two lines with slopes $(-X_1)$ and $(-X_5)$ respectively. The sample set used to generate Figure 5.18 is, in a (X_i, Y_i) manner, $[(-1.4, -0.4), (0.6, 8.3), (1.2, 0.5), (-0.7, -0.9), (0.8, 2.6)]$.

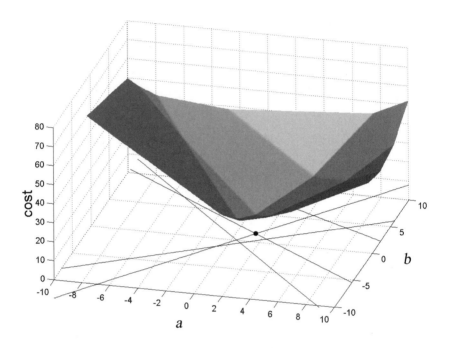

Figure 5.19 The cost surface of the LAD regression problem. The dot at an intersection on the a-b plane represents the global minimum. To better illustrate the inner topology of the function, the half surface that towards the viewers is cut off.

The structure of the objective function $F(a, b)$ is well defined as a polyhedron sitting on top of the a-b plane, as seen in Figure 5.19. The projections of the polyhedron edges onto the plane are exactly the lines defined by sample pairs (X_i, Y_i), which is why the term "edge line" is used. In other words, every sample pair (X_i, Y_i) has a corresponding edge line in the parameter space. Moreover, the projections of the polyhedron corners are those locations on the a-b plane where two or more of the edge lines intersect. Most importantly, the minimum of this convex, linearly segmented error surface occurs at one of these corners.

To describe the dynamics of this simple iterative method, consider Step 2 in the procedure, where a new estimate b_k is calculated based on a fixed, previously obtained a_{k-1} through a median operation. Since the median is of selection type, its output is always one of the inputs. Without loss of generality, assume $b_k = Y_j - a_{k-1}X_j$, which means that the newly estimated parameter pair (a_{k-1}, b_k) is on the edge line defined by $(-X_j)$ and Y_j. Thus, the geometrical interpretation of Step 2 can be derived as follows: draw a vertical line at $a = a_{k-1}$ in the parameter space, mark all the intersections of this line with N edge lines[4]. The intersection on the edge line defined by $(-X_j)$ and Y_j is vertically the median of all, thus its b-coordinate value is accepted as b_k, the new update for b. Similar interpretation can be made for Step 3, except that the chosen intersection is a weighted median output, and there may be some edge lines parallel to the a-axis.

The drawback of this algorithm is that the convergence dynamics depends on the geometry of the edge lines in the parameter space. As can be seen in Figure 5.20a, where the iteration is carried on between edge lines in an inefficient zigzag manner needing infinite steps to converge to the global minimum. Moreover, as illustrated in Figure 5.20b, it is possible that vertical optimization and horizonal optimization on the edge lines can both give the same results in each iteration. Thus the algorithm gets stuck in a nonoptimal solution. The sample set used for Figure 5.20a is $[(-0.1, -3.2),$ $(-0.9, -2.2), (0.4, 5.7), (-2.4, -2.1), (-0.4, -1.0)]$, the initial values for a and b are 5 and 6. The sample set used for Figure 5.20b is $[(0.3, -1.0), (-0.4, -0.1),$ $(-2.0, -2.9), (-0.9, -2.4), (-1.1, 2.2)]$, the initial values for a and b are -1 and 3.5.

5.4.2 LAD Regression with Weighted Medians

To overcome these limitations, the iterative algorithm must be modified exploiting the fact that the optimal solution is at an intersection of edge lines. Thus, if the search is directed along the edge lines, then a more accurate and more efficient algorithm can be formulated. The approach described here is through coordinates transformation. The basic idea is as follows: In the parameter space, if the coordinates are transformed so that the edge line containing the previous estimate (a_{k-1}, b_{k-1}) is parallel to the a'-axis at height b'_{k-1}, then the horizontal optimization based upon b'_{k-1} is essentially an optimization along this edge line. The resultant (a'_k, b'_k) will be one of the intersections that this line has with all other edge lines, thus avoiding possible zigzag dynamics during the iterations. Transforming the obtained parameter pair back to the original coordinates results in (a_k, b_k). This is illustrated in Figure 5.21. The only requirement for this method is that the shape of the cost surface must be preserved upon transformation, thus the same optimization result can be achieved. Notice that if an edge line is horizontal, its slope $(-X_j)$ has to be 0. We will show shortly that a simple shifting in the sample space can satisfy the requirement.

The following is the resultant algorithm for LAD regression.

(1) Set $k = 0$. Initialize b to be b_0 using the LS solution

$$b_0 = \frac{\sum_{i=1}^{N}(X_i - \bar{X})(\bar{Y}X_i - \bar{X}Y_i)}{\sum_{i=1}^{N}(X_i - \bar{X})^2}. \qquad (5.86)$$

---

[4]Since all meaningful samples are finite, no edge lines will be parallel to the b-axis, hence there must be N intersections.

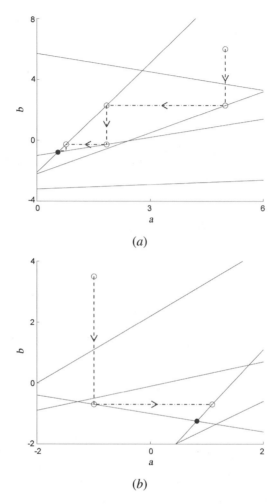

Figure 5.20 The parameters' trajectories during the iterations. Vertical dashed lines represent b updates, horizontal dash-dotted lines represent a updates. (*a*) zigzag case, (*b*) non-optimal case. The marked dots represent the global minima. To better illustrate, the initial values for a and b are not set from the LS solution.

Calculate a_0 by a weighted median

$$a_0 = \text{MED} \left(|X_i| \diamond \left. \frac{Y_i - b_0}{X_i} \right|_{i=1}^{N} \right). \tag{5.87}$$

Keep the index j which satisfies $a_0 = \frac{Y_j - b_0}{X_j}$. In the parameter space, (a_0, b_0) is on the edge line with slope $(-X_j)$ and intercept Y_j.

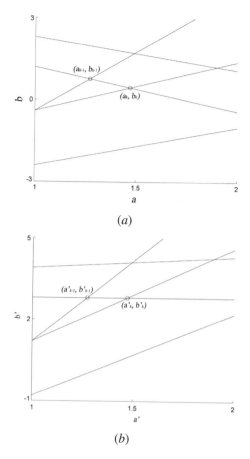

Figure 5.21 The illustration of one iteration. The previous estimate (a_{k-1}, b_{k-1}) is mapped into the transformed coordinates as (a'_{k-1}, b'_{k-1}); (a'_k, b'_k) is obtained through ML estimation in the transformed coordinates; The new estimate (a_k, b_k) is formed by mapping (a'_k, b'_k) back into the original coordinates. The sample set is $[(1.6, 2.8), (-1.4, -3.8), (1.2, 3.5), (-4.3, -4.7), (-1.8, -2.2)]$.

(2) Set $k = k+1$. In the sample space, right shift the coordinates by X_j, so that the newly formed y'-axis goes through the original (X_j, Y_j). The transformations in the sample space are

$$X'_i = X_i - X_j, \quad Y'_i = Y_i, \tag{5.88}$$

and the transformations in the parameter space

$$a'_{k-1} = a_{k-1}, \quad b'_k = b'_{k-1} = b_{k-1} + a_{k-1}X_j. \tag{5.89}$$

The shifted sample space (X', Y') corresponds to a new parameter space (a', b'), where $(-X'_j, Y'_j)$ represents a horizontal line.

(3) Perform a weighted median to get a new estimate of a'

$$a'_k = \text{MED}\left(|X'_i| \diamond \left.\frac{Y'_i - b'_k}{X'_i}\right|_{i=1}^N\right). \tag{5.90}$$

Keep the new index t which gives $a'_k = \frac{Y'_t - b'_k}{X'_t}$.

(4) Transform back to the original coordinates

$$a_k = a'_k, \quad b_k = b'_k - a'_k X_j \tag{5.91}$$

(5) Set $j = t$. If a_k is identical to a_{k-1} within the tolerance, end the program. Otherwise, go back to step 2.

It is simple to verify that the transformed cost function is the same as the original one using the relations in (5.88) and (5.89). For fixed b_k,

$$\begin{aligned}
F'(a') &= \sum_{i=1}^N |Y'_i - a'X'_i - b'_k| \\
&= \sum_{i=1}^N |Y_i - a(X_i - X_j) - (aX_j + b_k)| \\
&= \sum_{i=1}^N |Y_i - aX_i - b_k| = F(a).
\end{aligned} \tag{5.92}$$

This relationship guarantees that the new update in each iteration is correct.

5.4.3   Simulation

The computational complexity in Li and Arce's algorithm resides in the weighted median operation used at each iteration. Essentially, it is a sorting problem with complexity proportional to the order of $N \log N$, where N is the sample size. In this particular application, a speed-up can be achieved by not carrying out a full sorting operation every time. In [196], a short cut is used to circumvent the time-consuming full-sorting procedure. In essence, the previous estimate can be considered close enough to the true value, thus fine-tuning can be executed around this point.

Two criteria are often used to compare LAD algorithms: speed of convergence and complexity. Most of the efficient algorithms, in terms of convergence speed (except for Wesolowsky's and its variations), are derived from Linear Programming (LP) perspectives, such as simplex and interior point. Take Barrodale and Roberts' (BR) algorithm[5] (1973) for example. Its basic idea is to apply row and column operations

---

[5]which can be considered as the basic form of the other two best simplex-type algorithms, namely, Bloomfield and Steiger's (1983) (BS), and Armstrong, Frome and Kung's (1979) (AFK), according to [60].

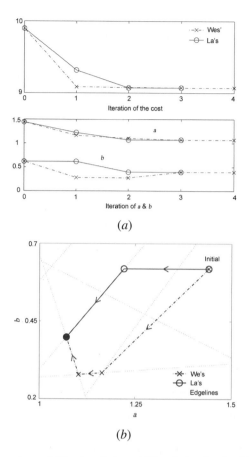

Figure 5.22 Comparison of Wesolowsky's and Li and Arce's algorithms. (*a*) shows the iterations of the parameters and the costs. (*b*) the convergence of the algorithms on the parameter space. Two algorithms choose the same LS solution as the initial point. The marked dot in (*b*) represents the global minimum. Notice that not all the edgelines are plotted.

on a constructed $(N + K) \times (K + 1)$ matrix $\mathbf{A}$. The initial value of $\mathbf{A}$ is

$$\mathbf{A} = \begin{bmatrix} \mathbf{X} & \mathbf{Y} \\ \mathbf{I} & \mathbf{0} \end{bmatrix}, \qquad (5.93)$$

where $\mathbf{Y}$ is an $N \times 1$ vector of observations of the dependent variable and $\mathbf{X}$ is an $N \times K$ matrix of the independent variables. For the simple regression case, $K = 2$. BR-like algorithms usually consist of two phases: Phase I forms a set of independent edge direction vectors, Phase II updates the variable basis until it converges. In general, BR-like algorithms are slightly faster than other algorithms with simpler structures. Their computational complexity, however, is significantly higher. The complicated variable definition and logical branches used in BR-like algorithms cause tremendous efforts in their hardware implementations and are thus less attractive in such cases. Focusing on efficient algorithms that have a simple structure for ease of implementation, Wesolowsky's direct descent algorithm stands out. The algorithm is summarized below.

Step 1 Set $k = 0$. Choose the initial values a_0, b_0. Choose j so that $|Y_j - a_0 X_j - b_0|$ is a minimum.

Step 2 Set $k = k + 1$. Using the weighted median structure to get the update for b

$$b_k = \text{MED} \left(\left| 1 - \frac{X_i}{X_j} \right| \diamond \frac{Y_i - \frac{Y_j X_i}{X_j}}{1 - \frac{X_i}{X_j}} \right|^N_{i=1} \right). \tag{5.94}$$

Recording the index i at which the term $(Y_i - Y_j X_i / X_j)/(1 - X_i / X_j)$ is the weighted median output.

Step 3   a) If $b_k - b_{k-1} = 0$: if $k \geq 3$, go to **Step 4**; if not, set $j = i$ and go to **Step 2**.

b) If $b_k - b_{k-1} \neq 0$: set $j = i$ and go to **Step 2**.

Step 4 Let $b^* = b_k$, $a^* = Y_j / X_j - b^* / X_j$.

The major difference between Wesolowsky's algorithm and that of Li and Arce's (LA) is that the weighted median operations in their case are used for intercept b updates, while in LA algorithm they are used for slope a updates. Also as depicted in Figure 5.22b, the first iterations of the two algorithms are different. LA algorithm picks the first a update horizontally, whereas Wesolowsky's algorithm chooses a nearby intersection based on the criterion described in **Step 1**.

Since the realization of the weighted median in both algorithms can benefit from the partial sorting scheme stated above, the computational complexity of both methods is comparable. Li and Arce's algorithm, however, is slightly more efficient, reaching convergence in less iterations as depicted in Figure 5.23, in terms of number of iterations. It can be observed in Figure 5.23 that for large sample sets, Li and Arce's LAD regression method requires 5% less iterations, and about 15% less for small sample sets.

Problems

5.1   Let $X_1, X_2, \ldots, X_N$ be N i.i.d. Cauchy observations with zero median and $k = 1$.

(a) Show that $E(X_{(k)}^2) < \infty$ if and only if $3 \leq k \leq N - 2$.

(b) Show that the median of N Cauchy i.i.d. samples has a finite second-order central moment only if $N \geq 5$.

5.2   Prove that the superposition property and the impulse response analysis do not apply to the running median.

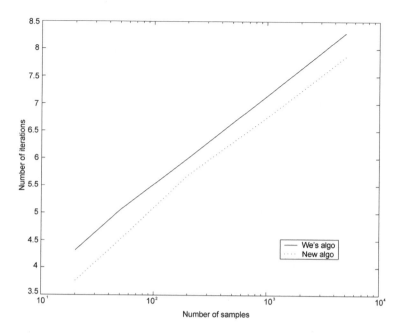

Figure 5.23 Comparison of the number of iterations of Wesolowsky's and LA algorithms. The dimensions of the sample sets are chosen as [20, 50, 200, 1000, 5000], each having 1000 averaging runs.

5.3

(a) Prove that the rth-order statistic of N samples can be found by the maxima of set of minimums:

$$x_{(r)} = MAX \left[\text{set of } \begin{pmatrix} N \\ N-r+1 \end{pmatrix} \text{ minima of } (x_{j_1}, x_{j_2}, \dots, x_{j_{N-r+1}}) \right]$$
$$(5.95)$$

where $(j_1, j_2, \dots, j_a)$ index all possible combinations of a samples in a set of N samples.

(b) Derive an expression for the rth-order statistic of N samples in terms of the minimum of a set of maximums.

5.4 Explain the problems encountered with the WM estimation if some of the weights in 5.16 are allowed to be negative.

5.5 Prove that the procedure described in (5.24) minimizes the sum of weighted deviations required to compute the WM in (5.15).

5.6 Recall that for a median filter of window size $N = 2N_1 + 1$, a necessary and sufficient condition for the signal to be invariant (a root) under median smoothing is that the extended (beginning and end appended) signal be LOMO($N_1 + 2$), where a sequence $\{X(\cdot)\}$ is said to be locally monotonic of length m, denoted LOMO(m), if the subsequence $X(n), X(n+1), \ldots, X(n+m-1)$ is monotonic for all $n > 1$.

Give the necessary and sufficient (local monotonic) conditions for a signal to be invariant under center-weighted smoothing for the center weight being equal to 3, for $N_1 = 1, 2, 3, \ldots$.

5.7   Prove that the center-weighted Median in (5.35) can be obtained by:

$$Y(n) = \text{MEDIAN}\left[X_{(k)},\ X_c,\ X_{(N+1-k)}\right]$$

as described in (5.36).

6

Weighted Median Filters

Weighted median smoothers admit only positive weights. This is a limitation as WM smoothers are, in essence, limited to have low-pass type filtering characteristics. Although WM smoothers have some analogies with linear FIR filters, they are equivalent to the normalized weighted average with non-negative weights – a severely constrained subset of linear FIR filter. A number of engineering applications require band-pass or high-pass frequency filtering characteristics. Equalization, deconvolution, prediction, beamforming, and system identification are example applications where filters having band-pass or high-pass characteristics are of fundamental importance. Linear FIR equalizers admitting only positive filter weights, for instance, would lead to inadequate results. Thus, it is not surprising that weighted median smoothers admitting only positive weights lead to inadequate results in some applications.

This Chapter focuses on weighted median filters admitting positive as well as negative weights. WM filters thus overcome the limitations of WM smoothers. As would be expected, weighted median filters reduce to weighted median smoothers whenever the filter coefficients in the WM filter structure are constrained to be positive.

6.1 WEIGHTED MEDIAN FILTERS WITH REAL-VALUED WEIGHTS

Weighted median smoothers emerge as the solution to the Maximum Likelihood estimation of location for a set of independently distributed Laplacian samples with unequal variance. For Gaussian-distributed samples, the equivalent estimate is the normalized weighted average with non-negative weights. In order to formulate the general weighted median filter structure, it is logical to ask how linear FIR filters arise within the location estimation problem. The answer provides the key to the formulation of the general WM filter. To this end, consider N samples $X_1, X_2, \ldots, X_N$ obeying a multivariate Gaussian distributiondistribution,multivariate Gaussian

$$f(\mathbf{X}) = \frac{1}{(2\pi)^{N/2}[\det(\mathbf{R})]^{1/2}} \exp[-\frac{1}{2}(\mathbf{X} - \mathbf{e}\beta)^T \mathbf{R}^{-1}(\mathbf{X} - \mathbf{e}\beta)] \qquad (6.1)$$

where $\mathbf{X} = [X_1, X_2, \ldots, X_N]^T$ is the observation vector, $\mathbf{e} = [1, 1, \ldots, 1]^T$, β is the location parameter, $\mathbf{R}$ is the covariance matrix, and $\det(\mathbf{R})$ is the determinant of $\mathbf{R}$. The Maximum Likelihood estimate of the location parameter β results in

$$\hat{\beta} = \frac{\mathbf{e}^T \mathbf{R}^T \mathbf{X}}{\mathbf{e}^T \mathbf{R}\mathbf{e}} = \mathbf{W}^T \mathbf{X} \qquad (6.2)$$

where $\mathbf{e}^T \mathbf{R}\mathbf{e} > 0$, due to the positive definite nature of the covariance matrix, and where elements in the vector $\mathbf{e}^T \mathbf{R}^T$ can take on positive as well as negative values. Thus, (6.2) takes the structure of a linear FIR filter whose weights W_i may take on negative values depending on the mutual correlation of the observation samples.

The extension of the above to the case of Laplacian distributed samples, and in general to other nonGaussian distributions, unfortunately becomes too cumbersome. The multivariate Laplacian distribution, and in general all nonGaussian multivariate distributions, do not lead to simple ML location estimates. The complexity in these solutions has hindered the development of nonlinear filters having attributes comparable to that of linear FIR filters. Notably, however, a simple approach was discovered that can overcome these limitations [6]. In this approach, a generalization of the sample mean leads to the class of linear FIR filters. This generalization is, in turn, applied to the sample median forming the class of weighted median filters that admits both positive and negative weights. The extension, turns out, not only to be natural, leading to a significantly richer filter class, but it is simple as well.

The sample mean MEAN $(X_1, X_2, \ldots, X_N)$ can be generalized to the class of linear FIR filters as

$$\bar{\beta} = \text{MEAN}(W_1 \cdot X_1, W_2 \cdot X_2, \ldots, W_N \cdot X_N) \qquad (6.3)$$

where $W_i \in R$. In order to apply the analogy to the median filter structure, (6.3) must be written as

$$\bar{\beta} = \text{MEAN}(|W_1| \cdot \text{sgn}(W_1)X_1, |W_2| \cdot \text{sgn}(W_2)X_2, \ldots, |W_N| \cdot \text{sgn}(W_n)X_N), \qquad (6.4)$$

where the sign of the weight affects the corresponding input sample and the weighting is constrained to be non-negative. By analogy, the class of weighted median filters admitting real-valued weights emerges as defined next.

Figure 6.45 Multivariate medians for color images in salt-and-pepper noise, $\mu = 0.001$ for the WVM, $\mu_v, \mu_w = 0.05$ for the marginal WMM. From left to right and top to bottom: noiseless image, contaminated image, WVM with 3×3 window, marginal WMM with 3×3 window.

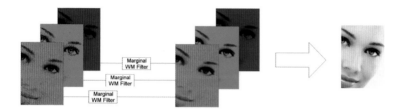

Figure 6.43 Center WM filter applied to each component independently.

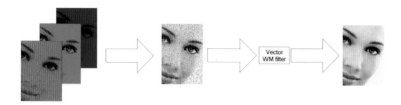

Figure 6.44 Center vector WM filter applied in the 3-dimensional space.

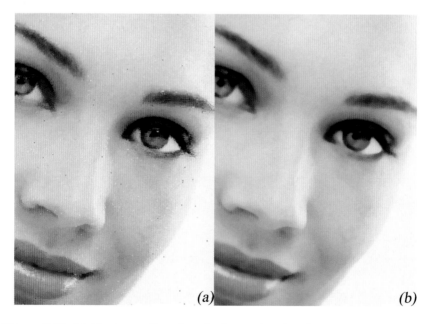

Figure 6.46 Multivariate medians for color images in salt-and-pepper noise, $\mu = 0.001$ for the WVM, $\mu_v, \mu_w = 0.05$ for the marginal WMM (continued). From left to right: WVM with 5×5 window, marginal WMM with 5×5 window.

DEFINITION 6.1 (WEIGHTED MEDIAN FILTERS) *Given a set of N real valued weights* $\langle W_1, W_2, \ldots, W_N \rangle$ *and the observation vector* $\mathbf{X} = [X_1, X_2, \ldots, X_N]^T$, *the weighted median filter output is defined as*

$$\tilde{\beta} = \text{MEDIAN}(|W_1| \diamond sgn(W_1)X_1, \ldots, |W_N| \diamond sgn(W_n)X_N), \qquad (6.5)$$

with $W_i \in R$ *for* $i = 1, 2, \ldots, N$, *and where* $\diamond$ *is the replication operator.*

Note that the weight signs are uncoupled from the weight magnitude values and are merged with the observation samples. The weight magnitudes play the equivalent role of positive weights in the framework of weighted median smoothers.

EXAMPLE 6.1 (WEIGHTED MEDIAN FILTER COMPUTATION)

Consider first the case where the weights are integer-valued and where these add up to an odd integer number. Let the window size be 5 defined by the symmetric weight vector $\mathbf{W} = \langle 1, -2, 3, -2, 1 \rangle$. For the observation vector $\mathbf{X}(n) = [2, -6, 9, 1, 12]$, the weighted median filter output is found as

$$
\begin{aligned}
Y(n) &= \text{MEDIAN}[\, 1 \diamond 2, -2 \diamond -6, 3 \diamond 9, -2 \diamond 1, 1 \diamond 12 \,] \\
&= \text{MEDIAN}[\, 1 \diamond 2, 2 \diamond 6, 3 \diamond 9, 2 \diamond -1, 1 \diamond 12 \,] \\
&= \text{MEDIAN}[\, 2, 6, 6, 9, 9, 9, -1, -1, 12 \,] \qquad (6.6) \\
&= \text{MEDIAN}[\, -1, -1, 2, 6, \underline{6}, 9, 9, 9, 12 \,] \\
&= 6
\end{aligned}
$$

where the median filter output value is underlined in equation (6.6). ∎

Note that the output in the example above is a signed sample whose value is not equal to that of any of the input samples. Note also that as a result of the negative weights, the computation of the weighted median filter is not shift invariant. Consider a shift of 2 on the samples of $\mathbf{X}$ such that $X_i' = X_i + 2$. The weighted median filtering of $\mathbf{X}' = [4, -4, 11, 3, 15]$ with the weight vector $\mathbf{W} = \langle 1, -2, 3, -2, 1 \rangle$ leads to the output $Y'(n) = 4$, which does not equal the previous output in (6.6) of 6 plus the appropriate shift.

EXAMPLE 6.2

Consider the case where the WM filter weights add up to an even integer with $\mathbf{W} = \langle 1, -2, 2, -2, 1 \rangle$. Furthermore, assume the observation vector consists of a set of constant valued samples $\mathbf{X}(n) = [5, 5, 5, 5, 5]$. The weighted median filter output in this case is found as

$$Y(n) = \text{MEDIAN}[\, 1 \diamond 5, -2 \diamond 5, 2 \diamond 5, -2 \diamond 5, 1 \diamond 5 \,]$$

$$= \text{MEDIAN}[\, 1 \diamond 5, 2 \diamond -5, 2 \diamond 5, 2 \diamond -5, 1 \diamond 5 \,]$$

$$= \text{MEDIAN}[\, 5, -5, -5, 5, 5, -5, -5, 5 \,] \tag{6.7}$$

$$= \text{MEDIAN}[\, -5, -5, -5, \underline{-5, 5}, 5, 5, 5 \,]$$

$$= 0,$$

where the median filter output is the average of the underlined samples in equation (6.7).   ∎

Note that in order for the WM filter to have band- or high-pass frequency characteristics where constant signals are annihilated, the weights must add to an even number such that averaging of the middle rank samples occurs. When the WM filter weights add to an odd number, the output is one of the signed input samples, and consequently the filter is unable to suppress constant-valued signals.

In general, the WM filter output can be computed without replicating the sample data according to the corresponding weights, as this increases the computational complexity. A more efficient method to find the WM is shown next, which not only is attractive from a computational perspective but it also admits real-valued weights. The weighted median filter output for noninteger weights can be determined as follows:

(1) Calculate the threshold $T_0 = \frac{1}{2} \sum_{i=1}^{N} |W_i|$.

(2) Sort the signed observation samples $\text{sgn}(W_i)X_i$.

(3) Sum the magnitude of the weights corresponding to the sorted "signed" samples beginning with the maximum and continuing down in order.

(4) The output is the signed sample whose weight magnitude causes the sum to become $\geq T_0$. For band- and high-pass characteristics, the output is the average between the signed sample whose weight magnitude causes the sum to become $\geq T_0$ and the next smaller signed sample.

EXAMPLE 6.3
___

Consider the window size 5 WM filter defined by the real valued weights $\langle W_1, W_2, W_3, W_4, W_5 \rangle = \langle 0.1, 0.2, 0.3, -0.2, 0.1 \rangle$. The output for this filter operating on the observation set $[X_1, X_2, X_3, X_4, X_5] = [-2, 2, -1, 3, 6]$ is found as follows. Summing the weights' magnitude gives the threshold $T_0 = \frac{1}{2} \sum_{i=1}^{5} |W_i| = 0.45$. The signed observation samples, sorted observation samples, their corresponding weight, and the partial sum of weights (from each ordered sample to the maximum) are:

observation samples	$-2, \quad 2, \quad -1, \quad 3, \quad 6$
corresponding weights	$0.1, 0.2, 0.3, -0.2, 0.1$
sorted signed observation samples	$-3, -2, -1, \quad 2, \quad 6$
corresponding weights' magnitude	$0.2, 0.1, 0.3, \quad 0.2, \quad 0.1$
partial weight sums	$0.9, 0.7, \underline{0.6}, \quad 0.3, \quad 0.1$

Thus, the output is -1 since when starting from the right (maximum sample) and summing the weights, the threshold $T_0 = 0.45$ is not reached until the weight associated with -1 is added. The underlined sum value above indicates that this is the first sum which meets or exceeds the threshold. To guarantee high- or band-pass characteristics, the WM filter output would be modified to compute the average between -1 and -2, leading to -1.5 as the output value. ∎

Although the four-step procedure described above to compute the weighted median filter is straightforward, the weighted median filter computation can be expressed more succinctly as follows. Let the signed samples $\text{sgn}(W_i)X_i$ and their corresponding absolute valued weights be denoted as S_i and $|W_i|$, respectively. The sorted "signed" samples are then denoted as $S_{(i)}$ where $S_{(1)} \leq S_{(2)} \leq \ldots \leq S_{(N)}$. The absolute valued weights corresponding to the sorted signed samples are denoted as $|W_{[k]}|$, that is, the absolute value of the concomitant of the kth order statistic. In the previous example, the weight associated with the fourth-order statistic $S_{(4)}$ is, for instance, $|W_{[4]}| = |W_2| = 0.2$. With this notation, the selection-weighted median filter output can be written as

$$\hat{\beta} = \{S_{(k)} : \min_k \text{ for which } \sum_{i=0}^{k} |W_{[N-i]}| \geq T_0\}. \tag{6.8}$$

The weighted median filter output when the average of the middle rank samples is used can be written as

$$\hat{\beta} = \{(S_{(k)} + S_{(k-1)})/2 : \min_k \text{ for which } \sum_{i=0}^{k} |W_{[N-i]}| \geq T_0\}. \tag{6.9}$$

Cost Function Interpretation The effect that negative weights have on the weighted median operation is similar to the effect that negative weights have on linear FIR filter outputs. This is illustrated by the effect that negative weights have on the cost function minimized by the WM filter. To this end, it is simple to show that the weighted mean and the weighted median operations, shown in (6.4) and (6.5) respectively, minimize

$$G_2(\beta) = \sum_{i=1}^{N} |W_i| \left(\mathrm{sgn}(W_i)X_i - \beta\right)^2$$

and

$$G_1(\beta) = \sum_{i=1}^{N} |W_i||\mathrm{sgn}(W_i)X_i - \beta|. \tag{6.10}$$

While $G_2(\beta)$ is a convex continuous function, $G_1(\beta)$ is a convex but piecewise linear function whose minima is guaranteed to be one of the signed input samples (i.e., $\mathrm{sgn}(W_i) X_i$).

As an example consider the observation vector $[X_1, X_2, X_3, X_4, X_5] = [-2, 2, -1, 3, 6]$ applied to a WM filter where two sets of weights are used. The first set is $\langle W_1, W_2, W_3, W_4, W_5 \rangle = \langle 0.1, 0.2, 0.3, 0.2, 0.1 \rangle$ where all the coefficients are positive, and the second set being $\langle 0.1, 0.2, 0.3, -0.2, 0.1 \rangle$ where W_4 has been changed, with respect to the first set of weights, from 0.2 to -0.2. Recall that the linear FIR and WM filter outputs respectively minimize the cost functions $G_2(\beta)$ and $G_1(\beta)$.

Figure 6.1a shows the cost functions $G_2(\beta)$, as a function of β, corresponding to the linear FIR filter for the two sets of filter weights. Notice that by changing the sign of W_4, we are effectively moving X_4 to its new location $\mathrm{sgn}(W_4)X_4 = -3$. This, in turn, pulls the minimum of the cost function towards the relocated sample $\mathrm{sgn}(W_4)X_4$. Negatively weighting X_4 on $G_1(\beta)$ has a similar effect as shown in Figure 6.1(b). In this case, the minimum is pulled towards the new location of $\mathrm{sgn}(W_4)X_4$. The minimum, however, occurs at one of the signed samples $\mathrm{sgn}(W_i)X_i$.

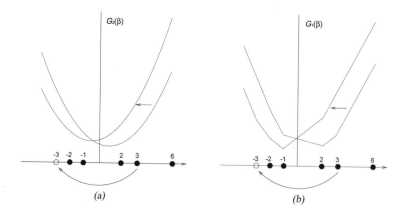

(a)   (b)

Figure 6.1 Effects of negative weighting on the cost functions $G_2(\beta)$ and $G_1(\beta)$. The input samples are $[X_1, X_2, X_3, X_4, X_5] = [-2, 2, -1, 3, 6]$, which are filtered by the two set of weights $\langle 0.1, 0.2, 0.3, 0.2, 0.1 \rangle$ and $\langle 0.1, 0.2, 0.3, -0.2, 0.1 \rangle$, respectively.

To illustrate the characteristics of WM filters and their advantages over WM smoothers, several examples are described next. The first example shows that WM filters, like their linear FIR filter counterparts, can be designed to have frequency selection characteristics.

EXAMPLE 6.4 (BANDPASS FILTERING)

Since linear FIR filters output the mean of a set of weighted samples, the median of an equivalently weighted sample set ought to provide a similar output. Notably, this is the case even when the same set of weights as those designed for a linear FIR filter is used in the weighted median filter structure. This approach to assign the WM filter weights, however, is not optimal and will lead to undesirable artifacts on the output. Nonetheless, the frequency response characteristics of the attained weighted median filter follows that of the equivalent linear FIR filter, but more importantly, these are significantly more robust in the processing of signals embedded in noise.

Figure 6.2a depicts a linearly swept-frequency cosine signal spanning instantaneous frequencies ranging from 0 to 400 Hz. Figure 6.2b shows the chirp signal filtered by a 120-tap linear FIR filter designed by MATLAB's fir1 function with pass band $0.075 \leq \omega \leq 0.125$ (normalized frequency with Nyquist=1). Figure 6.2c shows the best WM smoother output when the coefficients are constrained to positive values only. The positive coefficients are found by the method described in [156]. The WM smoother clearly fails to delete the low frequency components and it also introduces artifacts at higher frequencies. Figure 6.2d depicts the WM filter output where real-valued weights are allowed. The 120 median filter weights are given values identical to that of the linear FIR filter weights. Although the weight values for the weighted median filter are suboptimal, Figure 6.2d shows the significant attenuation obtained in the low-frequency components. The high-frequency terms are cancelled almost completely as well. The small amplitude artifacts exhibited at low-frequencies arise from the fact that the WM filter output is constrained to be equal to the average of only two of the signed input samples. Methods to optimally design the weighted median filter weights will be described shortly. ∎

EXAMPLE 6.5 (BANDPASS FILTERING IN NOISE)

Consider the case where the observed signals are noisy. Figure 6.3a depicts the chirp test signal with added α-stable noise. The parameter $\alpha = 1.4$ was used, simulating noise with impulsive characteristics. Figure 6.3a is truncated so that the same scale is used in all plots. Figure 6.3b shows the noisy chirp signal filtered by the 120-tap linear FIR filter. The output is affected severely by the noise components. Ringing artifacts emerge with each impulse fed into the filter. Figure 6.3c shows the WM filter output when the coefficients are constrained to positive values only. In this case, the noise does not deteriorate the response significantly, but the response is not

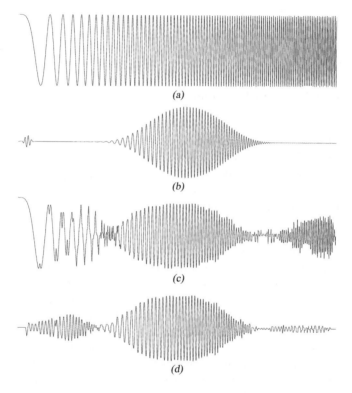

(a)

(b)

(c)

(d)

Figure 6.2   Frequency selective filter outputs: (*a*) chirp test signal, (*b*) linear FIR filter output, (*c*) weighted median smoother output, (*d*) weighted median filter output with suboptimal real-valued weights.

satisfactory due to the low-pass characteristics of the WM smoother. Figure 6.3*d* depicts the output of the WM filter with real-valued weights which shows a considerable improvement.   ■

EXAMPLE 6.6 (IMAGE SHARPENING WITH WM FILTERS)

In principle, image sharpening consists in adding to the original image a signal that is proportional to a high-pass filtered version of the original image. Figure 6.4 illustrates this procedure often referred to as unsharp masking [107] on a 1-dimensional signal. As shown in Figure 6.4, the original image is first filtered by a high-pass filter that extracts the high frequency components, and then a scaled version of the high-pass filter output is added to the original image, thus producing a sharpened image of the original. Note that the homogeneous regions of the signal, where the signal is constant, remain unchanged. The sharpening operation can be represented by

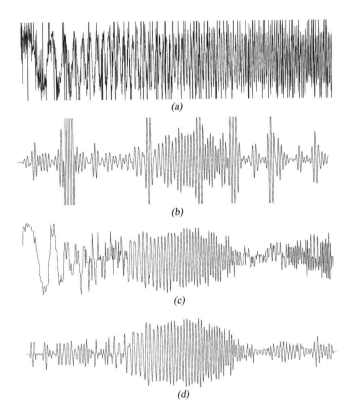

Figure 6.3   Frequency selective filter outputs in noise: (a) chirp test signal in stable noise, (b) linear FIR filter output, (c) weighted median smoother output, (d) weighted median filter output with suboptimal real-valued weights.

$$Y(m, n) = X(m, n) + \lambda \cdot \mathcal{F}(X(m, n)) \tag{6.11}$$

where $X(m, n)$ is the original pixel value at the coordinates (m, n), $\mathcal{F}(\cdot)$ is the high-pass filter, λ is a tuning parameter greater than or equal to zero, and $Y(m, n)$ is the sharpened pixel at the coordinates (m, n). The value taken by λ depends on the grade of sharpness desired. Increasing λ yields a more sharpened image. If background noise is present, however, increasing λ will rapidly amplify the noise.

The key point in the effective sharpening process lies in the choice of the high-pass filtering operation. Traditionally, linear filters have been used to implement the high-pass filter, however, linear techniques can lead to rapid performance degradation should the input image be corrupted with noise. A tradeoff between noise attenuation and edge highlighting can be obtained if a weighted median filter with appropriate weights is used. To illustrate this, consider a WM filter applied to a grayscale image where the following filter mask is used

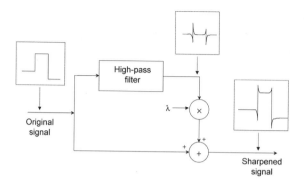

Figure 6.4 Image sharpening by high frequency emphasis.

$$W = \left\langle \begin{array}{ccc} -1 & -1 & -1 \\ -1 & 8 & -1 \\ -1 & -1 & -1 \end{array} \right\rangle . \tag{6.12}$$

Because of the weight coefficients in (6.12), for each position of the moving window, the output is proportional to the difference between the center pixel and the smallest pixel around the center pixel. Thus, the filter output takes relatively large values for prominent edges in an image, and small values in regions that are fairly smooth, being zero only in regions that have constant gray level.

Although this filter can effectively extract the edges contained in an image, the effect that this filtering operation has over negative-slope edges is different from that obtained for positive-slope edges[1]. Since the filter output is proportional to the difference between the center pixel and the smallest pixel around the center, for negative-slope edges, the center pixel takes small values producing small values at the filter output. Moreover, the filter output is zero if the smallest pixel around the center pixel and the center pixel have the same values. This implies that negative-slope edges are not extracted in the same way as positive-slope edges. To overcome this limitation, the basic image sharpening structure shown in Figure 6.4 must be modified such that positive-slope edges as well as negative-slope edges are highlighted in the same proportion. A simple way to accomplish that is: (a) extract the positive-slope edges by filtering the original image with the filter mask described above; (b) extract the negative-slope edges by first preprocessing the original image such that the negative-slope edges become positive-slope edges, and then filter the preprocessed image with the filter described above; (c) combine appropriately the original image, the filtered version of the original image and the filtered version of the pre-processed image to form the sharpened image.

---

[1]A change from a gray level to a lower gray level is referred to as a negative-slope edge, whereas a change from a gray level to a higher gray level is referred to as a positive-slope edge.

Thus both positive-slope edges and negative-slope edges are equally highlighted. This procedure is illustrated in Figure 6.5, where the top branch extracts the positive-slope edges and the middle branch extracts the negative-slope edges. In order to understand the effects of edge sharpening, a row of a test image is plotted in Figure 6.6 together with a row of the sharpened image when only the positive-slope edges are highlighted Figure 6.6*a*, only the negative-slope edges are highlighted Figure 6.6*b*, and both positive-slope and negative-slope edges are jointly highlighted Figure 6.6*c*.

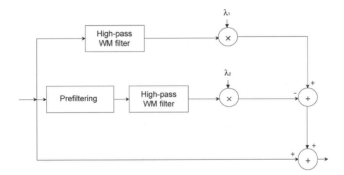

Figure 6.5 Image sharpening based on the weighted median filter.

In Figure 6.5, λ_1 and λ_2 are tuning parameters that control the amount of sharpness desired in the positive-slope direction and in the negative-slope direction respectively. The values of λ_1 and λ_2 are generally selected to be equal. The output of the prefiltering operation is defined as

$$X(m,n)' = M - X(m,n) \qquad (6.13)$$

with M equal to the maximum pixel value of the original image. This prefiltering operation can be thought of as a flipping and a shifting operation of the values of the original image such that the negative-slope edges are converted to positive-slope edges. Since the original image and the prefiltered image are filtered by the same WM filter, the positive-slope edges and negative-slope edges are sharpened in the same way.

In Figure 6.7, the performance of the WM filter image sharpening is compared with that of traditional image sharpening based on linear FIR filters. For the linear sharpener, the method shown in Figure 6.4 was used. The parameter λ was set to 1 for the clean image and to 0.75 for the noisy image. For the WM sharpener, the method of Figure 6.5 was used with $\lambda_1 = \lambda_2 = 2$ for the clean image, and $\lambda_1 = \lambda_2 = 1.5$ for the noisy image. The filter mask given by (6.12) was used in both linear and median image sharpening. Sharpening with WM filters does not suffer from noise amplification to the extent that sharpening with FIR filters do.

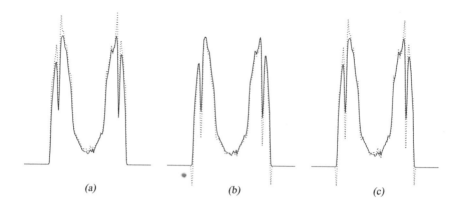

Figure 6.6 Original row of a test image (solid line) and row sharpened (dotted line) with (*a*) only positive-slope edges, (*b*) only negative-slope edges, and (*c*) both positive-slope and negative-slope edges.

■

EXAMPLE 6.7 (EDGE DETECTION WITH WM FILTERS)

The most common approach used for edge detection is illustrated in Figure 6.8. A high-pass filter is applied to the image to obtain the amount of change present in the image at every pixel. The output of the filter is thresholded to determine those pixels that have a high enough rate of change to be considered lying on an edge, that is all pixels with filter output greater than some value T are taken as edge pixels. The value of T is a tunable parameter that can be adjusted to give the best visual results. High thresholds lose some of the real edges, while low values result in many false edges, thus a tradeoff needs to be made to get the best results. Other techniques such as edge thinning can be applied to further pinpoint the location of the edges in an image.

The most common linear filter used for the initial high-pass filtering is the Sobel operator, which uses the following 3×3 masks:

$$\mathbf{W}_V = \begin{bmatrix} -1 & -2 & -1 \\ 0 & 0 & 0 \\ 1 & 2 & 1 \end{bmatrix} \qquad \mathbf{W}_H = \begin{bmatrix} -1 & 0 & 1 \\ -2 & 0 & 2 \\ -1 & 0 & 1 \end{bmatrix}.$$

These two masks are convolved with the image separately to measure the strength of horizontal edges and vertical edges, respectively, present at each pixel. Thus if the amount to which a horizontal edge is present at the pixel in the ith row and jth column is represented as $E_{i,j}^h$, and if the vertical edge indicator is $E_{i,j}^v$, then the values are:

Figure 6.7 (*a*: top-left) Original image sharpened with (*b*: top-right) the FIR-sharpener, and (*c*: middle-left) with the WM-sharpener. (*d*: middle-right) Image with added Gaussian noise sharpened with (*e*: bottom-left) the FIR-sharpener, and (*f*: bottom-right) the WM-sharpener.

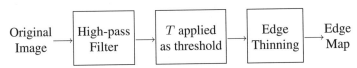

Figure 6.8 The process of edge detection

$$E_{i,j}^h = -X_{i-1,j-1} - 2X_{i-1,j} - X_{i-1,j+1} + X_{i+1,j-1} + 2X_{i+1,j} + X_{i+1,j+1}$$

$$E_{i,j}^v = -X_{i-1,j-1} - 2X_{i,j-1} - X_{i+1,j-1} + X_{i-1,j+1} + 2X_{i,j+1} + X_{i+1,j+1}$$

The two strengths are combined to find the total amount to which any edge exists at the pixel: $E_{i,j}^{total} = \sqrt{E_{i,j}^h{}^2 + E_{i,j}^v{}^2}$. This value is then compared to the threshold T to determine the existence of an edge.

Instead of using linear high-pass filters, weighted median filters can be used. To apply weighted medians to the high-pass filtering, the weights from the Sobel masks can be used. The Sobel linear high-pass filters take a weighted difference between the pixels on either side of $X_{i,j}$. On the other hand, if the same weights are used in a weighted median filter, the value returned is the difference between the lowest-valued pixels on either side of $X_{i,j}$. If the pixel values are then flipped about some middle value, the difference between the *highest* pixels on either side can also be obtained. The flipping is found as in (6.13) yielding the "flipped" sample $X'_{i,j} = M - X_{i,j}$ where M is the maximum pixel value in the image. The lower of the two differences across the pixel can then be used as the indicator of the presence of an edge. If there is a true edge present, then both differences should be high in magnitude, while if noise causes one of the differences to be too high, the other difference is not necessarily affected. Thus the horizontal and vertical edge indicators are:

$$E_{i,j}^h = \min \left(\begin{array}{l} \mathrm{MED} \left[\begin{array}{ccc} -1 \diamond X_{i-1,j-1}, & -2 \diamond X_{i-1,j}, & -1 \diamond X_{i-1,j+1}, \\ 1 \diamond X_{i+1,j-1}, & 2 \diamond X_{i+1,j}, & 1 \diamond X_{i+1,j+1} \end{array} \right], \\ \mathrm{MED} \left[\begin{array}{ccc} -1 \diamond X'_{i-1,j-1}, & -2 \diamond X'_{i-1,j}, & -1 \diamond X'_{i-1,j+1}, \\ 1 \diamond X'_{i+1,j-1}, & 2 \diamond X'_{i+1,j}, & 1 \diamond X'_{i+1,j+1} \end{array} \right] \end{array} \right)$$

$$E_{i,j}^v = \min \left(\begin{array}{l} \mathrm{MED} \left[\begin{array}{cc} -1 \diamond X_{i-1,j-1}, & 1 \diamond X_{i-1,j+1}, \\ -2 \diamond X_{i,j-1}, & 2 \diamond X_{i,j+1}, \\ -1 \diamond X_{i+1,j-1}, & 1 \diamond X_{i+1,j+1} \end{array} \right], \\ \mathrm{MED} \left[\begin{array}{cc} -1 \diamond X'_{i-1,j-1}, & 1 \diamond X'_{i-1,j+1}, \\ -2 \diamond X'_{i,j-1}, & 2 \diamond X'_{i,j+1}, \\ -1 \diamond X'_{i+1,j-1}, & 1 \diamond X'_{i+1,j+1} \end{array} \right] \end{array} \right)$$

and the strength of horizontal and vertical edges $E_{(i,j)}^{h,v}$ is determined in the same way as the linear case: $E_{i,j}^{h,v} = \sqrt{E_{i,j}^h{}^2 + E_{i,j}^v{}^2}$.

Another addition to the weighted median method is necessary in order to detect diagonal edges. Horizontal and vertical indicators are not sufficient to register diagonal edges, so the following two masks must also be used:

$$\begin{bmatrix} -2 & -1 & 0 \\ -1 & 0 & 1 \\ 0 & 1 & 2 \end{bmatrix} \qquad \begin{bmatrix} 0 & 1 & 2 \\ -1 & 0 & 1 \\ -2 & -1 & 0 \end{bmatrix}.$$

These masks can be applied to the image just as the Sobel masks above. Thus the strengths of the two types of diagonal edges are $E_{i,j}^{d1}$ for diagonal edges going from the bottom left of the image to the top right (using the mask on the left above) and $E_{i,j}^{d2}$ for diagonal edges from top left to bottom right (the mask on the right), and the values are given by:

$$E_{i,j}^{d1} = \min \left(\begin{matrix} \mathrm{MED} \begin{bmatrix} -2 \diamond X_{i-1,j-1}, & -1 \diamond X_{i-1,j}, \\ -1 \diamond X_{i,j-1}, & 1 \diamond X_{i,j+1}, \\ 1 \diamond X_{i+1,j}, & 2 \diamond X_{i+1,j+1} \end{bmatrix}, \\ \\ \mathrm{MED} \begin{bmatrix} -2 \diamond X'_{i-1,j-1}, & -1 \diamond X'_{i-1,j}, \\ -1 \diamond X'_{i,j-1}, & 1 \diamond X'_{i,j+1}, \\ 1 \diamond X'_{i+1,j}, & 2 \diamond X'_{i+1,j+1} \end{bmatrix} \end{matrix} \right)$$

$$E_{i,j}^{d2} = \min \left(\begin{matrix} \mathrm{MED} \begin{bmatrix} 1 \diamond X_{i-1,j}, & 2 \diamond X_{i-1,j+1}, \\ -1 \diamond X_{i,j-1}, & 1 \diamond X_{i,j+1}, \\ -2 \diamond X_{i+1,j-1}, & -1 \diamond X_{i+1,j} \end{bmatrix}, \\ \\ \mathrm{MED} \begin{bmatrix} 1 \diamond X'_{i-1,j}, & 2 \diamond X'_{i-1,j+1}, \\ -1 \diamond X'_{i,j-1}, & 1 \diamond X'_{i,j+1}, \\ -2 \diamond X'_{i+1,j-1}, & -1 \diamond X'_{i+1,j} \end{bmatrix} \end{matrix} \right).$$

A diagonal edge strength is determined in the same way as the horizontal and vertical edge strength above: $E_{i,j}^{d1,d2} = \sqrt{E_{i,j}^{d1^2} + E_{i,j}^{d2^2}}$. The indicator of all edges in any direction is the maximum of the two strengths $E_{i,j}^{h,v}$ and $E_{i,j}^{d1,d2}$: $E_{i,j}^{total} = \max \left(E_{i,j}^{h,v}, E_{i,j}^{d1,d2} \right)$. As in the linear case, this value is compared to the threshold T to determine whether a pixel lies on an edge. Figure 6.9 shows the results of calculating $E_{i,j}^{total}$ for an image. The results of the median edge detection are similar to the results of using the Sobel linear operator. Similar approaches to edge detection with generalized median filters have been proposed in [18, 17], where a new differential filter is implemented via negative weights. The generalization in [18, 17] is refered to as RONDO: rank-order based nonlinear differential operator.

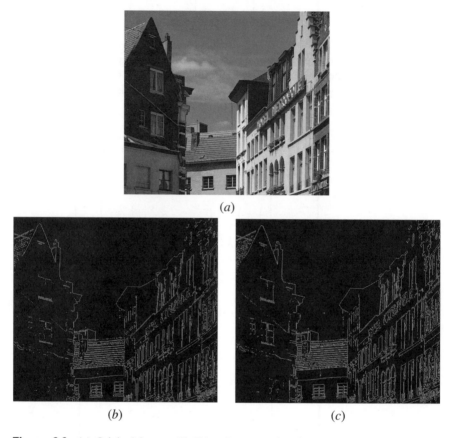

(a)

(b)   *(c)*

Figure 6.9   (*a*) Original image, (*b*) Edge detector using linear method, and (*c*) median method.

6.1.1   Permutation-Weighted Median Filters

Permutation WM filters closely resemble permutation WM smoothers with the exception that the weights are not only data dependent but can also take on negative values [10].

DEFINITION 6.2 (PERMUTATION WM FILTERS) *Let* $\langle W_{1(R_1)}, W_{2(R_2)}, \ldots, W_{N(R_N)} \rangle$ *be rank-order dependent weights assigned to the input observation samples. The output of the permutation WM filter is found as*

$$Y = \text{MEDIAN}[|W_{1(R_1)}| \diamond \text{sgn}(W_{1(R_1)})X_1), \ldots, |W_{N(R_N)}| \diamond \text{sgn}(W_{N(R_N)})X_N)].$$
(6.14)

where $W_{i(R_i)}$ is the weight assigned to X_i and selected according to the sample's rank R_i.

Note that the signs of the weights are decoupled from the replication operator and applied to the data sample. The weight assigned to X_i is drawn from the weight set $\{W_{i(1)}, W_{i(2)}, \ldots, W_{i(N)}\}$. Having N weights per sample, a total of N^2 weights need to be stored for the computation of (6.14). In general, an optimization algorithm is needed to design the set of weights although in some cases only a few rank-order dependent weights are required and their design is simple. Permutation WM filters can provide significant improvement in performance at the higher cost of memory cells.

To illustrate the versatility of permutation WM filters, consider again the image sharpening example. Recall that linear high-pass filters are inadequate in unsharp masking whenever background noise is present. Although WM high-pass filters ameliorate the problem, the goal is to improve their performance by allowing the WM filter weights to take on rank-dependent values. The unsharp WM filter structure shown in Figure 6.5 is used with the exception that permutation WM filters are now used to synthesize the high-pass filter operation. The weights used for the WM high-pass filter in (6.12) were proportional to

$$W = \left\langle \begin{array}{ccc} -1 & -1 & -1 \\ -1 & \underline{8} & -1 \\ -1 & -1 & -1 \end{array} \right\rangle.$$
(6.15)

The weight mask for the permutation WM high-pass filter is

$$W = \left\langle \begin{array}{ccc} W_{1(R_1)} & W_{2(R_2)} & W_{3(R_3)} \\ W_{4(R_4)} & \underline{W_{c(R_c)}} & W_{6(R_6)} \\ W_{7(R_7)} & W_{8(R_8)} & W_{9(R_9)} \end{array} \right\rangle,$$
(6.16)

where $W_{i(R_i)} = -1$, for $i \neq 5$, with the following exceptions. The value of the center weight is given according to

$$W_{c(R_c)} = \begin{cases} 8 & \text{for } R_c = 2, 3, \ldots, 8 \\ -1 & \text{otherwise.} \end{cases}$$
(6.17)

That is, the value of the center weight is 8 if the center sample is not the smallest or largest in the observation window. If it happens to be the smallest or largest, its reliability is low and the weighting strategy must be altered such that the center weight is set to -1, and the weight of 8 is given to the sample that is closest in rank to the center sample leading to

$$W_{[8]} = \begin{cases} 8 & \text{if } X_c = X_{(9)} \\ \\ -1 & \text{otherwise,} \end{cases} \tag{6.18}$$

$$\tag{6.19}$$

$$W_{[2]} = \begin{cases} 8 & \text{if } X_c = X_{(1)} \\ \\ -1 & \text{otherwise,} \end{cases} \tag{6.20}$$

where $W_{[i]}$ is the concomitant weight of the ith order statistic of the input vector.

This weighting strategy can be extended to the case where the L smallest and L largest samples in the window are considered unreliable and the weighting strategy applied in (6.18) and (6.20) now applies to the weights $W_{[L+1]}$ and $W_{[N-L]}$.

Figure 6.10 illustrates the image sharpening performance when permutation WM filters are used. The "Saturn image" with added Gaussian background noise is shown in Figure 6.10(a). Figures 6.10(b–f) show this image sharpened with (b) a LUM sharpener[2], (c) a linear FIR filter sharpener, (d) the WM filter sharpener, (e) the permutation WM filter sharpener with $L = 1$, and (f) the permutation WM filter sharpener with $L = 2$. The λ parameters were given a value of 1.5 for all weighted median-type sharpeners, and it was set to 1 for the linear sharpener. The linear sharpener introduces background noise amplification. The LUM sharpener does not amplify the background noise; however, it introduces severe edge distortion artifacts. The WM filter sharpener ameliorates the noise amplification and does not introduce edge artifacts. The permutation WM filter sharpeners perform best, with higher robustness attributes as L increases.

6.2   SPECTRAL DESIGN OF WEIGHTED MEDIAN FILTERS

A classical approach to filter design is to modify the filter weights so as to attain a desired spectral profile. Usually the specifications include the type of filter required, that is, low-pass, high-pass, band-pass or band-stop, and a set of cutoff frequencies and attenuation. There are a number of design strategies for the design of linear filters. See for instance Proakis and Manolakis (1996) [166] and Mitra (2001) [144]. These techniques, however, cannot be applied to the design of weighted medians since they lack an impulse response characterization. This section defines the concept of frequency response for weighted median filters and develops a closed form solution for their spectral design.

---

[2]The LUM sharpener algorithm will be described in a later chapter.

Figure 6.10   (*a*: top left) Image with background noise sharpened with (*b*:top right) LUM sharpener, (*c*: middle left) the FIR sharpener, (*d*: middle right) the WM sharpener, (*e*: bottom left) the permutation WM sharpener with $L = 1$, (*f*: bottom right) the permutation WM sharpener with $L = 2$.

6.2.1 Median Smoothers and Sample Selection Probabilities

Spectral analysis of nonlinear smoothers has been carried out based on the theory developed by Mallows (1980) [137]. This theory allows us to analyze some characteristics of nonlinear filters based on the characteristics of a corresponding linear filter. In particular, we are interested in designing a nonlinear filter based on some frequency-response requirements. In order to do that, the spectrum of a nonlinear smoother is defined as the spectral response of the corresponding linear filter. Mallows focused on the analysis of the smoothing of a nonGaussian sequence $\mathbf{X}$ by a nonlinear function $\mathbf{S}$ and how this process can be approximated by a well defined linear smoothing function, as stated on the following theorem:

THEOREM 6.1 (MALLOWS[137]) *Given a nonlinear smoothing function $\mathbf{S}$ operating on a random sequence $\mathbf{X} = \mathbf{Y} + \mathbf{Z}$, where $\mathbf{Y}$ is a zero mean Gaussian sequence and $\mathbf{Z}$ is independent of $\mathbf{Y}$, we have that if $\mathbf{S}$ is stationary, location invariant, centered (i.e., $\mathbf{S}(0) = 0$), it depends on a finite number of values of $\mathbf{X}$ and $Var(\mathbf{S}(\mathbf{X})) < \infty$, There exist a unique linear function $\mathbf{S}^\mathbf{L}$ such that the MSE function:*

$$\mathbf{E}\left\{\left(\mathbf{S}(\mathbf{X}) - \mathbf{S}^\mathbf{L}(\mathbf{X})\right)^2\right\} \tag{6.21}$$

is minimized. The function $\mathbf{S}^\mathbf{L}$ is the closest linear function to the nonlinear smoothing function $\mathbf{S}$ or its linear part.

In particular, median smoothers have all the characteristics required for this theorem and, in consequence, they can be approximated by a linear function. Median smoothers are also selection type and, referring again to Mallows' theory, there is an important corollary of the previous theorem that applies to selection type smoothers whose output is identical to one of their input samples:

COROLLARY 6.1 *[137] If $\mathbf{S}$ is a selection type smoother, the coefficients of $\mathbf{S}^\mathbf{L}$ are the sample selection probabilities of the smoother.*

The sample selection probabilities are defined next for a WM smoother described by the weight vector $\mathbf{W} = \langle W_1, W_2, \ldots, W_N \rangle$ and a vector of independent and identically distributed samples $\mathbf{X} = (X_1, X_2, \ldots, X_N)$.

DEFINITION 6.3 *The Sample Selection Probabilities (SSPs) of a WM smoother $\mathbf{W}$ are the set of numbers p_j defined by:*

$$p_j = P\left(X_j = \text{MEDIAN}[W_1 \diamond X_1, W_2 \diamond X_2, \ldots, W_N \diamond X_N]\right) \tag{6.22}$$

Thus, p_j is the probability that the output of a weighted median filter is equal to the jth input sample.

Mallows' results provide a link between the linear and nonlinear domains that allows the approximation of a WM smoother by its *linear part*. The *linear part* also provides an approximation of the frequency behavior of the smoother. Thus, in order to obtain a WM smoother with certain frequency characteristics, a linear filter with such characteristics should be designed. This linear filter can be approximated by a WM filter with the required frequency characteristics.

6.2.2 SSPs for Weighted Median Smoothers

In order to find the linear smoother closer in the mean square error sense to a given weighted median smoother, a method to calculate the SSPs of the WM smoother is needed. Some examples of algorithms to calculate the SSPs of a WM smoother can be found in Prasad and Lee (1994) [165] and Shmulevich and Arce (2001) [175]. The calculation is carried out here based on the calculation of the weighted median that is reproduced here for convenience:

Suppose that the WM filter described by the weight vector $\mathbf{W} = \langle W_1, W_2, \ldots, W_N \rangle$ is applied to the set of independent and identically distributed samples $\mathbf{X} = (X_1, X_2, \ldots, X_N)$, then the output is calculated through the steps in section 5.2 which are repeated here for convenience.

(1) Calculate the threshold $T_0 = \frac{1}{2} \sum_{i=1}^{N} W_i$;

(2) Sort the samples in the observation vector $\mathbf{X}$;

(3) Sum the concomitant weights of the sorted samples beginning with the maximum sample and continuing down in order;

(4) The output $\hat{\beta}$ is the first sample whose weight causes the sum to become $\geq T_0$.

The objective is to find a general closed form expression for the probability that the jth sample is chosen as the output of the WM filter, that is, to find the value $p_j = P(\hat{\beta} = X_j)$. The jth sample in the input vector can be ranked in N different, equally likely positions in its order statistics, since the samples are independent and identically distributed. For all i this probability is

$$P(X_{(i)} = X_j) = \frac{1}{N}. \tag{6.23}$$

Because of the different weight values applied to the input samples, each sample has a different probability of being the output of the median depending on where it lies in the set of ordered input samples. The final value of p_j is found as the sum of the probabilities of the sample X_j being the median for each one of the order statistics

$$p_j = \sum_{i=1}^{N} P(X_{(i)} = X_j) P(\hat{\beta} = X_{(i)} | X_{(i)} = X_j)$$

$$= \frac{1}{N} \sum_{i=1}^{N} P(\hat{\beta} = X_{(i)} | X_{(i)} = X_j) = \frac{1}{N} \sum_{i=1}^{N} \frac{K_{ij}}{\binom{N-1}{i-1}}. \tag{6.24}$$

The result in (6.24) can be explained as follows. After the sample X_j has been ranked in the ith order statistic, there are $N - 1$ samples left to occupy the remaining $N - 1$ order statistics: $i - 1$ before $X_j = X_{(i)}$ and $N - i$ after it. The total number of nonordered ways to distribute the remaining samples between the remaining order statistics is then equal to the number of ways in which we can distribute the set of $N - 1$ samples in two subsets of $i - 1$ and $N - i$ samples, leading to the denominator $\binom{N-1}{i-1}$ in (6.24). The order of the samples in each one of this subsets is not important since, as it will be shown shortly, only the sum of the associated weights is relevant. The term K_{ij} represents how many of these orderings will result in the output of the median being the sample X_j while it is ranked in the ith-order statistic, that is, the number of times that $\hat{\beta} = X_j = X_{(i)}$. K_{ij} is found as the number of subsets of $N - i$ elements of the vector $\mathbf{W}$ satisfying:

$$\sum_{m=i+1}^{N} W_{[m]} < T_0 \qquad (6.25)$$

$$\sum_{m=i}^{N} W_{[m]} \geq T_0, \qquad (6.26)$$

where $T_0 = \frac{1}{2} \sum_{m=1}^{N} W_{[m]}$ and where $W_{[m]}$ is the concomitant weight associated with the mth order statistic of the input vector. Conditions (6.25) and (6.26) are necessary and sufficient for $X_{(i)}$ to be the weighted median of the sample set. This was shown in Section 5.2, where it is stated that, in order to find the weighted median of a sample set, the samples are first ordered and then the concomitant weights of the ordered samples are added one by one beginning with the maximum sample and continuing down in order. The median of the set will be the value of the sample whose weight causes the sum to become greater or equal than the threshold T_0.

Conditions (6.25) and (6.26) can be rewritten in a more compact way as:

$$T_0 - W_j \leq \sum_{m=i+1}^{N} W_{[m]} < T_0 \qquad (6.27)$$

where $W_{[i]}$ has been replaced by W_j since it is assumed that the jth sample of the vector is the ith order statistic. In order to count the number of sets satisfying (6.27), a product of two step functions is used as follows: when the value $A = \sum_{m=i+1}^{N} W_{[m]}$ satisfies $T_0 - W_j \leq A < T_0$ the function:

$$u(A - (T_0 - W_j))u(T_0^- - A) \qquad (6.28)$$

will be equal to one. On the other hand, (6.28) will be equal to zero if A does not satisfy the inequalities. Here T_0^- represents a value approaching T_0 from the left in the real line and u is the unitary step function defined as: $u(x) = 1$ if $x \geq 0$, and 0 otherwise. On the other hand, (6.28) will be equal to zero if A does not satisfy the inequalities. Letting $T_1 = T_0 - W_j$ and adding the function in (6.28) over all the possible subsets of $i - 1$ elements of $\mathbf{W}$ excluding W_j the result is:

$$K_{ij} = \sum_{\substack{m_1=1 \\ m_1 \neq j}}^{N} \sum_{\substack{m_2=m_1+1 \\ m_2 \neq j}}^{N} \cdots \sum_{\substack{m_s=m_{s-1}+1 \\ m_s \neq j}}^{N} u(A - T_1)u(T_0^- - A) \qquad (6.29)$$

where $A = W_{m_1} + W_{m_2} + \ldots + W_{m_s}$ and $s = N - i$. The SSP vector is given by $\mathbf{P}(\mathbf{W}) = [p_1, p_2, \ldots, p_n]$, where p_j is defined as:

$$p_j = \frac{1}{N} \sum_{i=1}^{N} \frac{K_{ij}}{\binom{N-1}{i-1}}. \qquad (6.30)$$

This function calculates the sample selection probabilities of any WM smoother, that is, it leads to the linear smoother closest to a given WM smoother in the mean square error sense.

EXAMPLE 6.8 (SSPs for a four tap WM)

Given $\mathbf{W} = \langle 1, 3, 4, 1 \rangle$, find the sample selection probability of the third sample p_3.

T_1 and T_0 are found as:

$$T_0 = \frac{1}{2} \sum_{i=1}^{4} W_i = 4.5$$

$$T_1 = T_0 - W_3 = 0.5 \qquad (6.31)$$

Equation (6.30) reduces to

$$p_3(\mathbf{W}) = \frac{1}{4} \sum_{i=1}^{4} K_{i3} \frac{(i-1)!(4-i)!}{(4-1)!}. \qquad (6.32)$$

For $i = 1$, $W_{[1]} = 4$, thus:

$$A = \sum_{m=2}^{4} W_{[m]} = 1 + 3 + 1 = 5 \quad \text{then}$$

$$u(A - T_1)u(T_0^- - A) = u(5 - 0.5)u(4.5^- - 5) = 0, \qquad (6.33)$$

hence $K_{13} = 0$.

For $i = 2$, $W_{[2]} = 4$, then there are three possibilities for the ordering of the weights (the first weight can be either one of W_1, W_2 or W_4) and, in consequence, three different values for $A = \sum_{m=3}^{4} W_{[m]}$:

$$\begin{aligned}
A_1 &= 1 + 1 = 2 \\
u(A_1 - T_1)u(T_0^- - A_1) &= u(2 - 0.5)u(4.5^- - 2) = 1 \\
A_2 &= 1 + 3 = 4 \\
u(A_2 - T_1)u(T_0^- - A_2) &= u(4 - 0.5)u(4.5^- - 4) = 1 \\
A_3 &= 3 + 1 = 4 \\
u(A_3 - T_1)u(T_0^- - A_3) &= u(4 - 0.5)u(4.5^- - 4) = 1 \\
K_{23} &= 3.
\end{aligned} \tag{6.34}$$

Following the same procedure, the values of the remaining K_{i3} are found to be $K_{33} = 3$ and $K_{43} = 0$. Therefore, the sample selection probability results in:

$$\begin{aligned}
p_3(\mathbf{W}) &= \frac{1}{4}\left(0\frac{0!3!}{3!} + 3\frac{1!2!}{3!} + 3\frac{2!1!}{3!} + 0\frac{3!0!}{3!}\right) \\
&= \frac{1}{2}.
\end{aligned} \tag{6.35}$$

The full vector of SSPs is constructed as: $\mathbf{P}(\mathbf{W}) = \left[\frac{1}{6}, \frac{1}{6}, \frac{1}{2}, \frac{1}{6}\right]$ ∎

6.2.3 Synthesis of WM Smoothers

So far, this section has focused on the analysis of WM smoothers and the synthesis of linear smoothers having similar characteristics to a given WM. On the other hand, its final purpose is to present a spectral design method for WM smoothers. The approach is to find the closest WM filter to an FIR filter that has been carefully designed to attain a desired set of spectral characteristics. To attain this, the function obtained in (6.30) should be inverted; however, this nonlinear function is not invertible. Before studying other alternatives to solve this problem, certain properties of weighted median smoothers should be taken into account.

It has been demonstrated in Muroga (1971) [145] and Yli-Harja et al. (1991) [202] that weighted median smoothers of a given window size can be divided into a finite number of *classes*. Each one of the smoothers in a class produces the same output when they are fed with the same set of input samples. It has also been shown that each class contains at least one integer-valued weighted median smoother such that the sum of its coefficients is odd. Among these smoothers, the one with the minimum sum of components is called the representative of the class. Table 6.1 shows the representatives of the different classes of weighted median smoothers available for window sizes from one to five.

Weighted medians obtained as the permutation of the ones shown in Table 6.1 are also representatives of other classes. Additionally, a representative of a class can be padded with zeros to form a representative of another class with larger window size.

Table 6.1   Median weight vectors and their corresponding SSPs for window sizes 1 to 5

N	WM	SSP
1	$\langle 1 \rangle$	$[1]$
2	-	-
3	$\langle 111 \rangle$	$\left[\frac{1}{3}\frac{1}{3}\frac{1}{3}\right]$
4	$\langle 2111 \rangle$	$\left[\frac{1}{2}\frac{1}{6}\frac{1}{6}\frac{1}{6}\right]$
5	$\langle 11111 \rangle$	$\left[\frac{1}{5}\frac{1}{5}\frac{1}{5}\frac{1}{5}\frac{1}{5}\right]$
	$\langle 22111 \rangle$	$\left[\frac{3}{10}\frac{3}{10}\frac{2}{15}\frac{2}{15}\frac{2}{15}\right]$
	$\langle 31111 \rangle$	$\left[\frac{3}{5}\frac{1}{10}\frac{1}{10}\frac{1}{10}\frac{1}{10}\right]$
	$\langle 32211 \rangle$	$\left[\frac{2}{5}\frac{7}{30}\frac{7}{30}\frac{1}{15}\frac{1}{15}\right]$

For example, for window size three, we can construct four different weighted median vectors: $\langle 1,\ 1,\ 1 \rangle$ and the three permutations of $\langle 1,\ 0,\ 0 \rangle$.

It is also known that each weighted median filter has a corresponding equivalent self dual linearly separable positive boolean function (PBF) [145] and vice versa. This means that the number of different weighted medians of size N is the same as the number of self dual linearly separable PBFs of N variables. Equivalent WM vectors will correspond to the same PBF and they will also have the same vector of SSPs.

To illustrate the consequences of these properties, the case for smoothers of length three will be studied. Here the number of weighted median smoothers to be analyzed is reduced to include only normalized smoothers. These smoothers are included in the two dimensional simplex $W_1 + W_2 + W_3 = 1$. According to (6.27) and Table 6.1, there are four different classes of weighted medians for this window size. They will occupy regions in the simplex that are limited by lines of the form: $W_i + W_j = \frac{1}{2} = T_0$, where $i, j \in \{1, 2, 3\}, i \neq j$. Figure 6.11a shows the simplex with the four regions corresponding to the four classes of weighted medians and the representative of each class.

The weighted median closest to a given linear smoother in the mean square error sense is found by minimizing the mean square error cost function

$$J(\mathbf{W}) = \|\mathbf{P}(\mathbf{W}) - \mathbf{h}\|^2 = \sum_{j=1}^{N} (p_j(\mathbf{W}) - h_j)^2 \qquad (6.36)$$

where $\mathbf{h}$ is a normalized linear smoother. Since the number of SSP vectors $\mathbf{P}(\mathbf{W})$ for a given window size is finite, a valid option to solve this problem is to list all its possible values and find between them the one that minimizes the error measure

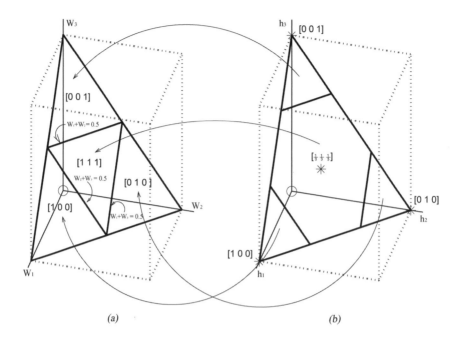

Figure 6.11   Illustrative example showing the mapping between the WM class regions and the linear smoother regions: (*a*) simplex containing the weighted median vectors for window size three. The simplex is divided in four regions, the representative of each region is also indicated; (*b*) correspondence between linear smoothers and SSP vectors of window size three. The SSP vectors are represented by '*'.

$J(\mathbf{W})$. This will lead to a division of the space of linear smoothers of window size N in regions, one for each SSP vector. Each point in the space is associated with the SSP vector that is the closest to it in Euclidean distance to conform the regions. This situation can be viewed as a quantization of the space of normalized linear smoothers where all the points in a quantization region are mapped to the (only) SSP included in that region. Figure 6.11*b* shows the case for window size three.

All vectors in the same WM class region are mapped into the linear domain as a single point, the corresponding SSP vector. Since all WM in a class are equivalent, the associated linear smoother to all of them is the same. Therefore, there is a unique solution to the problem of finding the linear smoother closest in the MSE sense to a given WM. On the other hand, the reverse problem, finding the WM smoother closest to a given linear smoother, has an infinite number of solutions. Since the linear smoother domain is quantized, a given vector h in a quantization region will be associated with the SSP vector contained in the region. This vector is mapped into the WM domain as a class of weighted medians instead of as a single WM smoother. Any set of weights in that class will result in the same value of the distance measure $J(\mathbf{W})$ and, in consequence, any of them can be chosen as the closest WM to the linear smoother represented by h. That is, the mapping in this case is established between a quantization region in the linear domain and a class region in the WM domain in such a way that any point in the latter can be associated with a given vector in the former. Figure 6.11 illustrates the mapping between quantization regions in the linear domain and class regions in the WM domain for window size three.

The procedure to transform a linear smoother into its associated weighted median reduces to finding the region in the linear space where it belongs, finding the corresponding SSP vector and then finding a corresponding WM vector. This is possible only if all the valid weighted median vectors and their corresponding SSPs for a certain window size are known. The problem of finding all the different weighted median vectors of size N has been subject to intensive research. However, a general closed-form solution that allows the generation of the list of PBFs, SSP vectors, or weighted median vectors has yet to be found. Partial solutions for the problem have been found for window sizes up to nine by Muroga (1971) [145]. Even if such a general form existed, the number of possibilities grows rapidly with the window size and the problem becomes cumbersome. For example, the number of different weighted medians grows from 2470 for window size eight to $175,428$ for window size nine according to Muroga (1971) [145]. There is no certainty about the number of vectors for window size ten and up.

Having all the possible sets of median weights for a certain window size will assure that the right solution of the problem can be found. As it was indicated before, this option becomes unmanageable for large window sizes. This does not disqualify the method for smoothers with small window size, but a faster, easier alternative is necessary to handle larger lengths. In the following section, an optimization algorithm for the function $J(\mathbf{W})$ is presented.

6.2.4 General Iterative Solution

The optimization process of the cost function in (6.36) is carried out with a gradient-based algorithm, and a series of approximations derived by Hoyos et al. (2003) [103]. The recursive equation for each of the median weights is:

$$
\begin{aligned}
W_l(n+1) &= W_l(n) + \mu\left(-\nabla_l \mathbf{J}(\mathbf{W})\right) \\
&= W_l(n) + \mu\left(-\frac{\partial}{\partial W_l}\mathbf{J}(\mathbf{W})\right).
\end{aligned}
\tag{6.37}
$$

The first step is to find the gradient of (6.36)

$$
\nabla \mathbf{J}(\mathbf{W}) = \begin{pmatrix} \frac{\partial}{\partial W_1}\mathbf{J}(\mathbf{W}) \\ \frac{\partial}{\partial W_2}\mathbf{J}(\mathbf{W}) \\ \vdots \\ \frac{\partial}{\partial W_N}\mathbf{J}(\mathbf{W}) \end{pmatrix}
\tag{6.38}
$$

where each of the terms in (6.38) is given by:

$$
\nabla_l \mathbf{J}(W) = \frac{\partial}{\partial W_l} \mathbf{J}(W) = \frac{\partial}{\partial W_l} \|\mathbf{P}(\mathbf{W}) - \mathbf{h}\|^2
$$

$$
= \sum_{j=1}^{N} \frac{\partial}{\partial W_l} (p_j(\mathbf{W}) - h_j)^2
$$

$$
= \sum_{j=1}^{N} 2 (p_j(\mathbf{W}) - h_j) \frac{\partial}{\partial W_l} p_j(W). \tag{6.39}
$$

The derivative of $p_j(W)$ is:

$$
\frac{\partial p_j(W)}{\partial W_l} = \frac{\partial}{\partial W_l} \frac{1}{N} \sum_{i=1}^{N} \frac{K_{ij}}{\binom{N-1}{i-1}}
$$

$$
= \frac{1}{N} \sum_{i=1}^{N} \frac{\frac{\partial K_{ij}}{\partial W_l}}{\binom{N-1}{i-1}}. \tag{6.40}
$$

The term K_{ij} given in (6.29) is not differentiable because of the discontinuities of the step functions. To overcome this situation, $u(x)$ is approximated by a smooth differentiable function: $u(x) \approx \frac{1}{2}(\tanh(x) + 1)$. The derivative on the right hand side of (6.40) can be computed as:

$$
\frac{\partial K_{ij}}{\partial W_l} = \frac{1}{4} \sum_{\substack{m_1=1 \\ m_1 \neq j}}^{N} \sum_{\substack{m_2=m_1+1 \\ m_2 \neq j}}^{N} \cdots \sum_{\substack{m_s=m_{s-1}+1 \\ m_s \neq j}}^{N} \frac{\partial}{\partial W_l} B,
$$

$$
\tag{6.41}
$$

where $B = (\tanh(A - T_1) + 1) (\tanh(T_0^- - A) + 1)$ and

$$
\frac{\partial B}{\partial W_l} = C_1(W_l)\mathrm{sech}^2(A - T_1) (\tanh(T_0^- - A) + 1)
$$

$$
- C_2(W_l) (\tanh(A - T_1) + 1) \mathrm{sech}^2(T_0^- - A). \tag{6.42}
$$

The coefficients $C_1(W_l)$ and $C_2(W_l)$ above are defined by:

$$
C_1(W_l) = \begin{cases} \frac{1}{2} & l = j \\ \frac{1}{2} & \text{if } i \text{ exists s.t. } m_i = l \\ -\frac{1}{2} & \text{else} \end{cases}
$$

$$
C_2(W_l) = \begin{cases} -\frac{1}{2} & \text{if } i \text{ exists s.t. } m_i = l \\ \frac{1}{2} & \text{else.} \end{cases} \tag{6.43}
$$

That is, the coefficient $C_1(W_l)$ will be equal to $\frac{1}{2}$ if the term of the sum in (6.39) whose derivative is being found is the lth or when W_l is one of the weights included in the sum A in (6.29). Otherwise, $C_1(W_l) = -\frac{1}{2}$. On the other hand, $C_2(W_l)$ will be equal to $-\frac{1}{2}$ if W_l is one of the weights included in the sum A in (6.29) and $\frac{1}{2}$ otherwise. The iterative algorithm shown above approximates linear smoothers by weighted median smoothers.

The cost function in (6.36) is stepwise and, in consequence, it has an infinite number of local minima. Based on our simulation results, a smooth approximation is obtained when replacing the step functions with hyperbolic tangents, which allows the implementation of a steepest descent algorithm to minimize it. However, no formal proof of the uniqueness of the minimum of the approximated cost function has been found, and remains an open mathematical problem. Experimental results show that the steepest descent algorithm converges to the global minimum of this function.

Figure 6.12a illustrates the cost function with respect to the optimization of W_4 of a WM filter of size six while the other weights remain constant. Both the original cost function (solid line) and the one obtained after using the hyperbolic tangent approximations (dashed line) are shown. Figure 6.12b and 6.12c show the contours of the same cost function with respect to W_1 and W_5 for the original and the approximated cost function, respectively. Notice the staircase shape of the original cost function. It is also noticeable that the minimum of the approximation falls in the region where the original function reaches its minimum. In this case, the approximation is convex and in consequence it lacks local minima that can disrupt the performance of the iterative algorithm.

This procedure is generalized next to the design of weighted median filters admitting real valued weights.

6.2.5 Spectral Design of Weighted Median Filters Admitting Real-Valued Weights

Weighted median filters admitting real-valued weights are obtained by properly modifying the input samples according to the sign of the associated weights and then using the magnitude of the weights for the calculation of a weighted median smoother. It was stated in Theorem 6.1 that a nonlinear function needs to satisfy certain properties in order to be best approximated under the mean squared error sense by a linear filter. Unfortunately, the real-valued medians do not satisfy the location invariance property. However, Mallows results can be extended to cover medians like (6.5) in the case of an independent, zero mean, Gaussian input sequence.

THEOREM 6.2 *If the input series is Gaussian, independent, and zero centered, the coefficients of the linear part of the weighted median defined in (6.5) are defined as:* $h_i = \mathrm{sgn}(W_i)p_i$, where p_i *are the SSPs of the WM smoother* $|W_i|$.

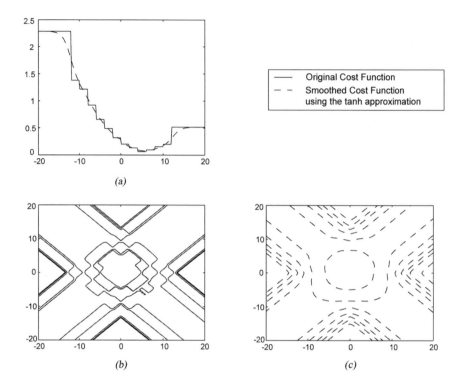

Figure 6.12 Cost functions in 6.36 with input FIR filter $h = [-0.0078, 0.0645, 0.4433,$ $0.4433, 0.0645, -0.0078]$. (*a*) Cost functions with respect to one weight for both the original (solid line) and the approximated cost function (dashed line), (*b*) contours with respect to two weights for the original cost function, (*c*) contours with respect to the same weights for the approximated cost function.

To show this theorem define $Y_i = \text{sgn}(W_i)X_i$. In this case, the Y_i will have the same distribution as the X_i. In consequence:

$$\mathbf{E}\left\{(\text{MEDIAN}(W_i \diamond X_i) - \sum h_i X_i)^2\right\} \tag{6.44}$$

$$= \mathbf{E}\left\{\left(\text{MEDIAN}(|W_i| \diamond Y_i) - \sum q_i Y_i\right)^2\right\}$$

where $q_i = h_i/\text{sgn}(W_i)$. From Theorem 6.1, (6.45) is minimized when the q_i equal the SSPs of the smoother $|W_i|$, say p_i. In consequence:

$$q_i = h_i/\text{sgn}(W_i) = p_i \rightarrow h_i = \text{sgn}(W_i)p_i \tag{6.45}$$

According to this theorem, the proposed algorithm can be used to design WM filters using the following procedure [103, 175]:

(1) Given the desired impulse response, design the linear FIR filter $\mathbf{h} = (h_1, h_2, \ldots, h_N)$ using one of the traditional design tools for linear filters.

(2) Decouple the signs of the coefficients to form the vectors $|\mathbf{h}| = (|h_1|, |h_2|, \ldots, |h_N|)$ and $\text{sgn}(\mathbf{h}) = (\text{sgn}(h_1), \text{sgn}(h_2), \ldots, \text{sgn}(h_N))$.

(3) After normalizing the vector $|\mathbf{h}|$, use the algorithm in Section 6.2.4 to find the closest WM filter to it, say $\mathbf{W}' = \langle W_1', W_2', \ldots, W_N' \rangle$.

(4) The WM filter weights are given by $\mathbf{W} = \left\langle \text{sgn}(h_i) W_i' \big|_{i=1}^N \right\rangle$

EXAMPLE 6.9

Design 11 tap (a) low-pass, (b) band-pass, (c) high-pass, and (d) band-stop WM filters with the cutoff frequencies shown in Table 6.2.

Table 6.2 Characteristics of the WM filters to be designed

Filter	Cut-off frequencies
Low pass	0.25
Band pass	0.35–0.65
High pass	0.75
Band stop	0.35–0.65

Initially, 11 tap linear filters with the spectral characteristics required were designed using MATLAB's function **fir1**. These filters were used as a reference for the design of the weighted median filters. The spectra of the WM and linear filters were approximated using the Welch method [192]. The results are shown in Figure 6.13. The median and linear weights are shown in Table 6.3. ∎

The plots show that WM filters are able to attain arbitrary frequency responses. The characteristics of the WM filters and the linear filters are very similar in the pass band, whereas the major difference is the lower attenuation provided by the WM in the stop band.

6.3 THE OPTIMAL WEIGHTED MEDIAN FILTERING PROBLEM

The spectral design of weighted median filters is only one approach to the design of this class of filters. Much like linear filters can be optimized in an statistical sense

Table 6.3 Weights of the median filters designed using the algorithm in Section 6.2.5 and the linear filters used as reference.

Low-pass		Band-pass		High-pass		Band-stop	
Linear	Median	Linear	Median	Linear	Median	Linear	Median
-0.0039	-0.0223	-0.0000	-0.0092	0.0039	0.0223	0.0000	0.0261
0.0000	0.0211	0.0362	0.0384	-0.0000	-0.0211	-0.0254	-0.0468
0.0321	0.0472	0.0000	0.0092	-0.0321	-0.0472	0.0000	0.0261
0.1167	0.1094	-0.2502	-0.2311	0.1167	0.1094	0.1756	0.1610
0.2207	0.1898	0.0000	0.0092	-0.2207	-0.1898	-0.0000	-0.0261
0.2687	0.2205	0.4273	0.4056	0.2687	0.2205	0.6996	0.4278
0.2207	0.1898	0.0000	0.0092	-0.2207	-0.1898	-0.0000	-0.0261
0.1167	0.1094	-0.2502	-0.2311	0.1167	0.1094	0.1756	0.1610
0.0321	0.0472	0.0000	0.0092	-0.0321	-0.0472	0.0000	0.0261
0.0000	0.0211	0.0362	0.0384	-0.0000	-0.0211	-0.0254	-0.0468
-0.0039	-0.0223	0.0000	-0.0092	0.0039	0.0223	0.0000	0.0261

using the Wiener filter theory, weighted median filters enjoy an equivalent theory for optimization. The theory to be described below emerged from the concepts developed in Coyle and Lin (1998) [55], Lin et al. (1990) [132], Kim and Lin (1994) [131], Yin and Neuvo (1994) [200], and Arce (1998) [6]. In order to develop the various optimization algorithms, threshold decomposition is first extended to admit real-valued inputs. The generalized form of threshold decomposition plays a critical role in the optimization of WM filters.

6.3.1 Threshold Decomposition For Real-Valued Signals

This far, threshold decomposition has been defined for input sequences with a finite size input alphabet. In order to use the properties of threshold decomposition for the optimization of WM filters, this framework must first be generalized to admit real-valued input signals. This decomposition, in turn, can be used to analyze weighted median filters having real-valued weights.

Consider the set of real-valued samples $X_1, X_2, \ldots, X_N$ and define a weighted median filter by the corresponding real-valued weights $W_1, W_2, \ldots, W_N$. Decompose each sample X_i as

$$x_i^q = \operatorname{sgn}(X_i - q) \tag{6.46}$$

where $-\infty < q < \infty$, and

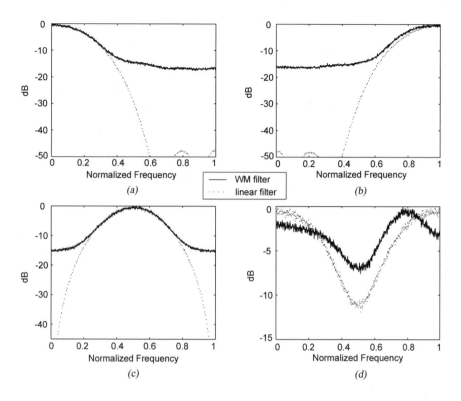

Figure 6.13 Approximated frequency response of WM filters designed with Mallows iterative algorithm: (*a*) low-pass, (*b*) high-pass, (*c*) band-pass, (*d*) band-stop.

$$\text{sgn}\,(X_i - q) = \begin{cases} 1 & \text{if } X_i \geq q; \\ -1 & \text{if } X_i < q. \end{cases} \tag{6.47}$$

Thus, each sample X_i is decomposed into an infinite set of binary points taking values in $\{-1, 1\}$. Figure 6.14 depicts the decomposition of X_i as a function of q.

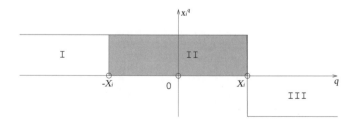

Figure 6.14 Decomposition of X_i into the binary x_i^q signal.

Threshold decomposition is reversible since the original real-valued sample X_i can be perfectly reconstructed from the infinite set of thresholded signals. To show this, let $\hat{X}_i = \lim_{T \to \infty} X_i^{<T>}$ where

$$X_i^{<T>} = \frac{1}{2} \int_{-T}^{-|X_i|} x_i^q \, dq \;+\; \frac{1}{2} \int_{-|X_i|}^{|X_i|} x_i^q \, dq \;+\; \frac{1}{2} \int_{|X_i|}^{T} x_i^q \, dq. \tag{6.48}$$

Since the first and last integrals in (6.48) cancel each other and since

$$\int_{-|X_i|}^{|X_i|} x_i^q \, dq = 2X_i, \tag{6.49}$$

it follows that $X_i^{<T>} = \hat{X}_i = X_i$. Hence, the original signal can be reconstructed from the infinite set of thresholded signals as

$$\begin{aligned}
X_i &= \frac{1}{2} \int_{-\infty}^{\infty} x_i^q \, dq \\
&= \frac{1}{2} \int_{-\infty}^{\infty} \mathrm{sgn}\,(X_i - q) \; dq.
\end{aligned} \tag{6.50}$$

The sample X_i can be reconstructed from its corresponding set of decomposed signals and consequently X_i has a unique threshold signal representation, and vice versa:

$$X_i \xleftrightarrow{T.D.} \{x_i^q\}, \tag{6.51}$$

where $\xleftrightarrow{T.D.}$ denotes the one-to-one mapping provided by the threshold decomposition operation. Since q can take any real value, the infinite set of binary samples $\{x_i^q\}$ seems redundant in representing X_i. Letting $\mathbf{x}^q = [x_1^q, \ldots, x_N^q]^T$, some of the binary vectors $\{\mathbf{x}^q\}$ are infinitely repeated. For $X_{(1)} < q \leq X_{(2)}$, for instance, all the binary vectors $\{\mathbf{x}^q\}$ are identical. Note however that the threshold signal representation can be simplified based on the fact that there are at most $L + 1$ different binary vectors $\{\mathbf{x}^q\}$ for each observation vector $\mathbf{X}$. Using this fact, (6.51) reduces to

$$\mathbf{X} \xleftrightarrow{T.D.} \{\mathbf{x}^q\} = \begin{cases} [1, 1, \cdots, 1]^T & \text{for } -\infty < q \leq X_{(1)} \\[2mm] [x_1^{X_{(i)}^+}, x_2^{X_{(i)}^+}, \cdots, x_L^{X_{(i)}^+}]^T & \text{for } X_{(i)} < q \leq X_{(i+1)}, \\[2mm] & \qquad 1 \leq i \leq L - 1 \\[2mm] [-1, -1, \cdots, -1]^T & \text{for } X_{(L)} < q < +\infty \end{cases} \tag{6.52}$$

where $X_{(i)}^+$ denotes a value on the real line approaching $X_{(i)}$ from the right. The simplified representation in (6.52) will be used shortly.

Threshold decomposition in the real-valued sample domain also allows the order of the median and threshold decomposition operations to be interchanged without affecting the end result. To illustrate this concept, consider three samples X_1, X_2, X_3 and their threshold decomposition representations x_1^q, x_2^q, x_3^q shown in Figure 6.15a. The plots of x_1^q and x_3^q are slightly shifted in the vertical axis for illustrative purposes. As it is shown in the figure, assume that $X_3 = X_{(3)}, X_2 = X_{(1)}$, and $X_1 = X_{(2)}$. Next, for each value of q, the median of the decomposed signals is defined as

$$y^q = \text{MEDIAN}(x_1^q, x_2^q, x_3^q). \tag{6.53}$$

Referring to Figure 6.15a, note that for $q \leq X_{(2)}$ two of the three x_i^q samples have values equal to 1, and for $q > X_{(2)}$ two of these have values equal to -1. Thus,

$$y^q = \begin{cases} 1 & \text{for } q \leq X_{(2)}; \\ -1 & \text{for } q > X_{(2)}. \end{cases} \tag{6.54}$$

A plot of y^q as a function of q is shown in Figure 6.15b. Reversing the decomposition using y^q in (6.50), it follows that

$$\begin{aligned} Y &= \frac{1}{2} \int_{-\infty}^{\infty} y^q \, dq \\ &= \frac{1}{2} \int_{-\infty}^{\infty} \text{sgn}\left(X_{(2)} - q\right) \, dq \\ &= X_{(2)}. \end{aligned} \tag{6.55}$$

Thus, in this example, the reconstructed output is the second order-statistic namely the median.

In the general case, we consider N samples $X_1, X_2, \ldots, X_N$ and their corresponding threshold decomposition representations $x_1^q, x_2^q, \ldots, x_N^q$. The median of the decomposed signals at a fixed value of q is

$$y^q = \text{MEDIAN}(x_1^q, x_2^q, \ldots, x_N^q) = \begin{cases} 1 & \text{for } q \leq X_{\left(\frac{N+1}{2}\right)}; \\ -1 & \text{for } q > X_{\left(\frac{N+1}{2}\right)}. \end{cases} \tag{6.56}$$

Reversing the threshold decomposition, Y is obtained as

$$\begin{aligned} Y &= \frac{1}{2} \int_{-\infty}^{\infty} y^q \, dq \\ &= \frac{1}{2} \int_{-\infty}^{\infty} \text{sgn}\left(X_{\left(\frac{N+1}{2}\right)} - q\right) \, dq \\ &= X_{\left(\frac{N+1}{2}\right)}. \end{aligned} \tag{6.57}$$

Thus, applying the median operation on a set of samples and applying the median operation on a set threshold decomposed set of samples and reversing the decomposition give exactly the same result.

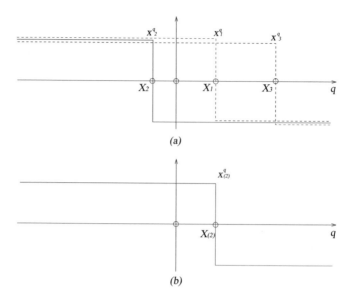

(a)

(b)

Figure 6.15 The decomposed signal $x^q_{(2)}$ as the median of x^q_1, x^q_2, x^q_3. The reconstructed signal results in $X_{(2)}$.

With this threshold decomposition, the weighted median filter operation can be implemented as

$$
\hat{\beta} = \text{MEDIAN}\left(|W_i| \diamond \text{sgn}\,(W_i)\, X_i|_{i=1}^{N} \right)
$$

$$
= \text{MEDIAN}\left(|W_i| \diamond \frac{1}{2} \int_{-\infty}^{\infty} \text{sgn}\,[\text{sgn}\,(W_i)\, X_i - q]\ dq|_{i=1}^{N} \right). \quad (6.58)
$$

The expression in (6.58) represents the median operation of a set of weighted integrals, each synthesizing a signed sample. Note that the same result is obtained if the weighted median of these functions, at each value of q, is taken first and the resultant signal is integrated over its domain. Thus, the order of the integral and the median operator can be interchanged without affecting the result leading to

$$
\hat{\beta} = \frac{1}{2} \int_{-\infty}^{\infty} \text{MEDIAN}\left(|W_i| \diamond \text{sgn}\,[\text{sgn}\,(W_i)\, X_i - q]\,|_{i=1}^{N} \right)\ dq. \quad (6.59)
$$

In this representation, the signed samples play a fundamental role; thus, we define the signed observation vector **S** as

$$
\mathbf{S} = [\text{sgn}(W_1)X_1, \text{sgn}(W_2)X_2, \ldots, \text{sgn}(W_N)X_N]^T
$$

$$
= [S_1, S_2, \ldots, S_N]^T. \quad (6.60)
$$

The threshold decomposed signed samples, in turn, form the vector $\mathbf{s}^q$ defined as

$$\mathbf{s}^q = [\text{sgn}\,[\text{sgn}(W_1)X_1 - q]\,,\ \text{sgn}\,[\text{sgn}(W_2)X_2 - q]\,,\dots$$
$$\dots,\ \text{sgn}\,[\text{sgn}(W_N)X_N - q]]^T$$
$$= [s_1^q,\ s_2^q,\dots,\ s_N^q]^T. \tag{6.61}$$

Letting $\mathbf{W}_a$ be the vector whose elements are the weight's magnitudes, $\mathbf{W}_a = \langle|W_1|,\ |W_2|,\cdots,\ |W_N|\rangle^T$, the WM filter operation can be expressed as

$$\hat{\beta} = \frac{1}{2}\int_{-\infty}^{\infty} \text{sgn}\left(\mathbf{W}_a^T \mathbf{s}^q\right)\,dq. \tag{6.62}$$

The WM filter representation using threshold decomposition is compact although it may seem that the integral term may be difficult to implement in practice. Equation (6.62), however, is used for the purposes of analysis and not implementation. In addition, if desired, it can be simplified, based on the fact that there are at most $N + 1$ different binary signals for each observation vector $\mathbf{s}^q$. Let $S_{(i)}$ be the ith smallest signed sample, then the $N + 1$ different vectors $\mathbf{s}^q$ are

$$\mathbf{s}^q = \begin{cases} [1, 1,\dots, 1] & \text{for } -\infty < q < S_{(1)} \\ \left[s_1^{S_{(i)}^+}, s_2^{S_{(i)}^+},\dots, s_N^{S_{(i)}^+}\right] & \text{for } S_{(i)} < q < S_{(i+1)} \\ [-1, -1,\dots, -1] & \text{for } S_{(N)} < q < \infty \end{cases} \tag{6.63}$$

where $S_{(i)}^+$ denotes a value on the real line approaching $S_{(i)}$ from the right. Using these vectors in (6.62) we have

$$\hat{\beta} = \frac{1}{2}\int_{-\infty}^{S_{(1)}} \text{sgn}\left(\mathbf{W}_a^T \mathbf{s}^{S_{(1)}^-}\right)\,dq$$
$$+ \frac{1}{2}\sum_{i=1}^{N-1}\int_{S_{(i)}}^{S_{(i+1)}} \text{sgn}\left(\mathbf{W}_a^T \mathbf{s}^{S_{(i)}^+}\right)\,dq \tag{6.64}$$
$$+ \frac{1}{2}\int_{S_{(N)}}^{\infty} \text{sgn}\left(\mathbf{W}_a^T \mathbf{s}^{\infty}\right)\,dq.$$

The above equation reduces to

$$\hat{\beta} = \tfrac{1}{2}\lim_{Q\to\infty}\left(S_{(1)} + Q\right) + \frac{1}{2}\sum_{i=1}^{N-1}\left(S_{(i+1)} - S_{(i)}\right)\text{sgn}\left(\mathbf{W}_a^T \mathbf{s}^{S_{(i)}^+}\right)$$
$$- \frac{1}{2}\lim_{Q\to\infty}\left(Q - S_{(N)}\right), \tag{6.65}$$

which simplifies to

$$\hat{\beta} = \frac{S_{(1)} + S_{(N)}}{2} + \frac{1}{2}\sum_{i=1}^{N-1}\left(S_{(i+1)} - S_{(i)}\right)\text{sgn}\left(\mathbf{W}_a^T \mathbf{s}^{S_{(i)}^+}\right). \tag{6.66}$$

The computation of weighted median filters with the new threshold decomposition architecture is efficient requiring only $N - 1$ threshold logic (sign) operators, it allows the input signals to be arbitrary real-valued signals, and it allows positive and negative filter weights.

The filter representation in (6.66) also provide us with a useful interpretation of WM filters. The output $\hat{\beta}$ is computed by the sum of the midrange of the signed-samples $V = (S_{(1)} + S_{(N)})/2$, which provides a coarse estimate of location, and by a linear combination of the $(i, i+1)$th spacing $V_i = S_{(i)} - S_{(i-1)}$, for $i = 1, 2, \ldots, N$. Hence

$$\hat{\beta} = V + \sum_{i=2}^{N} C(\mathbf{W}_a, \mathbf{s}^{S^+_{(i)}}) V_i. \tag{6.67}$$

The coefficients $C(\cdot)$ take on values $-1/2$ or $1/2$ depending on the values of the observation samples and filter weights.

6.3.2 The Least Mean Absolute (LMA) Algorithm

The least mean square algorithm (LMS) is perhaps one of the most widely used algorithms for the optimization of linear FIR filters in a broad range of applications [197]. In the following, a similar adaptive algorithm for optimizing WM filters is described – namely the *Least Mean Absolute* (LMA) adaptive algorithm. The LMA algorithm shares many of the desirable attributes of the LMS algorithm including simplicity and efficiency [6].

Assume that the observed process $\{X(n)\}$ is statistically related to some desired process $\{D(n)\}$ of interest. $\{X(n)\}$ is typically a transformed or corrupted version of $\{D(n)\}$. Furthermore, it is assumed that these processes are jointly stationary. A window of width N slides across the input process pointwise estimating the desired sequence. The vector containing the N samples in the window at time n is

$$\begin{aligned}
\mathbf{X}(n) &= [X(n - N_1), \ldots, X(n), \ldots, X(n + N_2)]^T \\
&= [X_1(n), X_2(n), \ldots, X_N(n)]^T, \tag{6.68}
\end{aligned}$$

with $N = N_1 + N_2 + 1$. The running weighted median filter output estimates the desired signal as

$$\hat{D}(n) = \text{MEDIAN}\left[|W_i| \diamond \text{sgn}(W_i)X_i(n)|_{i=1}^{N}\right],$$

where both the weights W_i and samples $X_i(n)$ take on real values. The goal is to determine the weight values in $\mathbf{W} = \langle W_1, W_2, \ldots, W_N \rangle^T$ which will minimize the estimation error. Under the Mean Absolute Error (MAE) criterion, the cost to minimize is

$$J(\mathbf{W}) = E\left\{|D(n) - \hat{D}(n)|\right\} \tag{6.69}$$

$$= E\left\{\frac{1}{2}\left|\int_{-\infty}^{\infty} \text{sgn}(D-q) - \text{sgn}\left(\mathbf{W}_a^T \mathbf{s}^q\right) dq\right|\right\}, \qquad (6.70)$$

where the threshold decomposition representation of the signals was used. The absolute value and integral operators in (6.70) can be interchanged since the integral acts on a strictly positive or a strictly negative function. This results in

$$J(\mathbf{W}) = \frac{1}{2}\int_{-\infty}^{\infty} E\left\{\left|\text{sgn}(D-q) - \text{sgn}\left(\mathbf{W}_a^T \mathbf{s}^q\right)\right|\right\} dq. \qquad (6.71)$$

Furthermore, since the argument inside the absolute value operator in (6.71) can only take on values in the set $\{-2, 0, 2\}$, the absolute value operator can be replaced by a properly scaled second power operator. Thus

$$J(\mathbf{W}) = \frac{1}{4}\int_{-\infty}^{\infty} E\left\{\left(\text{sgn}(D-q) - \text{sgn}\left(\mathbf{W}_a^T \mathbf{s}^q\right)\right)^2\right\} dq. \qquad (6.72)$$

Taking the gradient of the above results in

$$\frac{\partial}{\partial \mathbf{W}} J(\mathbf{W}) = -\frac{1}{2}\int_{-\infty}^{\infty} E\left\{e^q(n)\frac{\partial}{\partial \mathbf{W}}\text{sgn}\left(\mathbf{W}_a^T \mathbf{s}^q\right)\right\} dq \qquad (6.73)$$

where $e^q(n) = \text{sgn}(D-q) - \text{sgn}\left(\mathbf{W}_a^T \mathbf{s}^q\right)$. Since the sign function is discontinuous at the origin, its derivative will introduce Dirac impulse terms that are inconvenient for further analysis. To overcome this difficulty, the sign function in (6.73) is approximated by a smoother differentiable function. A simple approximation is given by the hyperbolic tangent function

$$\text{sgn}(x) \approx \tanh(x) = \frac{e^x - e^{-x}}{e^x + e^{-x}}. \qquad (6.74)$$

Since $\frac{\partial}{\partial x}\tanh(x) = \text{sech}^2(x) = \frac{2}{e^x + e^{-x}}$, it follows that

$$\frac{\partial}{\partial \mathbf{W}}\text{sgn}\left(\mathbf{W}_a^T \mathbf{s}^q\right) \approx \text{sech}^2\left(\mathbf{W}_a^T \mathbf{s}^q\right)\frac{\partial}{\partial \mathbf{W}}\left(\mathbf{W}_a^T \mathbf{s}^q\right). \qquad (6.75)$$

Evaluating the derivative in (6.75) and after some simplifications leads to

$$\frac{\partial}{\partial \mathbf{W}}\text{sgn}\left(\mathbf{W}_a^T \mathbf{s}^q\right) \approx \text{sech}^2\left(\mathbf{W}_a^T \mathbf{s}^q\right)\begin{bmatrix} \text{sgn}(W_1)s_1^q \\ \text{sgn}(W_2)s_2^q \\ \vdots \\ \text{sgn}(W_N)s_N^q \end{bmatrix}. \qquad (6.76)$$

Using (6.76) in (6.73) yields

$$\frac{\partial}{\partial W_j}J(\mathbf{W}) = -\frac{1}{2}\int_{-\infty}^{\infty} E\left\{e^q(n)\text{sech}^2\left(\mathbf{W}_a^T \mathbf{s}^q\right)\text{sgn}(W_j)s_j^q\right\} dq. \qquad (6.77)$$

Using the gradient, the optimal coefficients can be found through the steepest descent recursive update

$$W_j(n+1) = W_j(n) + 2\mu \left[-\frac{\partial}{\partial W_j} J(\mathbf{W}) \right] \tag{6.78}$$

$$= W_j(n) + \mu \left[\int_{-\infty}^{\infty} E \left\{ e^q(n) \text{sech}^2 \left(\mathbf{W}_a^T(n) \mathbf{s}^q(n) \right) \right. \right.$$

$$\left. \left. \times \text{sgn}(W_j(n)) s_j^q(n) \right\} dq \right].$$

Using the instantaneous estimate for the gradient we can derive an adaptive optimization algorithm where

$$W_j(n+1) = W_j(n) + \mu \int_{-\infty}^{\infty} e^q(n) \text{sech}^2 \left(\mathbf{W}_a^T(n) \mathbf{s}^q(n) \right) \text{sgn}(W_j(n)) s_j^q(n) dq$$

$$= W_j(n) + \mu \int_{-\infty}^{S_{(1)}} \left[e^{S_{(1)}^-}(n) \text{sech}^2 \left(\mathbf{W}_a^T(n) \mathbf{s}^{S_{(1)}^-}(n) \right) \right.$$

$$\left. \times \text{sgn}(W_j(n)) s_j^{S_{(1)}^-}(n) \right] dq$$

$$+ \mu \sum_{i=1}^{N-1} \left[(S_{(i+1)} - S_{(i)}) s_j^{S_{(i)}^+}(n) e^{S_i^+}(n) \text{sgn}(W_j(n)) \right.$$

$$\left. \times \text{sech}^2 \left(\mathbf{W}_a^T(n) \mathbf{s}^{S_{(i)}^+}(n) \right) \right]$$

$$+ \mu \int_{S_{(N)}}^{\infty} \left[e^{S_{(N)}^+}(n) \text{sech}^2 \left(\mathbf{W}_a^T(n) \mathbf{s}^{S_{(N)}^+}(n) \right) \right.$$

$$\left. \times \text{sgn}(W_j(n)) s_j^{S_{(N)}^+}(n) \right] dq. \tag{6.79}$$

The error term $e^q(n)$ in the first and last integrals can be shown to be zero; thus, the adaptive algorithm reduces to

$$W_j(n+1) = W_j(n) + \mu \sum_{i=1}^{N-1} \left[(S_{(i+1)} - S_{(i)}) s_j^{S_{(i)}^+}(n) e^{S_{(i)}^+}(n) \text{sgn}(W_j(n)) \right.$$

$$\left. \times \text{sech}^2 \left(\mathbf{W}_a^T(n) \mathbf{s}^{S_{(i)}^+}(n) \right) \right], \tag{6.80}$$

for $j = 1, 2, \ldots, N$. Since the MAE criterion was used in the derivation, the recursion in (6.80) is referred to as the *Least Mean Absolute* (LMA) weighted median adaptive algorithm.

The contribution of most of the terms in (6.80), however, is negligible compared to that of the vector $\mathbf{s}^{\hat{D}(n)}(n)$ as it will be described here. Using this fact and following the arguments used in [147, 200], the algorithm in (6.80) can be simplified considerably leading to a fast LMA WM adaptive algorithm. The contribution of each term in (6.80) is, to a large extent, determined by $\text{sech}^2(\mathbf{W}_a^T \mathbf{s}^q)$, for $q \in \mathcal{S}$. The sech^2 function achieves its maximum value when its argument satisfies $\mathbf{W}_a^T \mathbf{s}^q = 0$. Its value decreases rapidly and monotonically to zero as the argument departs from zero. From the $N - 1$ vectors $\mathbf{s}^q$, $q \in \mathcal{S}$, there is one for which the inner product $\mathbf{W}_a^T \mathbf{s}^q$ is closest to zero. Consequently, the update term corresponding to this vector will provide the biggest contribution in the update. Among all vectors $\mathbf{s}^q$, $q \in \mathcal{S}$, the vector providing the largest update contribution can be found through the definition of the weighted median filter. Since $\hat{D}(n)$ is equal to one of the signed input samples, the output of the WM filter is given by

$$\hat{D}(n) = \left\{ S_{(k)}(n) : k = \max_j \text{ for which } \sum_{i=j}^N |W_{[i]}| \geq \frac{1}{2} \sum_{i=1}^N |W_i| = T_0 \right\} \quad (6.81)$$

The constraints can be rewritten as:

$$|W_{[k-1]}| + \ldots + |W_{[N]}| > T_0$$
$$|W_{[k]}| + \ldots + |W_{[N]}| \geq T_0$$
$$|W_{[k+1]}| + \ldots + |W_{[N]}| < T_0$$

Replacing T_0 and cancelling common terms in the summations leads to:

$$-|W_{[1]}| - \ldots - |W_{[k-2]}| + |W_{[k-1]}| + \ldots + |W_{[N]}| > 0$$
$$-|W_{[1]}| - \ldots - |W_{[k-1]}| + |W_{[k]}| + \ldots + |W_{[N]}| \geq 0$$
$$-|W_{[1]}| - \ldots - |W_{[k]}| + |W_{[k+1]}| + \ldots + |W_{[N]}| < 0 \quad (6.82)$$

The threshold decomposition of the signed input vector will be, according to (6.63):

$$\mathbf{s}^{S(k-1)} = [-1, \ldots, -1, \ 1, \underline{1}, 1, \ldots, 1]$$
$$\mathbf{s}^{S(k)} = [-1, \ldots, -1, -1, \underline{1}, 1, \ldots, 1]$$
$$\mathbf{s}^{S(k+1)} = [-1, \ldots, -1, -1, \underline{-1}, 1, \cdots, 1]$$

where the underlined element in each vector represents the kth component. Using these vectors (6.82) can be rewritten as:

$$\mathbf{W}_a^T \mathbf{s}^{S(k-1)} > 0$$
$$\mathbf{W}_a^T \mathbf{s}^{S(k)} \geq 0$$
$$\mathbf{W}_a^T \mathbf{s}^{S(k+1)} < 0.$$

This ensures that $\mathbf{s}^{S_{(k)}}$ is the vector whose inner product $\mathbf{W}_a^T \mathbf{s}^{S_{(k)}}$ is closest to zero. Accordingly $(S_{(k+1)} - S_{(k)}) \mathrm{sech}^2 \left(\mathbf{W}_a^T \mathbf{s}^{S_{(k)}} \right) S_j^{S_{(k)}}$ is the largest contributor in (6.80). In consequence, the derivative can be approximated as:

$$\frac{\partial}{\partial W_j} J(\mathbf{W}) \approx \frac{1}{2} \mathrm{sgn}(W_j)(S_{(k+1)} - S_{(k)}) \mathrm{sech}^2 \left(\mathbf{W}_a^T \mathbf{s}^{S_{(k)}} \right) S_j^{S_{(k)}} \qquad (6.83)$$

Removing scale factors and applying TD:

$$\begin{aligned} \frac{\partial}{\partial W_j} J(\mathbf{W}) &\approx \mathrm{sgn}(W_j) S_j^{S_{(k)}} \\ &= \mathrm{sgn}(W_j) \mathrm{sgn}(\mathrm{sgn}(W_j) X_j - S_{(k)}) \end{aligned} \qquad (6.84)$$

Using this as the principal contributor of the update, and since $S_{(k)}$ is the output of the weighted median at time n ($S_{(k)} = \hat{D}(n)$), the algorithm in (6.80) is simplified leading to the following recursion referred to as the fast LMA WM adaptive algorithm:

$$\begin{aligned} W_j(n+1) = W_j(n) &+ \mu \left(D(n) - \hat{D}(n) \right) \mathrm{sgn}(W_j(n)) \\ &\times \mathrm{sgn} \left(\mathrm{sgn}(W_j(n)) X_j(n) - \hat{D}(n) \right), \qquad (6.85) \end{aligned}$$

for $j = 1, 2, \ldots, N$.

The updates in (6.85) have an intuitive explanation described in Figure 6.16. When the output of the WM filter is smaller than the desired output, the magnitude of the weights corresponding to the signed samples which are larger than the actual output are increased. Thus, the weight for the signed sample $(-1)X_i$ is decreased (larger negative value) whereas the weight for signed sample $(+1)X_j$ is increased. Both cases will lead to updated weights that will push the estimate higher towards $D(n)$. Similarly, the weights corresponding to the signed samples which are smaller than the actual output are reduced. Thus, the weight for the signed sample $(-1)X_\ell$ is increased (smaller negative value) whereas the weight for signed sample $(+1)X_k$ is decreased. Figure 6.16b depicts the response of the algorithm when the WM filter output is larger than the desired output. The updates of the various samples follow similar intuitive rules as shown in Fig. 6.16b.

Since the updates only use the most significant update term in (6.80), it is expected that the fast algorithm requires a good initial weight vector. It has been experimentally shown that a good initial weight vector is that of the median filter. Because of the nonlinear nature of the adaptive algorithm, a convergence analysis cannot be derived. The fast algorithm, however, in practice works quite well. Since a convergence analysis is not available for the fast LMA WM adaptive algorithm, exact bounds on the stepsize μ are not available. A reliable guideline to select the step size of this algorithm is to select it on the order of that required for the standard LMS algorithm. The step size can then be further tuned according to the user's requirements and by evaluation of the response given by the initial step size choice.

Figure 6.16 Weight updates when: (a) $D(n) > \hat{D}(n)$, and (b) $D(n) < \hat{D}(n)$. The signed samples are denoted as either $(-1)X_i$ or $(1)X_i$.

An example of the contours of the cost function for the optimization of the WM filter is shown in Figure 6.17. The cost function is not continuous. It is composed of constant regions, represented in the figure by different levels of gray. These regions are separated by sharp transitions. The objective of the optimization algorithm is to find the region with the lowest value (displayed in the figure with color black). The plot shown represents the cost as a function of two out of seven weights. The white line represents the path followed by the LMA algorithm during the optimization process.

EXAMPLE 6.10 (DESIGN OF OPTIMAL HIGH-PASS WM FILTER)

Having the optimization framework at hand, consider next the design of a high-pass WM filter whose objective is to preserve a high frequency tone while remove all low frequency terms. Figure 6.18a depicts a two-tone signal with normalized frequencies of 0.04 and 0.4 Hz. Figure 6.18b shows the multi-tone signal filtered by a 28-tap linear FIR filter designed by MATLAB's fir1 function with a normalized cutoff frequency 0.2 Hz. The fast adaptive LMA algorithm was used to optimize a WM filter with 28 weights. These weights, in turn, were used to filter the multitone signal resulting in the estimate shown in Figure 6.18c. The low-frequency components have been clearly filtered out. There are, however, some minor artifacts present. Figure 6.18d depicts the WM filter output when the weights values of the linear FIR filter are used. Although the frequency content of the output signal is within the specifications, there is a significant distortion in the amplitude of the signal in Figure 6.18d. Next, Yin et. al's fast adaptive LMA algorithm was used to optimize a WM filter (smoother) with 28 (positive) weights. The filtered signal attained with the optimized weights is shown in Figure 6.18e. The weighted median smoother clearly fails to remove the low frequency components, as expected. The weighted median smoother output closely resembles the input signal as it is the closest output to the desired signal it can produce.

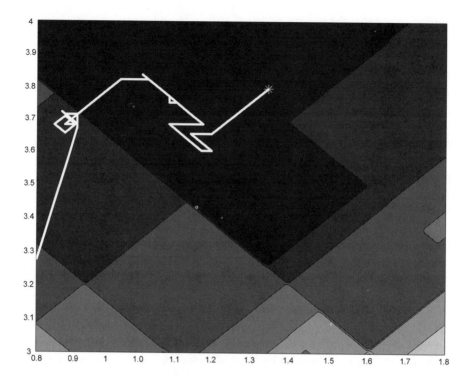

Figure 6.17 Contours of the cost function of the optimization of a WM filter and weight optimization trajectory for two filter weights

The step size used in all adaptive optimization experiments was 10^{-3}. The performance of the adaptive LMA algorithm in (6.80) and of the fast adaptive LMA algorithm in (6.85) were very similar. The algorithm in (6.80), however, converges somewhat faster than the algorithm in (6.85). This is not surprising as the fast algorithm uses the most important information available, but not all, for the update of the adaptive LMA algorithm. Figure 6.19 shows a single-realization learning curve for the fast adaptive LMA WM filter algorithm in (6.85) and the ensemble average of 1000 realizations of the same algorithm. It can be seen that 200 iterations were needed for the fast adaptive LMA algorithm to converge. The algorithm in (6.80) required only 120 iterations, however, due to its computational load, the fast LMA algorithm would be preferred in most applications. The mean absolute error (MAE) between the desired signal and the output of the various filters is summarized in Table 6.4. The advantage of allowing negative weights on the median filter structure is readily seen in Table 6.4. The performance of the LMA WM optimization and of the fast implementation are equivalent. The linear filter outperforms the median structures in the noise-free case, as expected.

Having designed the various high-pass filters in a noiseless environment, the performance on signals embedded in noise is tested next. Stable noise with $\alpha = 1.4$ was

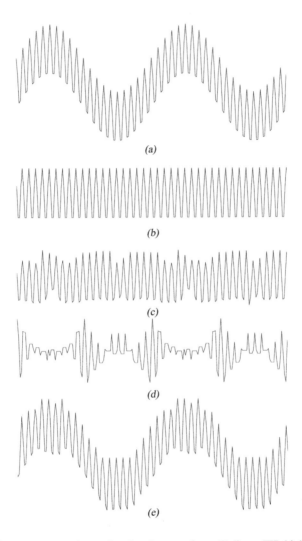

Figure 6.18 (*a*) Two-tone input signal and output from (*b*) linear FIR high-pass filter, (*c*) optimal WM filter, (*d*) WM filter using the linear FIR weight values, (*e*) optimal WM smoother with non-negative weights.

added to the two-tone signal. Rather than training the various filters to this noisy environment, we used the same filter coefficients as in the noise-free simulations. Figure 6.20*a–d* illustrates the results. The MAE for the linear, WM filter, and WM smoother were computed as 0.979, 0.209, and 0.692, respectively. As expected, the outputs of the weighted median filter and smoother are not affected, whereas the output of the linear filter is severely degraded as the linear high-pass filter amplifies the high fre-

Table 6.4 Mean Absolute Filtering Errors

Filter	noise free	with stable noise
Linear FIR	0.012	0.979
Optimal WM smoother	0.688	0.692
WMF with FIR weights	0.501	0.530
Optimal WMF (fast alg.)	0.191	0.209
Optimal WMF	0.190	0.205

quency noise. Table 6.4 summarizes the MAE values attained by the various filters. ■

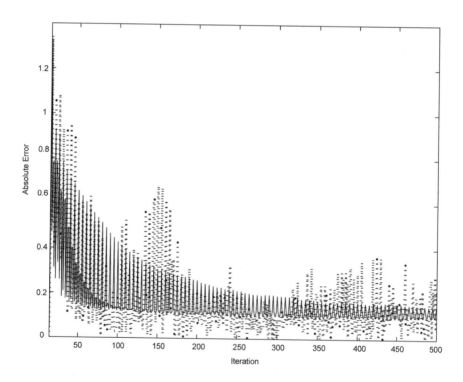

Figure 6.19 Learning characteristics of the fast LMA adaptive WM filter algorithm admitting real-valued weights, the dotted line represents a single realization, the solid line the average of 1000 realizations.

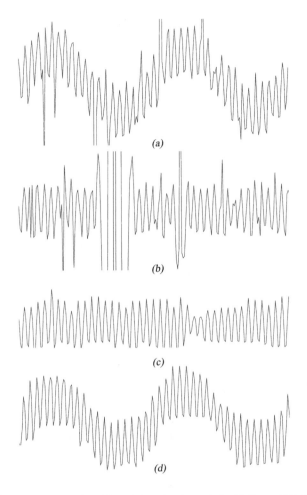

Figure 6.20   (*a*) Two-tone signal in stable noise ($\alpha = 1.4$), (*b*) linear FIR filter output, (*c*) WM filter output, (*d*) WM smoother output with positive weights.

6.4   RECURSIVE WEIGHTED MEDIAN FILTERS

Having the framework for weighted median filters, it is natural to extend it to other more general signal processing structures. Arce and Paredes (2000) [14] defined the class of recursive weighted median filters admitting real-valued weights. These filters are analogous to the class of infinite impulse response (IIR) linear filters. Recursive filter structures are particularly important because they can be used to model "resonances" that appear in many natural phenomena such as in speech. In fact, in the linear filtering framework, a large number of systems can be better characterized/modeled by a pole-zero transfer function than by a transfer function containing only zeros. In addition, IIR linear filters often lead to reduced computational complexity. Much like IIR linear filters provide these advantages over linear FIR filters, recursive WM filters also exhibit superior characteristics than nonrecursive WM filters. Recursive WM filters can synthesize nonrecursive WM filters of much larger window sizes. In terms of noise attenuation, recursive median smoothers have far superior characteristics than their nonrecursive counterparts [5, 8].

The general structure of linear IIR filters is defined by the difference equation

$$Y(n) = \sum_{\ell=1}^{N} A_\ell \, Y(n - \ell) + \sum_{k=-M_1}^{M_2} B_k \, X(n - k), \qquad (6.86)$$

where the output is formed not only from the input, but also from previously computed outputs. The filter weights consist of two sets: the feedback coefficients $\{A_\ell\}$, and the feed-forward coefficients $\{B_k\}$. In all, $N + M_1 + M_2 + 1$ coefficients are needed to define the recursive difference equation in (6.86).

The generalization of (6.86) to a RWM filter structure is straight-forward. Following a similar approach used in the optimization of nonrecursive WM filters, the summation operation is replaced with the *median* operation, and the *multiplication* weighting is replaced by weighting through *signed replication*:

$$Y(n) = \mathrm{MEDIAN} \left(|A_\ell| \; \diamond \; \mathrm{sgn}(A_\ell) Y(n - \ell)|_{\ell=1}^{N}, \right. \qquad (6.87)$$
$$\left. |B_k| \; \diamond \; \mathrm{sgn}(B_k) X(n - k)|_{k=-M_1}^{M_2} \right)$$

A noncausal implementation is assumed from now on where $M_2 = 0$ and $M_1 = M$ leading to the following definition:

DEFINITION 6.4 (RECURSIVE WEIGHTED MEDIAN FILTERS) *Given a set of N real-valued feed-back coefficients $A_i|_{i=1}^{N}$ and a set of $M + 1$ real-valued feed-forward coefficients $B_i|_{i=0}^{M}$, the $M + N + 1$ recursive WM filter output is defined as*

$$Y(n) = \mathrm{MEDIAN} \left(|A_N| \diamond \mathrm{sgn}(A_N) Y(n - N), \ldots, |A_1| \diamond \mathrm{sgn}(A_1) Y(n - 1), \right.$$
$$\left. |B_0| \diamond \mathrm{sgn}(B_0) X(n), \ldots, |B_M| \diamond \mathrm{sgn}(B_M) X(n + M) \right).$$
$$(6.88)$$

Note that if the weights A_ℓ and B_k are constrained to be positive, (6.88) reduces to the recursive WM smoother described in Chapter 5. For short notation, recursive WM filters will be denoted with double angle brackets where the center weight is underlined. The recursive WM filter in 6.88 is, for example, denoted by $\langle\langle A_N, \ldots, A_1, \underline{B_0}, B_1, \ldots, B_M \rangle\rangle$.

The recursive WM filter output for noninteger weights can be determined as follows:

(1) Calculate the threshold $T_0 = \frac{1}{2} \left(\sum_{\ell=1}^{N} |A_\ell| + \sum_{k=0}^{M} |B_k| \right)$.

(2) Jointly sort the signed past output samples $\mathrm{sgn}(A_\ell)Y(n-\ell)$ and the signed input observations $\mathrm{sgn}(B_k)X(n+k)$.

(3) Sum the magnitudes of the weights corresponding to the sorted signed samples beginning with the maximum and continuing down in order.

(4) If $2T_0$ is an even number, the output is the average between the signed sample whose weight magnitude causes the sum to become $\geq T_0$ and the next smaller signed sample, otherwise the output is the signed sample whose weight magnitude causes the sum to become $\geq T_0$.

EXAMPLE 6.11
___

Consider the window size 6 RWM filter defined by the real valued weights $\langle\langle A_2, A_1,$ $B_0, B_1, B_2, B_3\rangle\rangle = \langle\langle 0.2, 0.4, \underline{0.6}, -0.4, 0.2, 0.2\rangle\rangle$, where the use of double angle brackets is introduced to denote the recursive WM filter operation. The output for this filter operating on the observation set $[Y(n-2), Y(n-1), X(n), X(n+1), X(n+2), X(n+3)]^T = [-2, 2, -1, 3, 6, 8]^T$ is found as follows. Summing the absolute weights gives the threshold $T_0 = \frac{1}{2}(|A_1|+|A_2|+|B_0|+|B_1|+|B_2|) = 1$. The signed set of samples spanned by the filter's window, the sorted set, their corresponding weight, and the partial sum of weights (from each ordered sample to the maximum) are:

sample set in the window	$-2, \quad 2, \quad -1, \quad 3, \quad 6, \quad 8$
corresponding weights	$0.2, 0.4, 0.6, -0.4, 0.2, 0.2$
sorted signed samples	$-3, -2, -1, \quad 2, \quad 6, \quad 8$
corresponding absolute weights	$0.4, \ 0.2, \ 0.6, \quad 0.4, \quad 0.2 \ 0.2$
partial weight sums	$2.0, \ 1.6, \ \underline{1.4}, \quad 0.8, \quad 0.4 \ 0.2$

Thus, the output is $\frac{-1-2}{2} = -1.5$ since when starting from the right (maximum sample) summing the weights, the threshold $T_0 = 1$ is not reached until the weight associated with -1 is added. The underlined sum value above indicates that this is the first sum which meets or exceeds the threshold. ∎

Note that the definition of the weighted median operation with real-valued weights used here is consistent with both the definition of median operation when the window size is an even number and the definition of WM operation when the sum of the integer-valued weights adds up to an even number, in the sense that the filter's output is the average of two samples. The reason for using the average of two signed samples as the output of the recursive WM filter is that it allows the use of recursive WM filters with suitable weights in band-pass or high-pass applications where the filter should output zero when a DC component at the input is present. In addition, this overcomes the limitations of using a selection-type filter where the filter's output is constrained to be one of the input samples.

The signed samples in the window of the recursive WM filter at time n are denoted by the vector $\mathbf{S}(n) = [\mathbf{S}_Y^T(n), \ \mathbf{S}_X^T(n)]^T$ where

$$\mathbf{S}_Y(n) = [\text{sgn}(A_1)Y(n-1), \text{sgn}(A_2)Y(n-2), \ldots, \text{sgn}(A_N)Y(n-N)]^T$$

is the vector containing the signed past output samples, and

$$\mathbf{S}_X(n) = [\text{sgn}(B_0)X(n), \text{sgn}(B_1)X(n+1), \cdots, \text{sgn}(B_M)X(n+M)]^T$$

denotes the vector containing the signed input observation samples used to compute the filter's output at time n. The ith order statistic of $\mathbf{S}(n)$ is denoted as $S_{(i)}(n)$, $i = 1, \ldots, L$, where $S_{(1)}(n) \leq S_{(2)}(n) \leq \ldots \leq S_{(L)}(n)$ with $L = N + M + 1$ as the window size. Note that $S_{(i)}$ is the joint order statistic of the signed past output samples in $\mathbf{S}_Y$ and the signed input observation samples in $\mathbf{S}_X$. Furthermore, we let $\mathbf{A} = [A_1, A_2 \ldots, A_N]^T$ and $\mathbf{B} = [B_0, B_1, \ldots, B_M]^T$ be the vectors containing feedback and feed-forward filter coefficients respectively.

Stability of Recursive WM Filters   One of the main problems in the design of linear IIR filters is the stability under the bounded-input bounded-output (BIBO) criterion, which establishes certain constraints on the feedback filter coefficient values. In order to guarantee the BIBO stability of a linear IIR filter, the poles of its transfer function must lie within the unit circle in the complex plane [4]. Unlike linear IIR filters, recursive WM filters are guaranteed to be stable.

PROPERTY 6.1 *Recursive weighted median filters, as defined in (6.88), are stable under the bounded-input bounded-output criterion, regardless of the values taken by the feedback coefficients* $\{A_\ell\}$ *for* $\ell = 1, 2, \ldots, N$.

The proof of this property is left as an exercise. The importance of Property 6.1 cannot be overstated as the design of recursive WM filters is not as delicate as that of their linear counterparts.

6.4.1 Threshold Decomposition Representation of Recursive WM Filters

The threshold decomposition property states that a real-valued vector $\mathbf{X} = [X_1, X_2, \ldots, X_L]^T$ can be represented by a set of binary vectors $\mathbf{x}^q \in \{-1, 1\}^L$, $q \in (-\infty, \infty)$, where

$$\begin{aligned} \mathbf{x}^q &= [\text{sgn}(X_1 - q), \text{sgn}(X_2 - q), \ldots, \text{sgn}(X_L - q)]^T \\ &= [x_1^q, x_2^q, \ldots, x_L^q]^T, \end{aligned} \qquad (6.89)$$

where $\text{sgn}(\cdot)$ denotes the sign function. The original vector $\mathbf{X}$ can be exactly reconstructed from its binary representation through the inverse process as

$$X_i = \frac{1}{2} \int_{-\infty}^{+\infty} x_i^q dq, \tag{6.90}$$

for $i = 1, \ldots, L$.

Using the threshold signal decomposition in (6.89) and (6.90), the recursive WM operation in (6.88) can be expressed as

$$Y(n) = \text{MEDIAN} \left(|A_\ell| \diamond \frac{1}{2} \int_{-\infty}^{+\infty} \text{sgn}[\text{sgn}(A_\ell)Y(n-\ell) - q] \, dq|_{\ell=1}^{N}, \right.$$
$$\left. |B_k| \diamond \frac{1}{2} \int_{-\infty}^{+\infty} \text{sgn}[\text{sgn}(B_k)X(n+k) - q] \, dq|_{k=0}^{M} \right). \tag{6.91}$$

At this point, we resort to the weak superposition property of the nonlinear median operator described in Section 6.3.1, which states that applying a weighted median operator to a real-valued signal is equivalent to decomposing the real-valued signal using threshold decomposition, applying the median operator to each binary signal separately, and then adding the binary outputs to obtain the real-valued output. This superposition property leads to interchanging the integral and median operators in the above expression and thus, (6.91) becomes

$$Y(n) = \frac{1}{2} \int_{-\infty}^{+\infty} \text{MEDIAN} \left(|A_\ell| \diamond \text{sgn}[\text{sgn}(A_\ell)Y(n-\ell) - q]|_{\ell=1}^{N}, \right.$$
$$\left. |B_k| \diamond \text{sgn}[\text{sgn}(B_k)X(n+k) - q]|_{k=0}^{M} \right) dq. \tag{6.92}$$

To simplify the above expression, let $\{\mathbf{s}_Y^q\}$ and $\{\mathbf{s}_X^q\}$ denote the threshold decomposition of the signed past output samples and the signed input samples respectively, that is

$$\mathbf{S}_Y(n) \overset{T.D.}{\longleftrightarrow} \mathbf{s}_Y^q(n) = [\text{sgn}[\text{sgn}(A_1)Y(n-1) - q], \ldots$$
$$\ldots, \text{sgn}[\text{sgn}(A_N)Y(n-N) - q]]^T$$
$$\mathbf{S}_X(n) \overset{T.D.}{\longleftrightarrow} \mathbf{s}_X^q(n) = [\text{sgn}[\text{sgn}(B_0)X(n) - q], \ldots$$
$$\ldots, \text{sgn}[\text{sgn}(B_M)X(n+M) - q]]^T \tag{6.93}$$

where $q \in (-\infty, +\infty)$. Furthermore, we let $\mathbf{s}^q(n) = [[\mathbf{s}_Y^q(n)]^T, \ [\mathbf{s}_X^q(n)]^T]^T$ be the threshold decomposition representation of the vector $\mathbf{S}(n) = [\mathbf{S}_Y^T(n), \mathbf{S}_X^T(n)]^T$

containing the signed samples. With this notation and following a similar approach to that presented in Section 6.3. It can be shown that (6.92) reduces to

$$Y(n) = \frac{1}{2} \int_{-\infty}^{+\infty} \text{sgn} \left(\mathbf{A}_a^T \mathbf{s}_Y^q(n) + \mathbf{B}_a^T \mathbf{s}_X^q(n) \right) dq, \qquad (6.94)$$

where $\mathbf{A}_a$ is the vector whose elements are the magnitudes of the feedback coefficients: $\mathbf{A}_a = [|A_1|, |A_2|, \ldots, |A_N|]^T$, and $\mathbf{B}_a$ is the vector whose elements are the magnitudes of the feed-forward coefficients: $\mathbf{B}_a = [|B_0|, |B_1|, \ldots, |B_M|]^T$. Note in (6.94) that the filter's output depends on the signed past outputs, the signed input observations, and the feedback and feed-forward coefficients.

6.4.2   Optimal Recursive Weighted Median Filtering

The main objective here is to find the best filter coefficients, such that a performance cost criterion is minimized. Consider an observed process $\{X(n)\}$ that is statistically related to a desired process $\{D(n)\}$. Further, assume that both processes are jointly stationary. Under the mean absolute error (MAE) criterion the goal is to determine the weights $\{A_\ell\}|_{\ell=1}^N$ and $\{B_k\}|_{k=0}^M$ so as to minimize the cost function

$$J(A_1, \ldots, A_N, B_0, \ldots, B_M) = E\{|D(n) - Y(n)|\}, \qquad (6.95)$$

where $E\{\cdot\}$ denotes the statistical expectation and $Y(n)$ is the output of the recursive WM filter given in (6.88).

To form an iterative optimization algorithm, the steepest descent algorithm is used, in which the filter coefficients are updated according to

$$
\begin{aligned}
A_\ell(n+1) &= A_\ell(n) + 2\mu[-\frac{\partial}{\partial A_\ell} J(A_1, \ldots, A_N, B_0, \ldots, B_M)] \\
B_k(n+1) &= B_k(n) + 2\mu[-\frac{\partial}{\partial B_k} J(A_1, \ldots, A_N, B_0, \ldots, B_M)]
\end{aligned}
$$

$$(6.96)$$

for $\ell = 1, \ldots, N$ and $k = 0, \ldots, M$. Note that in (6.96), the gradient of the cost function (∇J) has to be previously computed to update the filter weights. Due to the feedback operation inherent in the recursive WM filter, however, the computation of ∇J becomes intractable.

To overcome this problem, the optimization framework referred to as *equation error formulation* is used [176]. Equation error formulation is used in the design of linear IIR filters and is based on the fact that ideally the filter's output is close to the desired response. The lagged values of $Y(n)$ in (6.88) can thus be replaced with the corresponding lagged values $D(n)$. Hence, the previous outputs $Y(n-\ell)|_{\ell=1}^N$ are replaced with the previous desired outputs $D(n-\ell)|_{\ell=1}^N$ to obtain a two-input, single-output filter that depends on the input samples $X(n+k)|_{k=0}^M$ and on delay samples of the desired response $D(n-l)|_{l=1}^N$, namely,

$$\hat{Y}(n) = \text{MEDIAN}\left(|A_N| \diamond \text{sgn}(A_N)D(n-N), \ldots, |A_1| \diamond \text{sgn}(A_1)D(n-1),\right.$$
$$\left.|B_0| \diamond \text{sgn}(B_0)X(n), \ldots, |B_M| \diamond \text{sgn}(B_M)X(n+M)\right).$$

$$(6.97)$$

The approximation leads to an output $\hat{Y}(n)$ that does not depend on delayed output samples and, therefore, the filter no longer introduces feedback reducing the output to a nonrecursive system. This recursive decoupling optimization approach provides the key to a gradient-based optimization algorithm for recursive WM filters.

According to the approximate filtering structure, the cost function to be minimized is

$$\hat{J}(A_1, \ldots, A_N, B_0, \ldots B_M) = E\{|D(n) - \hat{Y}(n)|\}, \quad (6.98)$$

where $\hat{Y}(n)$ is the nonrecursive filter output (6.97). Since $D(n)$ and $X(n)$ are not functions of the feedback coefficients, the derivative of $\hat{J}(A_1, \ldots, A_N, B_0, \ldots, B_M)$ with respect to the filter weights is nonrecursive and its computation is straightforward. The adaptive optimization algorithm using the steepest descent method (6.96), where $J(\cdot)$ is replaced by $\hat{J}(\cdot)$, is derived as follows. Define the vector $\mathbf{S}(n) = [\mathbf{S}_D^T(n), \quad \mathbf{S}_X^T(n)]^T$ as that containing the signed samples in the sliding window of the two-input, single-output nonrecursive filter (6.97) at time n, where

$$\mathbf{S}_D(n) = [\text{sgn}(A_1)D(n-1), \text{sgn}(A_2)D(n-2), \ldots, \text{sgn}(A_N)D(n-N)]^T$$

and $\mathbf{S}_X(n)$ is given by (6.93). With this notation and using threshold decomposition, (6.98) becomes

$$\hat{J}(A_1, \cdots, A_N, B_0, \cdots, B_M) = \frac{1}{2}E\left\{\left|\int_{-\infty}^{+\infty} [\text{sgn}(D(n) - q)\right.\right. \quad (6.99)$$
$$\left.\left.-\text{sgn}\left(\mathbf{A}_a^T \mathbf{s}_D^q(n) + \mathbf{B}_a^T \mathbf{s}_X^q(n)\right) dq\right]\right|\right\},$$

where $\{\mathbf{s}_D^q(n)\}$ is the corresponding threshold decomposition of the vector $\mathbf{S}_D(n)$.

Now, let $e^q(n)$ be the argument inside the integral operator, such that, $e^q(n) = \text{sgn}(D(n) - q) - \text{sgn}\left(\mathbf{A}_a^T \mathbf{s}_D^q(n) + \mathbf{B}_a^T \mathbf{s}_X^q(n)\right)$. Note that $e^q(n)$ can be thought of as the threshold decomposition of the error function $e(n) = D(n) - \hat{Y}(n)$ for a fixed n. Figure 6.21 shows $e^q(n)$ for two different cases. Figure 6.21a shows the case where the desired filter's output $D(n)$ is less than the filter's output $\hat{Y}(n)$. Figure 6.21b, shows the second case where the desired filter output $D(n)$ is greater than the filter output $\hat{Y}(n)$. The case where the desired response is equal to the filter's output is not shown in Figure 6.21. Note that for a fixed n, the integral operator in (6.99) acts on a strictly negative function (Figure 6.21a) or a strictly positive function (Figure 6.21b), therefore, the absolute value and integral operators in (6.99) can be interchanged leading to

$$\hat{J}(A_1, \ldots, A_N, B_0, B_1, \ldots, B_M) = \frac{1}{2} \int_{-\infty}^{+\infty} E\left[|e^q(n)|\right] dq, \qquad (6.100)$$

where we have used the linear property of the expectation.

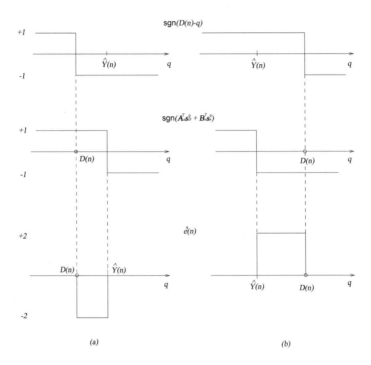

(a)   (b)

Figure 6.21 Threshold decompositions of the desired signal $D(n)$, filter output $\hat{Y}(n)$, and error function $e(n) = D(n) - \hat{Y}(n)$. (a) $D(n) < \hat{Y}(n)$, and (b) $D(n) > \hat{Y}(n)$.

Figure 6.21 also depicts that $e^q(n)$ can only take on values in the set $\{-2, 0, 2\}$, therefore, the absolute value operator can be replaced by a properly scaled second power operator. Thus

$$\hat{J}(A_1, \ldots, A_N, B_0, B_1, \ldots, B_M) = \frac{1}{4} \int_{-\infty}^{+\infty} E\left[\left(e^q(n)\right)^2\right] dq. \qquad (6.101)$$

Taking derivatives of the above expression with respect to the filter coefficients A_ℓ and B_k yields respectively

$$\frac{\partial}{\partial A_\ell} \hat{J}(A_1, \ldots, A_N, B_0, \ldots B_M) = -\frac{1}{2} \int_{-\infty}^{+\infty} E\left[e^q(n) \frac{\partial}{\partial A_\ell} \mathrm{sgn}(\mathbf{A}_a^T \mathbf{s}_D^q + \mathbf{B}_a^T \mathbf{s}_X^q)\right]$$

$$(6.102)$$

$$\frac{\partial}{\partial B_k} \hat{J}(A_1, \ldots, A_N, B_0, \ldots B_M) = -\frac{1}{2} \int_{-\infty}^{+\infty} E\left[e^q(n) \frac{\partial}{\partial B_k} \text{sgn}(\mathbf{A}_a^T \mathbf{s}_D^q + \mathbf{B}_a^T \mathbf{s}_X^q) \right].$$

Since the $\text{sgn}(\cdot)$ function has a discontinuity at the origin, it introduces the *Dirac* function in its derivative which is not convenient for further analysis. In order to overcome this difficulty the sgn function is approximated by the differentiable hyperbolic tangent function $\text{sgn}(x) \approx \tanh(x)$ whose derivative is $\frac{\partial}{\partial x} \tanh(x) = \text{sech}^2(x)$. Using this approximation and letting $\mathbf{W} = \langle\langle A_1, \ldots, A_N, B_0, \ldots B_M \rangle\rangle^T$, the derivation of the adaptive algorithm follows similar steps as that used in the derivation of the adaptive algorithm of nonrecursive WM filters. This leads to the following fast LMA adaptive algorithm for recursive WM filters

$$
\begin{aligned}
A_\ell(n+1) &= A_\ell(n) + \mu(D(n) - \hat{Y}(n))\text{sgn}(A_\ell(n))\text{sgn}(S_{D_\ell} - \hat{Y}(n)) \\
B_k(n+1) &= B_k(n) + \mu(D(n) - \hat{Y}(n))\text{sgn}(B_k(n))\text{sgn}(S_{X_k} - \hat{Y}(n)),
\end{aligned}
$$

$$(6.103)$$

for $\ell = 1, 2, \ldots, N$ and $k = 1, 2, \ldots, M$.

As with the nonrecursive case, this adaptive algorithm is nonlinear and a convergence analysis cannot be derived. Thus, the stepsize μ can not be easily bounded. On the other hand, experimentation has shown that selecting the step size of this algorithm in the same order as that required for the standard LMS algorithm gives reliable results. Another approach is to use a variable step size $\mu(n)$, where $\mu(n)$ decreases as the training progresses.

EXAMPLE 6.12 (IMAGE DENOISING)

The original portrait image used in the simulations is corrupted with impulsive noise. Each pixel in the image has a 10 percent probability of being contaminated with an impulse. The impulses occur randomly and were generated using MATLAB's imnoise function.

The noisy image is filtered by a 3×3 recursive center WM filter and by a 3×3 non-recursive center WM filter with the same set of weights [115]. Figures 6.22a and 6.22b show their respective filter outputs with a center weight $W_c = 5$. Note that the recursive WM filter is more effective than its nonrecursive counterpart. A small 60×60 pixel area in the upper left part of the original and noisy images are used to train the recursive WM filter using the Fast LMA algorithm. The same training data are used to train a nonrecursive WM filter. The initial conditions for the weights for both algorithms were the filter coefficients of the center WM filters described above. The step size used was 10^{-3} for both adaptive algorithms. The optimal weights found by the adaptive algorithms are

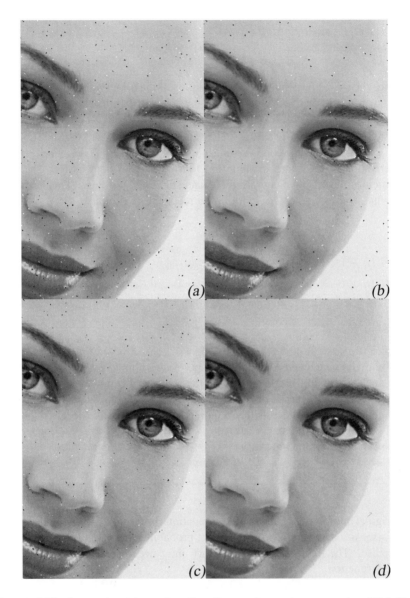

Figure 6.22 Image denoising using 3×3 recursive and nonrecursive WM filters: (*a*) nonrecursive center WM filter (PSNR=26.81dB), (*b*) recursive center WM filter (PSNR=28.33dB), (*c*) optimal nonrecursive WM filter (PSNR=29.91dB), (*d*) optimal RWM filter (PSNR=34.87dB).

$$\left\langle \begin{array}{ccc} 1.38 & 1.64 & 1.32 \\ 1.50 & \underline{5.87} & 2.17 \\ 0.63 & 1.36 & 2.24 \end{array} \right\rangle$$

Table 6.5 Results for impulsive noise removal

Image	Normalized MSE	Normalized MAE
Noisy image	2545.20	12.98
Nonrecursive center WM filter	243.83	1.92
Recursive center WM filter	189.44	1.69
Optimal nonrecursive WM filter	156.30	1.66
Optimal RWM filter	88.13	1.57

for the nonrecursive WM filter and

$$\left\langle\!\!\left\langle \begin{matrix} 1.24 & 1.52 & 2.34 \\ 2.07 & \underline{4.89} & 1.45 \\ 1.95 & 0.78 & 2.46 \end{matrix} \right\rangle\!\!\right\rangle$$

for the RWM filter, where the underlined weight is associated with the center sample of the 3×3 window. The optimal filters determined by the training algorithms were used to filter the entire image. Figures 6.22d and 6.22c show the output of the optimal RWM filter and the output of the non-recursive WM filter respectively. The normalized mean square errors and the normalized mean absolute errors produced by each of the filters are listed in Table 6.5. As can be seen by a visual comparison of the various images and by the error values, recursive WM filters outperform non-recursive WM filters.

Figures 6.23 and 6.24 repeat the denoising example, except that the image is now corrupted with stable noise ($\alpha = 1.2$). The set of weights for the previous example are used without further optimization. Similar conclusions to the example with "salt-and-pepper" noise can be drawn from Figs. 6.23 and 6.24. ∎

EXAMPLE 6.13 (DESIGN OF A BAND PASS RWM FILTER)

Here the LMA and fast LMA adaptive optimization algorithms are used to design a robust band-pass recursive WM filter. The performance of the designed recursive WM filter is compared with the performances of a linear FIR filter, a linear IIR filter,

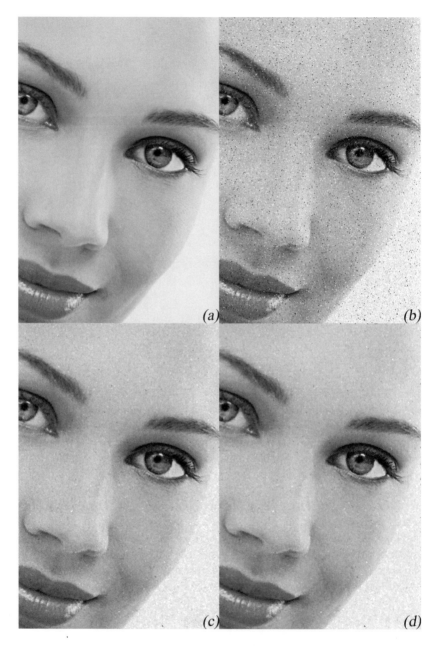

Figure 6.23 Image denoising using 3×3 recursive and nonrecursive WM filters: (a) original, (b) image with stable noise (PSNR=21.35dB), (c) nonrecursive center WM filter (PSNR=31.11dB), (d) recursive center WM filter (PSNR=31.41dB).

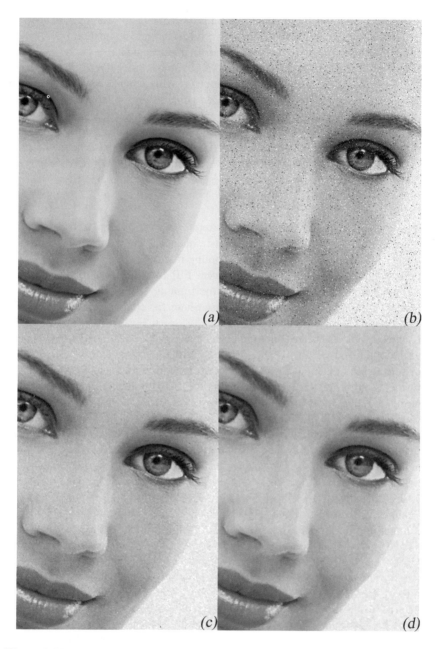

Figure 6.24 Image denoising using 3×3 recursive and nonrecursive WM filters (continued): (*a*) original, (*b*) image with stable noise, (*c*) optimal nonrecursive WM filter (PSNR=32.50dB), (*d*) optimal RWM filter (PSNR=33.99dB).

and a nonrecursive WM filter all designed for the same task. Moreover, to show the noise attenuation capability of the recursive WM filter and compare it with those of the other filters, an impulsive noisy signal is used as test signal.

The application at hand is the design of a 62-tap bandpass RWM filter with pass band $0.075 \leq \omega \leq 0.125$ (normalized frequency with Nyquist = 1). We use white Gaussian noise with zero mean and variance equal to one as input training signal. The desired signal is provided by the output of a large FIR filter (122-tap linear FIR filter) designed by MATLAB's fir1 function. The 31 feedback filter coefficients were initialized to small random numbers (on the order of 10^{-3}). The feed-forward filter coefficients were initialized to the values outputted by MATLAB's fir1 with 31 taps and the same pass band of interest. A variable step size $\mu(n)$ was used in both adaptive optimizations, where $\mu(n)$ changes according to $\mu_0 e^{-n/100}$ with $\mu_0 = 10^{-2}$.

A signal that spans all the range of frequencies of interest is used as a test signal. Figure 6.25a depicts a linear swept-frequency signal spanning instantaneous frequencies form 0 to 400 Hz, with a sampling rate of 2 kHz. Figure 6.25b shows the chirp signal filtered by the 122-tap linear FIR filter that was used as the filter that produced the desired signal during the training stage. Figure 6.25c shows the output of a 62-tap linear FIR filter used here for comparison purposes. The adaptive optimization algorithm described in Section 6.3 was used to optimize a 62-tap nonrecursive WM filter admitting negative weights. The filtered signal attained with the optimized weights is shown in Figure 6.25d. Note that the nonrecursive WM filter tracks the frequencies of interest but fails to attenuate completely the frequencies out of the desired pass band. MATLAB's yulewalk function was used to design a 62-tap linear IIR filter with pass band $0.075 \leq \omega \leq 0.125$. Figure 6.25e depicts the linear IIR filter's output. Finally, Figure 6.25f shows the output of the optimal recursive WM filter determined by the LMA training algorithm. Note that the frequency components of the test signal that are not in the pass band are attenuated completely. Moreover, the RWM filter generalizes very well on signals that were not used during the training stage. The optimal RWM filter determined by the fast LMA training algorithm yields similar performance to that of the optimal RWM filter determined by the LMA training algorithm and therefore, its output is not shown.

Comparing the different filtered signals in Figure 6.25, it can be seen that the recursive filtering operation outperforms its nonrecursive counterpart having the same number of coefficients. Alternatively, to achieve a specified level of performance, a recursive WM filter generally requires fewer filter coefficients than the corresponding non-recursive WM filter.

In order to test the robustness of the different filters, the test signal is contaminated with additive α-stable noise as shown in Figure 6.26a The parameter $\alpha = 1.4$ was used, simulating noise with impulsive characteristics. Figure 6.26a is truncated so that the same scale is used in all the plots. Figures 6.26b and 6.26d show the filter outputs of the linear FIR and the linear IIR filters respectively. Both outputs are severely affected by the noise. On the other hand, the non-recursive and recursive WM filters' outputs, shown in Figures 6.26c and 6.26e respectively, remain practically unaltered. Figure 6.26 clearly depicts the robust characteristics of median based filters.

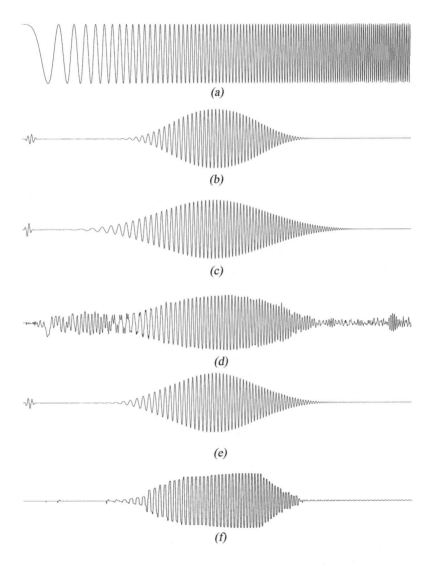

Figure 6.25 Band pass filter design: (*a*) input test signal, (*b*) desired signal, (*c*) linear FIR filter output, (*d*) nonrecursive WM filter output (*e*) linear IIR filter output, (*f*) RWM filter output.

To better evaluate the frequency response of the various filters, a frequency domain analysis is performed. Due to the nonlinearity inherent in the median operation, traditional linear tools, like transfer function-based analysis, cannot be applied. However, if the nonlinear filters are treated as a single-input single-output system, the magnitude of the frequency response can be experimentally obtained as follows. A single

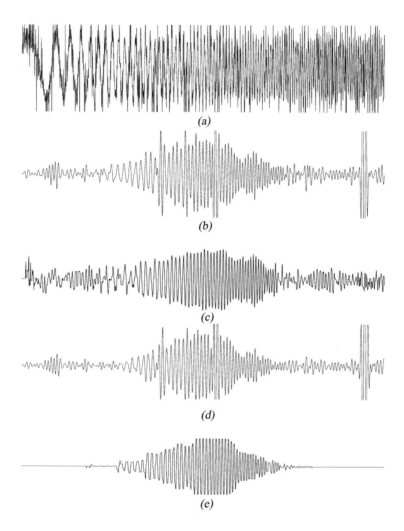

Figure 6.26 Performance of the band-pass filter in noise: (*a*) chirp test signal in stable noise, (*b*) linear FIR filter output, (*c*) nonrecursive WM filter output, (*d*) linear IIR filter output, (*e*) RWM filter output.

tone sinusoidal signal $\sin(2\pi f t)$ is given as the input to each filter, where f spans the complete range of possible frequencies. A sufficiently large number of frequencies spanning the interval $[0, 1]$ is chosen. For each frequency value, the mean power of each filter's output is computed. Figure 6.27*a* shows a plot of the normalized mean power versus frequency attained by the different filters. Upon closer examination of Figure 6.27*a*, it can be seen that the recursive WM filter yields the flattest response in the pass band of interest. A similar conclusion can be drawn from the time domain plots shown in Figure 6.25.

In order to see the effects that impulsive noise has over the magnitude of the frequency response, a contaminated sinusoidal signal, $\sin(2\pi ft) + \eta$, is given as the input to each filter, where η is α-stable noise with parameter $\alpha = 1.4$. Following the same procedure described above, the mean power versus frequency diagram is obtained and shown in Figure 6.27*b*. As expected, the magnitudes of the frequency responses for the linear filters are highly distorted; whereas the magnitudes of the frequency responses for the median based filters do not change significantly with noise. ∎

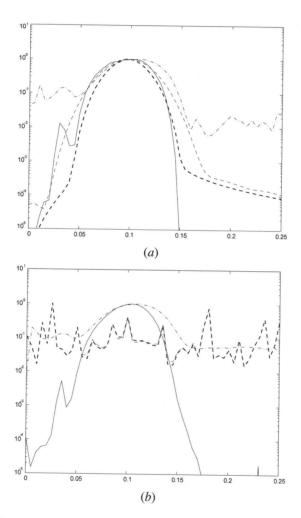

(a)

(b)

Figure 6.27 Frequency response (*a*) to a noiseless sinusoidal signal (*b*) to a noisy sinusoidal signal. (——) RWM, (— · — · —) non-recursive WM filter, (- - -) linear FIR filter, and (- - -) linear IIR filter.

6.5   MIRRORED THRESHOLD DECOMPOSITION AND STACK FILTERS

The threshold decomposition architecture provides the foundation needed for the definition of stack smoothers. The class of *stack filters* can be defined in a similar fashion provided that a more general threshold decomposition architecture, referred to as *mirrored threshold decomposition*, is defined by Paredes and Arce (1999) [158]. Unlike stack smoothers, stack filters can be designed to have arbitrary frequency selection characteristics.

Consider again the set of integer-valued samples $X_1, X_2, \ldots, X_N$ forming the vector $\mathbf{X}$. For simplicity purposes, the input signals are quantized into a finite set of values with $X_i \in \{-M, \ldots, -1, 0, \ldots, M\}$. Unlike threshold decomposition, *mirrored* threshold decomposition of $\mathbf{X}$ generates two sets of binary vectors, each consisting of $2M$ vectors. The first set consists of the $2M$ vectors associated with the traditional definition of threshold decomposition $\mathbf{x}^{-M+1}, \mathbf{x}^{-M+2}, \ldots, \mathbf{x}^0, \ldots, \mathbf{x}^M$. The second set of vectors is associated with the decomposition of the *mirrored* vector of $\mathbf{X}$, which is defined as

$$
\begin{aligned}
\mathbf{S} &= [\, S_1, \ S_2, \ldots, \ S_N]^T \\
&= [-X_1, -X_2, \ldots, -X_N]^T.
\end{aligned}
\tag{6.104}
$$

Since S_i take on symmetrical values about the origin from X_i, S_i is referred to as the *mirror* sample of X_i, or simply as the *signed* sample X_i. Threshold decomposition of $\mathbf{S}$ leads to the second set of $2M$ binary vectors $\mathbf{s}^{-M+1}, \mathbf{s}^{-M+2}, \ldots, \mathbf{s}^0, \ldots, \mathbf{s}^M$. The ith element of $\mathbf{x}^m$ is as before specified by

$$
x_i^m = T^m(X_i) = \begin{cases} 1 & \text{if } X_i \geq m; \\ -1 & \text{if } X_i < m, \end{cases}
\tag{6.105}
$$

whereas the ith element of $\mathbf{s}^m$ is defined by

$$
s_i^m = T^m(-X_i) = \begin{cases} 1 & \text{if } (-X_i) \geq m; \\ -1 & \text{if } (-X_i) < m. \end{cases}
\tag{6.106}
$$

The thresholded mirror signal can be written as $s_i^m = \mathrm{sgn}(-X_i - m^-) = -\mathrm{sgn}(X_i + m^- - 1)$.

X_i and S_i are both reversible from their corresponding set of decomposed signals and consequently, an integer-valued signal X_i has a unique *mirrored* threshold signal representation, and vice versa:

$$
X_i \overset{T.D.}{\longleftrightarrow} (\{x_i^m\}; \{s_i^m\})
$$

where $\overset{T.D.}{\longleftrightarrow}$ denotes the one-to-one mapping provided by the mirrored threshold decomposition operation. Each of the threshold decomposed signal sets possesses the stacking constraints, independently from the other set. In addition, since the vector $\mathbf{S}$ is the mirror of $\mathbf{X}$, a partial ordering relation exists between the two sets of thresholded signals. With $X_i = -S_i$, the thresholded samples satisfy $s^{M-\ell} = -x^{-M+1+\ell}$ and $x^{M-\ell} = -s^{-M+1+\ell}$ for $\ell = 0, 1, \ldots, 2M - 1$.

As an example, the representation of the vector $\mathbf{X} = [2, \ -1, \ 0, \ -2, \ 1, \ 2, \ 0]$ in the binary domain of mirrored threshold decomposition is

$$
\begin{aligned}
\mathbf{x}^2 &= [\ 1, -1, -1, -1, -1, \ 1, -1]^T & \mathbf{s}^2 &= [-1, -1, -1, \ 1, -1, -1, -1]^T \\
\mathbf{x}^1 &= [\ 1, -1, -1, -1, \ 1, \ 1, -1]^T & \mathbf{s}^1 &= [-1, \ 1, -1, \ 1, -1, -1, -1]^T \\
\mathbf{x}^0 &= [\ 1, -1, \ 1, -1, \ 1, \ 1, \ 1]^T & \mathbf{s}^0 &= [-1, \ 1, \ 1, \ 1, -1, -1, \ 1]^T \\
\mathbf{x}^{-1} &= [\ 1, \ 1, \ 1, -1, \ 1, \ 1, \ 1]^T & \mathbf{s}^{-1} &= [-1, \ 1, \ 1, \ 1, \ 1, -1, \ 1]^T.
\end{aligned}
$$

6.5.1 Stack Filters

Much like traditional threshold decomposition leads to the definition of stack smoothers, mirrored threshold decomposition leads to the definition of a richer class of nonlinear filters referred to as *stack filters*. The output of a stack filter is the result of a sum of a stack of binary operations acting on thresholded versions of the input samples and their corresponding mirrored samples. The stack filter output is defined by

$$
S_f(X_1, \ldots, X_N) = \frac{1}{2} \sum_{m=-M+1}^{M} f(x_1^m, \ldots, x_N^m; s_1^m, \ldots, s_N^m), \tag{6.107}
$$

where x_i^m and s_i^m, $i = 1, \ldots, N$, are the thresholded samples defined in (6.105) and (6.106), and where $f(\cdot)$ is a $2N - variable$ Positive Boolean Function (PBF) that, by definition, contains only uncomplemented input variables.

Given an input vector $\mathbf{X}$, its mirrored vector $\mathbf{S}$, and their set of thresholded binary vectors $\mathbf{x}^{-M+1}, \ldots, \mathbf{x}^0, \ldots, \mathbf{x}^M$; $\mathbf{s}^{-M+1}, \ldots, \mathbf{s}^0, \ldots, \mathbf{s}^M$, it follows from the definition of threshold decomposition that the set of thresholded binary vectors satisfy the partial ordering

$$
[\mathbf{x}^i; \mathbf{s}^i] \leq [\mathbf{x}^j; \mathbf{s}^j] \quad \text{if } i \geq j. \tag{6.108}
$$

Thus, $\mathbf{x}^i \in \{-1, 1\}^N$ and $\mathbf{s}^i \in \{-1, 1\}^N$ stack, that is, $x_k^i \leq x_k^j$ and $s_k^i \leq s_k^j$ if $i \geq j$, for all $k \in \{1, \ldots, N\}$. Consequently, the stack filtering of the thresholded binary vectors by the PBF $f(\cdot)$ also satisfy the partial ordering

$$
f(\mathbf{x}^i; \mathbf{s}^i) \leq f(\mathbf{x}^j; \mathbf{s}^j) \quad \text{if } i \geq j. \tag{6.109}
$$

The stacking property in (6.109) ensures that the decisions on different levels are consistent. Thus, if the filter at a given time location decides that the signal is less than j, then the filter outputs at levels $j + 1$ and greater must draw the same conclusion.

As defined in (6.107), stack filters input signals are assumed to be quantized to a finite number of signal levels. Following an approach similar to that with stack smoothers, the class of stack filters admitting real-valued input signals is defined next.

DEFINITION 6.5 (CONTINUOUS STACK FILTERS) *Given a set of N real-valued samples* $\mathbf{X} = (X_1, X_2, \ldots, X_N)$, *the output of a stack filter defined by a PBF* $f(\cdot)$ *is given by*

$$S_f(\mathbf{X}) = \max\{\ell \in \mathbf{R} : f(T^\ell(X_1), \ldots, T^\ell(X_N); T^\ell(-X_1), \ldots, T^\ell(-X_N)) = 1\}, \tag{6.110}$$

where the thresholding function $T^\ell(\cdot)$ *is defined in (6.105).*

The link between the continuous stack filter $S_f(\cdot)$ and the corresponding PBF $f(\cdot)$ is given by the following property.

PROPERTY 6.2 (MAX-MIN REPRESENTATION OF STACK FILTERS) *Let* $\mathbf{X} = (X_1, X_2, \ldots, X_N)$ *and* $\mathbf{S} = (-X_1, -X_2, \ldots, -X_N)$ *be a real-valued vector and its corresponding mirrored vector that are inputted to a stack filter* $S_f(\cdot)$ *defined by the positive Boolean function* $f(x_1, \ldots, x_N; s_1, \ldots, s_N)$. *The PBF with the sum of products expression*

$$f(x_1, \ldots, x_N; s_1, \ldots, s_N) = \sum_{i=1}^{K} \prod_{j \in P_i} x_j, \prod_{j \in Q_i} s_j, \tag{6.111}$$

where P_i *and* Q_i *are subsets of* $\{0, 1, 2, \ldots, N\}$, *has the stack filter representation*

$$S_f(\mathbf{X}) = \max\{\min\{X_j S_k : j \in P_1\ k \in Q_1\}, \ldots, \min\{X_j S_k : j \in P_K\ k \in Q_K\}\} \tag{6.112}$$

with $X_0 = S_0 = 1$ *and* P_i *and* Q_i *not having the 0th element at once.*

Thus, given a positive Boolean function $f(x_1, \ldots, x_N; s_1, \ldots, s_N,)$ that characterizes a stack filter, it is possible to find the equivalent filter in the real domain by replacing the binary AND and OR Boolean functions acting on the x_i's and s_i's with *max* and *min* operations acting on the real-valued X_i and S_i samples.

Integer Domain Filters of Linearly Separable Positive Boolean Functions

In general, stack filters can be implemented by $\max - \min$ networks in the integer domain. Although simple in concept, $\max - \min$ networks lack an intuitive interpretation. However, if the PBFs in the stack filter representation are further constrained, a number of more appealing filter structures emerge. These filter structures are more intuitive to understand and, in many ways, they are similar to linear FIR filters. Yli-Harja et al. [202] describe the various types of stack smoothers attained when the PBFs are constrained to be linearly separable. Weighted order statistic smoothers and weighted median smoothers are, for instance, obtained if the PBFs are restricted to be linearly separable and self-dual linearly separable, respectively. A Boolean function $f(\mathbf{x})$ is said to be linearly separable if and only if it can be expressed as

$$f(x_1, \cdots, x_N) = \text{sgn}\left(\sum_{i=1}^{N} W_i x_i - T\right) \qquad (6.113)$$

where x_i are binary variables, and the weights W_i and threshold T are nonnegative real-valued [174]. A self-dual linearly separable Boolean function is defined by further restricting (6.113) as

$$f(x_1, \ldots, x_N) = \text{sgn}\left(\sum_{i=1}^{N} W_i x_i\right). \qquad (6.114)$$

A Boolean function $f(x)$ is said to be self dual if and only if $f(x_1, x_2, \ldots, x_N) = 1$ implies $f(\bar{x}_1, \bar{x}_2, \ldots, \bar{x}_N) = 0$, and $f(x_1, x_2, \ldots, x_N) = 0$ implies $f(\bar{x}_1, \bar{x}_2, \ldots, \bar{x}_N) = 1$, where $\bar{x}$ denotes the Boolean complement of x [145].

Within the mirrored threshold decomposition representation, a similar strategy can be taken where the separable Boolean functions are progressively constrained leading to a series of stack filter structures that can be easily implemented in the integer domain. In particular, weighted order statistic filters, weighted median filters, and order statistic filters emerge by appropriately selecting the appropriate PBF structure in the binary domain. Figure 6.28 depicts the relationship among subclasses of stack filters and stack smoothers. As this figure shows, stack filters are much richer than stack smoothers. The class of WOS filters, for example, contains all WOS and OS smoothers, whereas even the simplest OS filter is not contained in the entire class of stack smoothers.

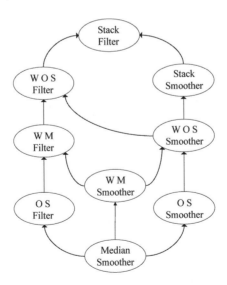

Figure 6.28 Relationship among subclasses of stack filters and stack smoothers.

Stack Filter Representation of Weighted Median Filters WM filters are generated if the positive Boolean function that defines the stack filter in (6.107) is constrained to be self dual and linearly separable. In the binary domain of mirrored threshold decomposition, *weighted median* filters are defined in terms of the thresholded vectors $\mathbf{x}^m = [x_1^m, \ldots, x_N^m]^T$ and the corresponding thresholded *mirror* vector $\mathbf{s}^m = [s_1^m, \ldots, s_N^m]^T$ as

$$S(X_1, X_2, \ldots, X_N) = \frac{1}{2} \sum_{m=-M+1}^{M} \text{sgn}\left(\mathbf{W}^T \mathbf{x}^m + |\mathbf{H}|^T \mathbf{s}^m - T_0\right) \quad (6.115)$$

where $\mathbf{W} = \langle W_1, W_2, \ldots, W_N \rangle^T$ and $|\mathbf{H}| = \langle |H_1|, |H_2|, \ldots, |H_N| \rangle^T$ are $2N$ positive-valued weights that uniquely characterize the WM filter. The constant T_0 is 0 or 1 if the weights are real-valued or integer-valued adding up to an odd integer, respectively. $|\cdot|$ represents the absolute value operator, and is used in the definition of binary domain WM filters for reasons that will become clear shortly. The role that W_i's and $|H_i|$'s play in WM filtering is very important as is described next.

Since the threshold logic gate $\text{sgn}(\cdot)$ in (6.115) is self-dual and linearly separable, and since the x_i^m and s_i^m respectively represent X_i and its mirror sample $S_i = -X_i$, the integer-domain representation of (6.115) is given by [145, 202]

$$Y = \text{MEDIAN}(W_1 \diamond X_1, |H_1| \diamond S_1, \ldots, W_N \diamond X_N, |H_N| \diamond S_N) \quad (6.116)$$

where $W_i \geq 0$ and $|H_i| \geq 0$. At this point it is convenient to associate the sign of the mirror sample S_i with the corresponding weight H_i as

$$Y = \text{MEDIAN}(W_1 \diamond X_1, H_1 \diamond X_1, \ldots, W_N \diamond X_N, H_N \diamond X_N) \quad (6.117)$$

leading to the following definition.

DEFINITION 6.6 (DOUBLE-WEIGHTED MEDIAN FILTER) *Given the N-long observation vector* $\mathbf{X} = [X_1, X_2, \ldots, X_N]^T$, *the set of* $2N$ *real valued weights* $\langle\langle W_1, H_1 \rangle, \langle W_2, H_2 \rangle, \ldots, \langle W_N, H_N \rangle\rangle$ *defines the double-weighted median filter output as*

$$Y = \text{MEDIAN}(\langle W_1, H_1 \rangle \diamond X_1, \ldots, \langle W_N, H_N \rangle \diamond X_N) \quad (6.118)$$

with $W_i \geq 0$ *and* $H_i \leq 0$, *where the equivalence* $H_i \diamond X_i = |H_i| \diamond \text{sgn}(H_i) X_i$ *is used.*

Thus, weighting in the WM filter structure is equivalent to uncoupling the weight sign from its magnitude, merging the sign with the observation sample, and replicating the signed sample according to the magnitude of the weight. Notice that each sample X_i in (6.117) is weighted twice — once positively by W_i and once negatively by H_i. In (6.118), the double weight $\langle W_i, H_i \rangle$ is defined to represent the positive and negative weighting of X_i. As expected, should one of the weight pairs is constrained to be zero, the double-weighted median filter reduces to the N weight median filter structure in Definition 6.1.

The general form of the weighted median structure contains $2N$ weights, N positive and N negative. Double weighting emerges through the analysis of mirrored threshold decomposition and the stack filter representation. In some applications, the simpler weighted median filter structure where a single real-valued weight is associated with each observation sample may be preferred in much the same way linear FIR filters only use N filter weights. At first, the WM filter structure in (6.118) seems redundant. After all, linear FIR filters only require a set of N weights, albeit real-valued. The reason for this is the associative property of the sample mean. As shown in [6], the linear filter structure analogous to (6.117) is

$$\bar{Y} = \text{MEAN}(W_1 \cdot X_1, |H_1| \cdot S_1, \ldots, W_N \cdot X_N, |H_N| \cdot S_N) \quad (6.119)$$
$$= \text{MEAN}((W_1 + H_1) \cdot X_1, \ldots, (W_N + H_N) \cdot X_N) \quad (6.120)$$

where $W_i \geq 0$ and $H_i \leq 0$ collapse to a single real-valued weight in (6.120). For the sample median, however,

$$\text{MEDIAN}(\langle W_i, H_i \rangle \diamond X_i|_{i=1}^N) \neq \text{MEDIAN}((W_i + H_i) \diamond X_i|_{i=1}^N), \quad (6.121)$$

thus the weight pair $\langle W_i, H_i \rangle$ is needed in general.

Weighted median filters have an alternate interpretation. Extending the concepts in the cost function representation of WM filters, it can be shown that the WM filter output in (6.118) is the value β minimizing the cost function

$$G_1(\beta) = \sum_{i=1}^N (W_i |X_i - \beta| + |H_i| |X_i + \beta|) \quad (6.122)$$

where β can only be one of the samples X_i or $-X_i$ since (6.122) is piecewise linear and convex. Figure 6.29 depicts the effects of double weighting in WM filtering where the absence of double weighting $\langle W_i, H_i \rangle$, distorts the shape of the cost function $G_1(\beta)$ and can lead to a distorted global minima.

6.5.2 Stack Filter Representation of Recursive WM Filters

The WM filtering characteristics can be significantly enriched, if the previous outputs are taken into account to compute future outputs. Recursive WM filters, taking advantage of prior outputs, exhibit significant advantage over their nonrecursive counterparts, particularly if negative as well as positive weights are used. Recursive WM filters can be thought of as the analogous of linear IIR filters with improved robustness and stability characteristics.

Given an N-input observation $\mathbf{X}_k = [X_{k-L}, \ldots, X_k, \ldots, X_{k+L}]$, the recursive counterpart of (6.115) is obtained by replacing the leftmost L samples of the input

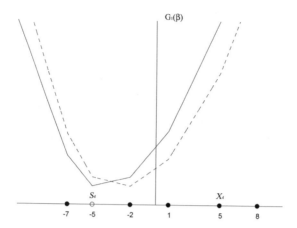

Figure 6.29 An observation vector $[X_1, X_2, X_3, X_4, X_5] = [-7, -2, 1, 5, 8]$ filtered by the two set of weights: $\langle 3, 2, 2, -3, 1\rangle$ (solid line) and $\langle 3, 2, 2, \langle 2, -3\rangle, 1\rangle$ (dashed line), respectively. Double weighting of X_4 shows the distinct cost function and minima attained.

vector $\mathbf{X}_k$ with the previous L output samples $Y_{k-L}, Y_{k-L+1}, \ldots Y_{k-1}$ leading to the definition:

DEFINITION 6.7 (RECURSIVE DOUBLE-WEIGHTED MEDIAN FILTER) *Given the input observation vector containing past output samples* $\mathbf{X}'_k = [Y_{k-L}, \ldots, Y_{k-1}, X_k, \ldots, X_{k+L}]^T$, *the set of positive weights* $\mathbf{W}_R = \langle W_{R_{-L}}, \ldots, W_{R_0}, \ldots, W_{R_L}\rangle^T$ *together with the set of negative weights* $\mathbf{H}_R = \langle H_{R_{-L}}, \ldots, H_{R_0}, \ldots, H_{R_L}\rangle^T$ *define the output of the recursive double-weighted median filter as*

$$Y_k = \text{MEDIAN}\left(\langle W_{R_{-L}}, H_{R_{-L}}\rangle \diamond Y_{k-L}, \ldots, \langle W_{R_0}, H_{R_0}\rangle \diamond X_k, \ldots \\ \ldots, \langle W_{R_L}, H_{R_L}\rangle \diamond X_{k+L}\right)$$

$$(6.123)$$

for $i \in [-L, \ldots, L]$ *and where the equivalence* $H_{R_i} \diamond X_{k+i} = |H_{R_i}| \diamond \text{sgn}(H_{R_i})X_{k+i}$ *is used.*

Clearly, Y_k in (6.123) is a function of previous outputs as well as the input signal.

Recursive WM filters have a number of desirable attributes. Unlike linear IIR filters, recursive WM filters are always stable under the bounded-input bounded-output (BIBO) criterion regardless of the values taken by the filter coefficients. Recursive WM filters can be used to synthesize a nonrecursive WM filter of much larger window size. To date, there is not a known method of computing the recursive WM filter equivalent to a nonrecursive one. However, a method, in the binary domain can be used to find nonrecursive WM filter approximations of a recursive WM filter [158]. For instance, the recursive 3-point WM filter given by

$$Y_k = \text{MEDIAN}\left(\langle 1,\ 0\rangle \diamond Y_{k-1},\ \langle 1,\ 0\rangle \diamond X_k,\ \langle 0,\ -1\rangle \diamond X_{k+1}\right) \qquad (6.124)$$

has the following first-order approximation

$$Y_k^1 = \text{MEDIAN}\left(\langle 1,\ 0\rangle \diamond X_{k-1},\ \langle 1, 0\rangle \diamond X_k,\ \langle 0,\ -1\rangle \diamond X_{k+1}\right), \qquad (6.125)$$

which, in turn, can be used to find the second-order approximation

$$Y_k^2 = \text{MEDIAN}\left(\langle 1, 0\rangle \diamond X_{k-2},\ \langle 1, 0\rangle \diamond X_{k-1},\ \langle 2, -1\rangle \diamond X_k,\ \langle 0, -2\rangle \diamond X_{k+1}\right). \qquad (6.126)$$

Note in (6.126) that sample X_k is weighted negatively and positively. This occurs naturally as a consequence of mirrored threshold decomposition. In a similar manner, the third-order approximation leads to the nonrecursive WM filter

$$Y_k^3 = \text{MEDIAN}\left(\langle 1,\ 0\rangle \diamond X_{k-3},\ \langle 1,\ 0\rangle \diamond X_{k-2},\ \langle 2,\ -1\rangle \diamond X_{k-1},\right.$$
$$\left.\langle 4,\ -2\rangle \diamond X_k,\ \langle 0,\ -4\rangle \diamond X_{k+1}\right). \quad (6.127)$$

In order to illustrate the effectiveness of the various nonrecursive WM approximations of the recursive WM filter, white Gaussian noise is inputted to the recursive WM filter and to the various nonrecursive approximation. The results are shown in Figure 6.30. Note that the approximation improves with the order as expected. Figure 6.30c shows that the output of the nonrecursive WM filter of length 5 is very close to the output of a RWM filter of length 3. This corroborates that recursive WM filters can synthesize a nonrecursive WM filter of much larger window size.

Notice also in expressions (6.126) and (6.127) that the nonrecursive realizations of the recursive 3-point WM filter given by (6.123) requires the use of weight pairs for some of the input samples. Indeed, binary representations having both x_i and s_i as part of the positive Boolean function will inevitably lead to having a weight pair $\langle W_i, H_i\rangle$ on X_i.

In order to illustrate the importance of the double weighting operation on the filter output, the same input signal used with the previous nonrecursive approximations is next fed into the nonrecursive WM filter given by (6.127), but with the positive weight related to X_{k-1} set to zero, that is $\langle W_{k-1}, H_{k-1}\rangle$ has been change from $\langle 2, -1\rangle$ to $\langle 0, -1\rangle$. The output of this filtering operation and the output of the recursive 3-point WM filter are shown in Figure 6.30d. Comparing Figures 6.30c and 6.30d, the strong influence of double-weighting on the filter output is easily seen.

Some interesting variants of the recursive three-point WM filters and their corresponding approximate nonrecursive WM filter are presented in Table 6.6.

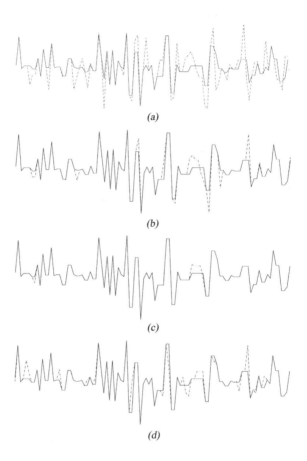

Figure 6.30 Output of the recursive 3-point WM filter $\langle\langle 1,\ \underline{1},\ -1\rangle\rangle$ (solid line), and its nonrecursive approximations (dashed line): (*a*) first order, (*b*) second order, (*c*) third order, (*d*) third-order approximation when W_{k-1} is set to 0.

6.6 COMPLEX-VALUED WEIGHTED MEDIAN FILTERS

Sorting and ordering a set of complex-valued samples is not uniquely defined, as with the sorting of any multivariate sample set. See Barnett (1976) [27]. The complex-valued median, however, is well defined from a statistical estimation framework.

The complex sample mean and sample median are two well known Maximum Likelihood (ML) estimators of location derived from sets of independent and identically distributed (i.i.d.) samples obeying the complex Gaussian and complex Laplacian distributions, respectively. Thus, if $X_i, i = 1, \ldots, N$ are i.i.d. complex Gaussian distributed samples with constant but unknown complex mean β, the ML estimate of

Table 6.6 Recursive three-point WM filters and their approximate non-recursive counterpart. The underline weight is related to the center sample of the window. For short notation only the nonzero weights are listed.

Recursive 3 pt WM filter	1st-order approx.	2nd-order approx.	3rd-order approx.
$\langle\langle 1, \underline{1}, 1\rangle\rangle$	$\langle 1, \underline{1}, 1\rangle$	$\langle 1, 1, \underline{2}, 1\rangle$	$\langle 1, 1, 2, \underline{3}, 2\rangle$
$\langle\langle 1, \underline{1}, -1\rangle\rangle$	$\langle 1, \underline{1}, -1\rangle$	$\langle 1, 1, \langle\underline{2}, -1\rangle, -2\rangle$	$\langle 1, 1, \langle 2, -1\rangle, \langle\underline{4}, -2\rangle, -4\rangle$
$\langle\langle 1, \underline{-1}, 1\rangle\rangle$	$\langle 1, \underline{-1}, 1\rangle$	$\langle 1, -1, \langle\underline{1}, -2\rangle, 2\rangle$	$\langle 1, -1, \langle 1, -2\rangle, \langle\underline{2}, -4\rangle, 4\rangle$
$\langle\langle -1, \underline{1}, 1\rangle\rangle$	$\langle -1, \underline{1}, 1\rangle$	$\langle 1, -1, \langle\underline{2}, -1\rangle, 2\rangle$	$\langle -1, 1, \langle 1, -2\rangle, \langle\underline{4}, -2\rangle, 4\rangle$
$\langle\langle 1, \underline{-1}, -1\rangle\rangle$	$\langle 1, \underline{-1}, -1\rangle$	$\langle 1, -1, \underline{-2}, -1\rangle$	$\langle 1, -1, -2, \underline{-3}, -2\rangle$
$\langle\langle -1, \underline{1}, -1\rangle\rangle$	$\langle -1, \underline{1}, -1\rangle$	$\langle 1, -1, \underline{2}, -1\rangle$	$\langle -1, 1, -2, \underline{3}, -2\rangle$
$\langle\langle -1, \underline{-1}, 1\rangle\rangle$	$\langle -1, \underline{-1}, 1\rangle$	$\langle 1, 1, \underline{-2}, 1\rangle$	$\langle -1, -1, 2, \underline{-3}, 2\rangle$
$\langle\langle -1, \underline{-1}, -1\rangle\rangle$	$\langle -1, \underline{-1}, -1\rangle$	$\langle 1, 1, \langle\underline{1}, -2\rangle, -2\rangle$	$\langle -1, -1, \langle 2, -1\rangle, \langle\underline{2}, -4\rangle, -4\rangle$

location is the value $\hat{\beta}$ that maximizes the likelihood function,

$$\hat{\beta} = \arg\max_{\beta}\left\{\left(\frac{1}{\pi\sigma^2}\right)^N exp\left(-\sum_{i=1}^{N}|X_i - \beta|^2/\sigma^2\right)\right\}.$$

This is equivalent to minimizing the sum of squares as

$$\hat{\beta} = \arg\min_{\beta}\sum_{i=1}^{N}|X_i - \beta|^2 = \text{MEAN}\,(X_1, X_2, \ldots, X_N).\qquad(6.128)$$

Letting each sample X_i be represented by its real and imaginary components $X_i = X_{R_i} + jX_{I_i}$[3], the minimization in (6.128) can be carried out marginally without losing optimality by minimizing real and imaginary parts independently as $\hat{\beta} = \hat{\beta}_R + j\hat{\beta}_I$, where

$$\hat{\beta}_R = \arg\min_{\beta_R}\sum_{i=1}^{N}(X_{R_i} - \beta_R)^2$$
$$= \text{MEAN}\,(X_{R_1}, X_{R_2}, \ldots, X_{R_N}),\qquad(6.129)$$

---

[3]The subindexes R and I represent real and imaginary part respectively

and

$$\hat{\beta}_I = \arg\min_{\beta_I} \sum_{i=1}^{N} (X_{I_i} - \beta_I)^2$$
$$= \text{MEAN}(X_{I_1}, X_{I_2}, \ldots, X_{I_N}). \tag{6.130}$$

When the set of i.i.d. complex samples obey the Laplacian distribution, it can be shown that the maximum likelihood estimate of location is the complex-valued estimate $\hat{\beta}$ that minimizes the sum of absolute deviations,

$$\hat{\beta} = \arg\min_{\beta} \sum_{i=1}^{N} |X_i - \beta|. \tag{6.131}$$

Unlike (6.128), the minimization in (6.131) cannot be computed marginally in the real and imaginary components and, in general, it does not have a closed-form solution, requiring a two-dimensional search over the complex space for the parameter $\hat{\beta}$. The suboptimal approach introduced by Astola et al. (1990) [19] referred to as the vector median, consists in assuming that the $\hat{\beta}$ that satisfies (6.131) is one of the input samples X_i. Thus, Astola's vector median outputs the input vector that minimizes the sum of Euclidean distances between the candidate vector and all the other vectors. Astola also suggested the marginal complex median, a fast but suboptimal approximation by considering the real and imaginary parts independent of each other, allowing to break up the complex-valued optimization into two real-valued optimizations leading to $\hat{\beta} \approx \tilde{\beta} = \tilde{\beta}_R + j\tilde{\beta}_I$ where $\tilde{\beta}_R = \text{MEDIAN}(X_{R_1}, X_{R_2}, \ldots, X_{R_N})$ and $\tilde{\beta}_I = \text{MEDIAN}(X_{I_1}, X_{I_2}, \ldots, X_{I_N})$.

When the complex samples are independent but not identically distributed, the ML estimate of location can be generalized. In particular, letting $X_1, X_2, \ldots, X_N$ be independent complex Gaussian variables with the same location parameter $\bar{\beta}$ but distinct variances $\sigma_1^2, \ldots, \sigma_N^2$, the location estimate becomes

$$\hat{\beta} = \arg\min_{\beta} \sum_{i=1}^{N} W_i |X_i - \beta|^2 = \frac{\sum_{i=1}^{N} W_i \cdot X_i}{\sum_{i=1}^{N} W_i}, \tag{6.132}$$

with $W_i = 1/\sigma_i^2$, a positive real-valued number. Likewise, under the Laplacian model, the maximum likelihood estimate of location minimizes the sum of weighted absolute deviations

$$\hat{\beta} = \arg\min_{\beta} \sum_{i=1}^{N} W_i |X_i - \beta|. \tag{6.133}$$

Once again, there is no closed-form solution to (6.133) in the complex-plane and a two-dimensional search must be used. Astola's approximations used for the identically distributed case can be used to solve (6.133), but in this case the effect of the weights W_i must be taken into account. These approaches, however, lead to severely constrained structures as only positive-valued weighting is allowed;

the attained complex medians are smoother operations where neither negative nor complex weights are admitted.

To overcome these limitations, the concept of *phase coupling* consisting in decoupling the phase of the complex-valued weight and merging it to the associated complex-valued input sample is used. This approach is an extension of the weighted median filter admitting negative weights described in Section 6.1, where the negative sign of the weight is uncoupled from its magnitude and is merged with the input sample to create a set of signed-input samples that constitute the output candidates. The phase coupling concept is used to define the *phase coupled* complex WM filter, which unlike the real-valued weighted median does not have a closed-form solution, thus requiring searching in the complex-plane. To avoid the high computational cost of the searching algorithm, a suboptimal implementation called *marginal phase coupled* complex WM was introduced in Hoyos et al. (2003) [104]. This definition leads to a set of complex weighted median filter structures that fully exploits the power of complex weighting and still keeps the advantages inherited from univariate medians.

The simplest approach to attain complex WM filtering is to perform marginal operations where the real component of the weights $W_R |_{i=1}^N$ affect the real part of the samples $X_R |_{i=1}^N$ and the imaginary component of the weights $W_I |_{i=1}^N$ affect the imaginary part of the samples $X_I |_{i=1}^N$. This approach, referred to as *marginal complex WM* filter, outputs:

$$\hat{\beta}_{marginal} = \text{MEDIAN}\left(|W_{R_i}|\lozenge\text{sgn}(W_{R_i})X_{R_i} |_{i=1}^N\right)$$
$$+ j\text{MEDIAN}\left(|W_{I_1}|\lozenge\text{sgn}(W_{I_1})X_{I_1} |_{i=1}^N\right), \qquad (6.134)$$

where the real and imaginary components are decoupled. The definition in (6.134) assumes that the real and imaginary components of the input samples are independent. On the other hand, if the real and imaginary domains are correlated, better performance is attained by mutually *coupling* the real and imaginary components of the signal and weights. This is shown in Section 6.6.1

In the context of filtering, weights are used to emphasize or deemphasize the input samples based on the temporal and ordinal correlation, or any other information contained in the signal. Consider the weighted mean operation with complex-valued weights,

$$\bar{\beta} = \text{MEAN}\left(|W_1| \cdot e^{-j\theta_1}X_1, |W_2| \cdot e^{-j\theta_2}X_2, \ldots, |W_N| \cdot e^{-j\theta_N}X_N\right)$$
$$= \frac{1}{N}\sum_{i=1}^N |W_i| \cdot e^{-j\theta_i}X_i. \qquad (6.135)$$

The simple manipulation used in (6.135) reveals that the weights have two roles in the complex weighted mean operation, first their phases are coupled into the samples changing them into a new group of *phased* samples, and then the magnitudes of the weights are applied. The process of decoupling the phase from the weight and merging it to the associated input sample is called *phase coupling*. The definition of the *phase coupled* complex WM filter follows by analogy.

6.6.1   Phase-Coupled Complex WM Filter

Given the complex valued samples $X_1, X_2, \ldots, X_N$ and the complex valued weights $W_i = |W_i|e^{j\theta_i}, i = 1, \ldots, N$, the output of the phase-coupled complex WM is defined as

$$\hat{\beta} = \arg\min_{\beta} \sum_{i=1}^{N} |W_i||e^{-j\theta_i}X_i - \beta|. \tag{6.136}$$

This definition of the complex weighted median delivers a rich class of complex median filtering structures. The solution to (6.136), however, suffers from computational complexity as the cost function must be searched for its minimum. Any one of the already mentioned suboptimal approximations, such as assuming that the output $\hat{\beta}$ is one of the phase-coupled input samples or, splitting the problem into real and imaginary parts, arise as effective ways to reduce the complexity. The following definition, from Hoyos et al. (2003) [104], provides efficient and fast complex-valued WM filter structures.

6.6.2   Marginal Phase-Coupled Complex WM Filter

DEFINITION 6.8 (MARGINAL PHASE-COUPLED COMPLEX WM FILTER)
Given a complex valued observation vector $\mathbf{X} = [X_1, X_2, \ldots, X_N]^T$ *and a set of complex valued weights* $\mathbf{W} = \langle W_1, W_2, \ldots, W_N \rangle$ *the marginal phase-coupled complex WM filter output is defined as:*

$$\begin{aligned}
\hat{\beta} &= \hat{\beta}_R + j\hat{\beta}_I \\
&= \text{MEDIAN}(|W_i|\diamond\text{Re}\{e^{-j\theta_i}X_i\} \,|_{i=1}^{N}), \\
&\quad + j\,\text{MEDIAN}(|W_i|\diamond\text{Im}\{e^{-j\theta_i}X_i\} \,|_{i=1}^{N}),
\end{aligned} \tag{6.137}$$

where $\diamond$ *is the replication operator,* $\text{Re}\{\cdot\}$ *and* $\text{Im}\{\cdot\}$ *denote real and imaginary part respectively. Thus,* $\hat{\beta}_R$ *is the weighted median of the real parts of the phase-coupled samples and* $\hat{\beta}_I$ *is the weighted median of the imaginary components of the phase-coupled samples.*

To help understand this definition better, a simple example is given in Figure 6.31. Three complex-valued samples X_1, X_2, X_3 and three complex-valued weights in the unit circle W_1, W_2, W_3 are arbitrarily chosen. The phase-coupled samples P_1, P_2, P_3 (where $P_i = e^{-j\theta_i}X_i$) are plotted to show the effect of phase coupling. The weights are not directly shown on the figure, but their phases $\theta_1, \theta_2, \theta_3$ are shown as the angles between original and altered samples. In addition, since the marginal phase-coupled complex WM filter outputs one of the real and imaginary parts of the phase-coupled samples, the filter does not necessarily select one of the phase-coupled inputs, which gives it more flexibility than the selection phase-coupled complex WM filter.

Because of the nonlinear nature of the median operations, direct optimization of the complex weighted median filter is not viable. To overcome this situation, threshold decomposition must be extended to the complex domain, and then used

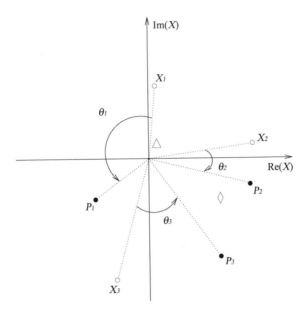

Figure 6.31 Marginal phase-coupled CWM illustration, "o" : original samples, "●" : phase-coupled samples, "△" : marginal median output, "◇" : marginal phase-coupled median output

to derive an adaptive algorithm for the marginal phase coupled complex weighted median filter in the minimum mean square error sense.

6.6.3 Complex threshold decomposition

For any real-valued signal X, its real threshold decomposition representation is given in equation (6.50), repeated here for convenience.

$$X = \frac{1}{2} \int_{-\infty}^{\infty} X^q dq, \tag{6.138}$$

where $-\infty < q < \infty$, and

$$X^q = \text{sgn}(X - q) = \begin{cases} 1 & \text{if } X \geq q; \\ -1 & \text{if } X < q. \end{cases} \tag{6.139}$$

Thus, given the samples $\{X_i \,|_{i=1}^N\}$ and the real-valued weights $\{W_i \,|_{i=1}^N\}$, the weighted median filter can be expressed as

$$Y = \frac{1}{2} \int_{-\infty}^{\infty} \text{MED}\left(|W_i| \diamond S_i^q \,|_{i=1}^N\right) dq, \tag{6.140}$$

where $S_i = \text{sgn}(W_i)X_i$, $\mathbf{S} = [S_1, S_2, \ldots, S_N]^T$, $S_i^q = \text{sgn}(S_i - q)$ and $\mathbf{S}^q = [S_1^q, S_2^q, \ldots S_N^q]^T$. Since the samples of the median filter in (6.140) are either 1 or

-1, this median operation can be efficiently calculated as $\text{sgn}(\mathbf{W}_a^T \mathbf{S}^q)$, where the elements of the new vector $\mathbf{W}_a^T$ are given by $\mathbf{W}_{a_i} = |W_i| \; |_{i=1}^N$. Equation (6.140) can be written as

$$Y = \frac{1}{2} \int_{-\infty}^{\infty} \text{sgn}(\mathbf{W}_a^T \mathbf{S}^q) dq. \tag{6.141}$$

Therefore, the extension of the threshold decomposition representation to the complex field can be naturally carried out as,

$$X = \frac{1}{2} \int_{-\infty}^{\infty} \text{sgn}(\text{Re}\{X\} - q) dq + j\frac{1}{2} \int_{-\infty}^{\infty} \text{sgn}(\text{Im}\{X\} - p) dp, \tag{6.142}$$

where real threshold decomposition is applied onto real and imaginary part of the complex signal X separately.

6.6.4 Optimal Marginal Phase-Coupled Complex WM

The real and imaginary parts of the output of the marginal phase-coupled complex WM in (6.137) are two separate real median operations, and thus the complex-valued threshold decomposition in (6.142) can be directly applied. Given the complex-valued samples $X_i \; |_{i=1}^N$, and the complex-valued weights $|W_i| e^{-j\theta_i} \; |_{i=1}^N$, define $P_i = e^{-j\theta_i} X_i \; |_{i=1}^N$ as the phase coupled input samples and its real and imaginary parts as $P_{R_i} = Re\{P_i\}$, $P_{I_i} = Im\{P_i\}$. Additionally define:

$$P_{R_i}^s = \text{sgn}(P_{R_i} - s) \;\; , \;\; \mathbf{P}_R^s = [P_{R_1}^s, P_{R_2}^s, \dots, P_{R_N}^s]^T$$

$$P_{I_i}^r = \text{sgn}(P_{I_i} - r) \;\; , \;\; \mathbf{P}_I^r = [P_{I_1}^r, P_{I_2}^r, \dots, P_{I_N}^r]^T.$$

Similar to the real threshold decomposition representation of WM in (6.141), the complex-valued threshold decomposition for the marginal phase-coupled complex WM can be implemented as

$$
\begin{aligned}
Y &= \text{MED}\left(|W_i| \diamond P_{R_i} \; |_{i=1}^N\right) + j\text{MED}\left(|W_i| \diamond P_{I_i} \; |_{i=1}^N\right) \\
&= \text{MED}\left(|W_i| \diamond \frac{1}{2} \int_{-\infty}^{\infty} P_{R_i}^s ds \; |_{i=1}^N\right) + j\text{MED}\left(|W_i| \diamond \frac{1}{2} \int_{-\infty}^{\infty} P_{I_i}^r dr \; |_{i=1}^N\right) \\
&= \frac{1}{2} \int_{-\infty}^{\infty} \text{MED}\left(|W_i| \diamond P_{R_i}^s \; |_{i=1}^N\right) ds + j\frac{1}{2} \int_{-\infty}^{\infty} \text{MED}\left(|W_i| \diamond P_{I_i}^r \; |_{i=1}^N\right) dr \\
&= \frac{1}{2} \left\{ \int_{-\infty}^{\infty} \text{sgn}(\mathbf{W}_a^T \mathbf{P}_R^s) ds + j \int_{-\infty}^{\infty} \text{sgn}(\mathbf{W}_a^T \mathbf{P}_I^r) dr \right\}. \tag{6.143}
\end{aligned}
$$

Assume the observed process $\{X(n)\}$ and the desired process $\{\beta(n)\}$ are jointly stationary. The filter output $\hat{\beta}(n)$ estimating the desired signal $\beta(n)$ is given in (6.143). Under the Mean Square Error (MSE) criterion, the cost function to minimize is

$$J(n) = \text{E}\{|\beta(n) - \hat{\beta}(n)|^2\}$$

$$= \mathrm{E}\left\{\left|\frac{1}{2}\int_{-\infty}^{\infty}(\mathrm{sgn}(\beta_R - s) - \mathrm{sgn}(\mathbf{W}_a^T\mathbf{P}_R^s))ds\right.\right.$$

$$\left.\left.+j\frac{1}{2}\int_{-\infty}^{\infty}(\mathrm{sgn}(\beta_I - r) - \mathrm{sgn}(\mathbf{W}_a^T\mathbf{P}_I^r))dr\right|^2\right\}$$

$$= \frac{1}{4}\mathrm{E}\left\{\left(\int_{-\infty}^{\infty}e_R^sds\right)^2 + \left(\int_{-\infty}^{\infty}e_I^rdr\right)^2\right\}, \qquad (6.144)$$

where $\beta_R = \mathrm{Re}\{\beta(n)\}$, $\beta_I = \mathrm{Im}\{\beta(n)\}$, $e_R = \mathrm{Re}\{\beta(n) - \hat{\beta}(n)\}$, $e_I = \mathrm{Im}\{\beta(n) - \hat{\beta}(n)\}$. Utilizing the relationship between the complex gradient vector ∇J and the conjugate derivative $\partial J/\partial \mathbf{W}^*$[99], results in

$$\nabla J(n) = 2\frac{\partial J(\mathbf{W})}{\partial \mathbf{W}^*}$$

$$= -\mathrm{E}\left\{\left(\int_{-\infty}^{\infty}e_R^sds\right)\left(\int_{-\infty}^{\infty}\frac{\partial}{\partial\mathbf{W}^*}\mathrm{sgn}(\mathbf{W}_a^T\mathbf{P}_R^s)ds\right)\right.$$

$$\left.+\left(\int_{-\infty}^{\infty}e_I^rdr\right)\left(\int_{-\infty}^{\infty}\frac{\partial}{\partial\mathbf{W}^*}\mathrm{sgn}(\mathbf{W}_a^T\mathbf{P}_I^r)dr\right)\right\}$$

$$= -2\mathrm{E}\left\{e_R\cdot\left(\int_{-\infty}^{\infty}\frac{\partial}{\partial\mathbf{W}^*}\mathrm{sgn}(\mathbf{W}_a^T\mathbf{P}_R^s)ds\right)\right.$$

$$\left.+e_I\cdot\left(\int_{-\infty}^{\infty}\frac{\partial}{\partial\mathbf{W}^*}\mathrm{sgn}(\mathbf{W}_a^T\mathbf{P}_I^r)dr\right)\right\}. \qquad (6.145)$$

To take the derivatives needed in (6.145), the sign function is approximated by a differentiable one to circumvent the inconvenience of having a Dirac impulse term in further analysis. The chosen substitute is the hyperbolic tangent function $\mathrm{sgn}(x) \approx \tanh(x) = \frac{e^x - e^{-x}}{e^x + e^{-x}}$ and its derivative $\frac{d}{dx}\tanh(x) = \mathrm{sech}^2(x) = \frac{4}{(e^x + e^{-x})^2}$. Thus, $\frac{\partial}{\partial\mathbf{W}^*}\mathrm{sgn}(\mathbf{W}_a^T\mathbf{P}_R^s) \approx \mathrm{sech}^2(\mathbf{W}_a^T\mathbf{P}_R^s)\frac{\partial}{\partial\mathbf{W}^*}(\mathbf{W}_a^T\mathbf{P}_R^s)$. Furthermore, the derivative with respect to only one weight is

$$\frac{\partial}{\partial W_i^*}\mathrm{sgn}(\mathbf{W}_a^T\mathbf{P}_R^s) \approx \mathrm{sech}^2(\mathbf{W}_a^T\mathbf{P}_R^s)\frac{\partial}{\partial W_i^*}(|W_i|P_{R_i}^s)$$

$$= \mathrm{sech}^2(\mathbf{W}_a^T\mathbf{P}_R^s)\left(\frac{\partial|W_i|}{\partial W_i^*}P_{R_i}^s + |W_i|\frac{\partial P_{R_i}^s}{\partial W_i^*}\right). \qquad (6.146)$$

Given the relationship

$$\frac{\partial P_{R_i}}{\partial W_i^*} = \frac{1}{2}\frac{\partial}{\partial W_i^*}(P_i + P_i^*)$$

$$= \frac{1}{2}\frac{1}{|W_i^*|}e^{j\theta_i}jP_{I_i},$$

equation (6.146) can be written as:

$$\frac{\partial}{\partial W_i^*}\mathrm{sgn}(\mathbf{W}_a^T\mathbf{P}_R^s) \approx \frac{1}{2}\mathrm{sech}^2(\mathbf{W}_a^T\mathbf{P}_R^s)e^{j\theta_i}\left(P_{R_i}^s + \mathrm{sech}^2(P_{R_i}-s)jP_{I_i}\right)$$

and similarly

$$\frac{\partial}{\partial W_i^*}\mathrm{sgn}(\mathbf{W}_a^T\mathbf{P}_I^r) \approx \frac{1}{2}\mathrm{sech}^2(\mathbf{W}_a^T\mathbf{P}_I^r)e^{j\theta_i}\left(P_{I_i}^r - \mathrm{sech}^2(P_{I_i}-r)jP_{R_i}\right).$$

Integrating both sides

$$\int_{-\infty}^{\infty}\frac{\partial}{\partial W_i^*}\mathrm{sgn}(\mathbf{W}_a^T\mathbf{P}_R^s)ds \approx \frac{1}{2}e^{j\theta_i}\int_{-\infty}^{\infty}\mathrm{sech}^2(\mathbf{W}_a^T\mathbf{P}_R^s)P_{R_i}^s ds$$
$$+ \frac{1}{2}e^{j\theta_i}jP_{I_i}\int_{-\infty}^{\infty}\mathrm{sech}^2(\mathbf{W}_a^T\mathbf{P}_R^s)\mathrm{sech}^2(P_{R_i}-s)ds.$$

$$(6.147)$$

The second integral in (6.147) can be expanded as follows

$$\int_{-\infty}^{\infty}\mathrm{sech}^2(\mathbf{W}_a^T\mathbf{P}_R^s)\mathrm{sech}^2(P_{R_i}-s)ds =$$

$$\mathrm{sech}^2\left(\mathbf{W}_a^T\mathbf{P}_R^{P_{R(1)}^-}\right)\int_{-\infty}^{P_{R(1)}}\mathrm{sech}^2(P_{R_i}-s)ds$$

$$+ \sum_{k=1}^{N-1}\mathrm{sech}^2\left(\mathbf{W}_a^T\mathbf{P}_R^{P_{R(k+1)}^-}\right)\int_{P_{R(k)}}^{P_{R(k+1)}}\mathrm{sech}^2(P_{R_i}-s)ds$$

$$+ \mathrm{sech}^2\left(\mathbf{W}_a^T\mathbf{P}_R^{P_{R(N)}^+}\right)\int_{P_{R(N)}}^{\infty}\mathrm{sech}^2(P_{R_i}-s)ds$$

$$(6.148)$$

and recalling that $\int\mathrm{sech}^2(x)dx = \int d\tanh(x)$

$$\int_{-\infty}^{\infty}\mathrm{sech}^2(\mathbf{W}_a^T\mathbf{P}_R^s)\mathrm{sech}^2(P_{R_i}-s)ds =$$

$$\mathrm{sech}^2\left(\mathbf{W}_a^T\mathbf{P}_R^{P_{R(1)}^-}\right)\tanh(P_{R_i}-s)\Big|_{-\infty}^{P_{R(1)}}$$

$$+ \sum_{k=1}^{N-1}\mathrm{sech}^2\left(\mathbf{W}_a^T\mathbf{P}_R^{P_{R(k+1)}^-}\right)\tanh(P_{R_i}-s)\Big|_{P_{R(k)}}^{P_{R(k+1)}}$$

$$+ \mathrm{sech}^2\left(\mathbf{W}_a^T\mathbf{P}_R^{P_{R(N)}^+}\right)\tanh(P_{R_i}-s)\Big|_{P_{R(N)}}^{\infty}.$$

$$(6.149)$$

At this time $\tanh(x)$ can be replaced again with $\text{sgn}(x)$. As a result, all terms involving $\text{sign}(P_{R_i} - s)$ in the previous equation will be zero, except the one when $P_{R_i} = P_{R_{(k)}}$. In this case: $\text{sgn}(P_{R_i} - s)\Big|_{P_{R_{(k)}}}^{P_{R_{(k+1)}}} = 2$.

On the other hand, when $P_{R_i} = \hat{\beta}_R$, the product $\mathbf{W}_a^T \mathbf{P}_R^s$ is approximately zero. In this case $\text{sech}^2(\mathbf{W}_a^T \mathbf{P}_R^s) \approx 1$, and since this is the largest contributor to the sum in (6.149) all the other terms can be omitted. All these approximations result in:

$$\int_{-\infty}^{\infty} \frac{\partial}{\partial W_i^*} \text{sgn}(\mathbf{W}_a^T \mathbf{P}_R^s) ds \approx \frac{1}{2} e^{j\theta_i} \left(\text{sgn}(P_{R_i} - \hat{\beta}_R) + 2jP_{I_i}\delta(P_{R_i} - \hat{\beta}_R) \right)$$

$$\int_{-\infty}^{\infty} \frac{\partial}{\partial W_i^*} \text{sgn}(\mathbf{W}_a^T \mathbf{P}_I^r) dr \approx \frac{1}{2} e^{j\theta_i} \left(\text{sgn}(P_{I_i} - \hat{\beta}_I) + 2jP_{R_i}\delta(P_{I_i} - \hat{\beta}_I) \right),$$

$$(6.150)$$

leading to the following weight update equation:

$$\begin{aligned}
W_i(n+1) &= W_i(n) + \mu\{-\nabla J(n)\} \\
&\approx W_i(n) + \mu e^{j\theta_i} \Big\{ e_R(n)\text{sgn}(P_{R_i}(n) - \hat{\beta}_R(n)) \\
&\quad + e_I(n)\text{sgn}(P_{I_i}(n) - \hat{\beta}_I(n)) \\
&\quad + 2je_R(n)(P_{I_i}(n)\delta(P_{R_i}(n) - \hat{\beta}_R(n)) \\
&\quad + 2je_I(n)(P_{R_i}(n)\delta(P_{I_i}(n) - \hat{\beta}_I(n)) \Big\}.
\end{aligned}$$

$$(6.151)$$

EXAMPLE 6.14 (LINE ENHANCEMENT)

Adaptive line enhancement consists of an adaptive filter driven with a delayed version of the input signal, which uses the noisy signal itself as the reference. The goal is to exploit the signal correlation and the noise uncorrelation between the received signal and its shifted version to filter out the noise. The algorithm also tunes the weights to correct the phase introduced between the filter input and the reference signal. A basic block diagram of a line-enhancer implemented with the complex WM filter is shown in Figure 6.32.

In the first experiment, the input of an 11-tap line enhancer is a complex exponential contaminated with α-stable noise with dispersion $\gamma = 0.2$, α running from 1.3 to 2 (Gaussian noise) to show different levels of noise impulsiveness. The weights of the marginal phase-coupled complex WM filter are designed using the previously developed LMS algorithm. In addition, LMS algorithms are implemented to design a marginal complex weighted median and a linear complex-valued filter. The same noisy signal will be filtered using these three schemes to compare the results obtained

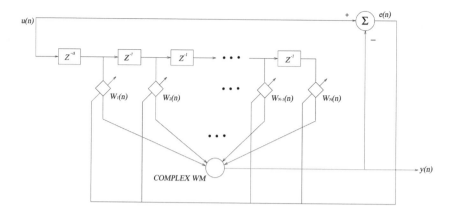

Figure 6.32 Block diagram for line enhancer implemented with complex WM filter.

with each one of them. To analyze the convergence properties of the algorithms, the learning curves calculated as the average MSE of 1000 realizations of the experiment are plotted. Figure 6.33 shows the results for two values of α: (a) α=1.3 and (b) α=1.7 where the LMS algorithm for the linear filter diverges for $\alpha < 2$. On the other hand, the robustness of the marginal phase-coupled complex WM is clearly seen. For the values of α shown, the plot of the MSE remains almost unaltered, that is, the impulsiveness of the noise does not have a major effect in the performance of the algorithm for $\alpha < 2$. Table 6.7 summarizes the average of 2000 values of the MSE after the convergence of the LMS algorithm for the complex filters. These results show the reliability of the complex WM filters in α-stable environments. For this particular application and noise conditions, the marginal phase-coupled outperforms the marginal complex WM filter.

Unlike linear filters, the step size has a small effect on the floor error of the complex WM filter as it is illustrated in Figure 6.34 where the learning curves of the LMS algorithm for $\mu = 0.1$ and $\mu = 0.001$ are shown. The plot shows how a higher value of the step size improves the convergence rate of the algorithm without harming the robustness of the filter or modifying significantly the value of the floor error.

Table 6.7 LMS average MSE for line enhancement. ($\mu = 0.001$, $\gamma = 0.2$)

Filter	$\alpha = 1.3$	$\alpha = 1.5$	$\alpha = 1.7$	$\alpha = 2$
Noisy signal	40.3174	7.3888	2.4300	0.8258
Linear filter	∞	∞	∞	0.0804
Marginal complex WM	0.3621	0.3728	0.3975	0.4297
Marginal phase coupled complex WM	0.1455	0.1011	0.1047	0.1162

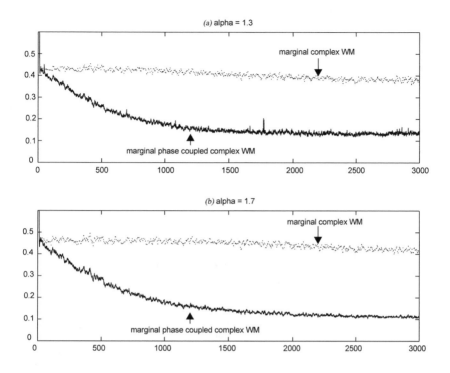

Figure 6.33 Learning curves of the LMS algorithm of a linear filter, marginal complex WM and marginal phase-coupled complex WM (μ=0.001) for line enhancement in α-stable noise with dispersion $\gamma = 0.2$ (ensemble average of 1000 realizations): (*a*) α=1.3, (*b*) α=1.7

For illustrative purposes the real part and the phase of the filter outputs are shown in Figure 6.35 and Figure 6.36, respectively. The plot shows 2000 samples of the filter output taken after the LMS algorithm has converged. As it can be seen, the linear filter is not successful at filtering the impulsive noise, while the complex WM filters are able to recover the original shape of the signal, being the output of the marginal phase-coupled complex WM the one that resembles the best the original signal. ∎

EXAMPLE 6.15 (COMPLEX WM FILTER DESIGN BY FREQUENCY RESPONSE)

In this example, the complex weighted median filter is designed to approximate the frequency response of a complex linear filter. To obtain this, the system shown in Figure 6.37 is used.

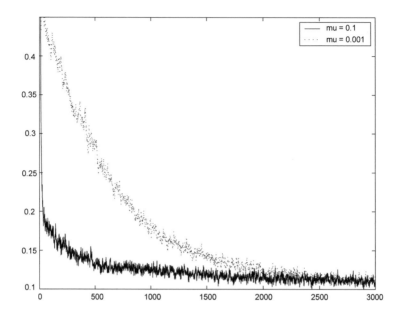

Figure 6.34 Learning curves of the LMS algorithm of the marginal phase-coupled complex WM (ensemble average of 1000 realizations) with $\mu = 0.1$ and $\mu = 0.001$ for Line enhancement in α-stable noise ($\gamma = 0.2$).

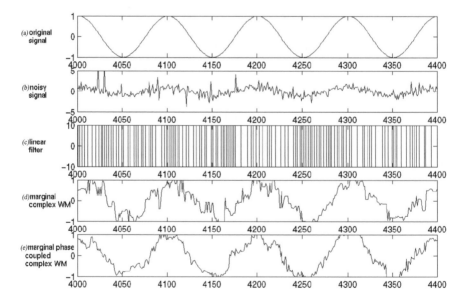

Figure 6.35 Real part of the output of the filters for $\alpha = 1.7$, $\gamma = 0.2$ and $\mu = 0.1$.

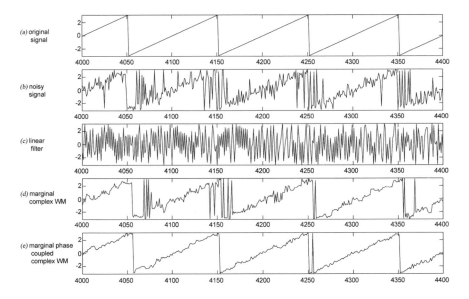

Figure 6.36 Phase of the output of the filters for $\alpha = 1.7$, $\gamma = 0.2$ and $\mu = 0.1$.

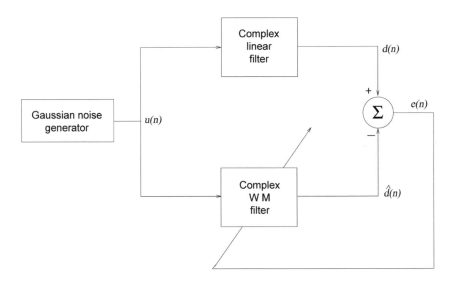

Figure 6.37 Block diagram of the frequency response design experiment.

The Gaussian noise generator provides a complex Gaussian sequence that is fed to both a complex linear filter and to the complex weighted median filter being tuned. The difference between the outputs of the two filters is the error parameter used in an LMS algorithm that calculates the optimum weights for the complex-weighted median filter. After convergence, the complex-weighted median filter should be a close approximation of the original linear filter in the mean square error sense. Figure 6.38 shows the ensemble average learning curves for 1000 realizations of the experiment. The learning curve of a linear filter calculated in the same way has also been included. For this experiment the complex linear filter:

$$h = \begin{bmatrix} -0.0123 - 0.0123i \\ -0.0420 - 0.1293i \\ -0.0108 + 0.0687i \\ -0.0541 + 0.0744i \\ 0.2693 - 0.1372i \\ 0.5998 \\ 0.2693 + 0.1372i \\ -0.0541 - 0.0744i \\ -0.0108 - 0.0687i \\ -0.0420 + 0.1293i \\ -0.0123 + 0.0123i \end{bmatrix}$$

was used. This is a complex low pass filter with normalized cut off frequencies $\omega_1 = -0.4$ and $\omega_2 = 0.7$. The designed complex WM filters have the same number of taps (11).

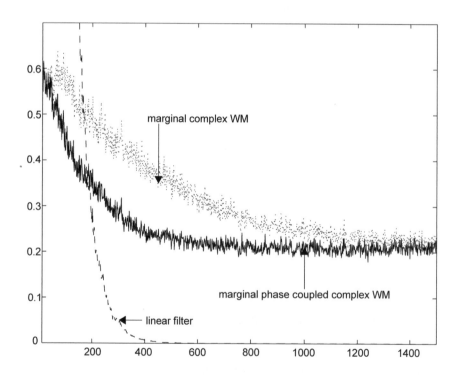

Figure 6.38 Learning curves of the LMS algorithm of the marginal phase coupled complex WM, the marginal complex WM and a linear filter with $\mu = 0.01$ for the frequency response design problem.

As expected, the MSE for the linear filter reaches a minimum of zero. In addition, the floor error for both complex WM filters is similar. The frequency response of the complex weighted median filters, as well as the one of the original linear filter, were calculated as follows: 10,000 samples of complex Gaussian noise were fed to the filters and the spectra of the outputs were calculated using the Welch method [192], the experiment was repeated 50 times to get an ensemble average, the results are shown in Figure 6.39.

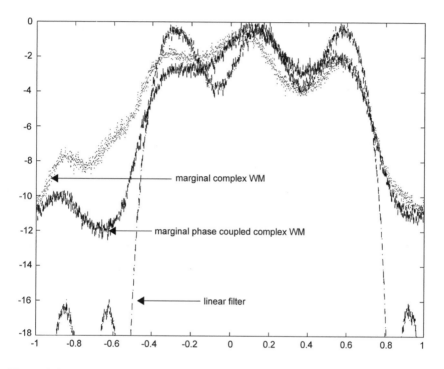

Figure 6.39 Approximated frequency response of the complex WM filters for the frequency response design problem.

The filters have approximately the same band-pass gain and even though the rejection in the stop band is not as good as the obtained with the linear filter, the levels reached for the complex WM filters are acceptable.

All the previous merit figures applied to this experiment had been designed for linear filters in a Gaussian environment. The real strength of the weighted median filter comes out when the classical Gaussian model is abandoned and heavy-tailed random processes are included. In order to show the power of the filters designed with the adaptive LMS algorithm, the sum of two complex exponentials of magnitude one and two different normalized frequencies, one in the pass band (0.2) and one in the stop band (-0.74) of the filters is contaminated with α-stable noise with α values 1, 1.3, 1.7, 2 and $\gamma = 0.1$. If a linear filter were used to filter the clean signal, the output will be a complex exponential of normalized frequency 0.2. This signal is used as a reference to calculate the MSE of the outputs of the complex WM filters and a linear filter in the presence of the noise. Table 6.8 shows the average of 100 realizations of the filtering of 200 samples of the noisy signal for each case. As it can be seen, for this application, the marginal phase-coupled complex WM obtains the best results. On the other hand, the linear filter is unable to remove the impulsive noise, as is shown in the high values of the MSE of its output. In the Gaussian case the linear filter shows its superiority.

Table 6.8 Average MSE of the output of the complex WM filters and the linear filter in presence of α-stable noise

Filter	$\alpha = 1$	$\alpha = 1.3$	$\alpha = 1.7$	$\alpha = 2$
Linear filter	555.1305	13.4456	2.8428	0.2369
Marginal complex WM	0.7352	0.7230	0.7139	0.7099
Marginal phase coupled complex WM	0.7660	0.7192	0.6937	0.6642

An example of the real part of the original signal, the noisy signal and the output of the filters is shown in Figure 6.40. The plots show the presence of only one sinusoidal in the outputs, still showing some artifacts from the remaining noise after filtering. The other exponential has been eliminated from the signal, showing the frequency selection capabilities of the complex WM filters.

■

6.6.5 Spectral Design of Complex-Valued Weighted Medians

Equation (6.137) shows that the complex-valued weighted median filter operation consists of properly modifying the input samples according to the associated weights and then using the magnitude of the weights for the calculation of positive weighted medians. It was stated in Theorem 6.1 that a nonlinear function needs to satisfy certain properties in order to be best approximated under the mean squared error sense by a linear filter. Unfortunately, the complex-valued medians do not satisfy the location invariance property. A similar procedure to the one in Section 6.2.5 can be used to extend Mallows results to the complex domain.

THEOREM 6.3 *If the real and imaginary parts of the input series are Gaussian, independent, and zero centered, the coefficients of the linear part of the weighted median defined in (6.137) are defined as:* $h_i = e^{-j\theta_i} p_i$, *where* p_i *are the SSPs of the WM smoother* $|W_i|$.

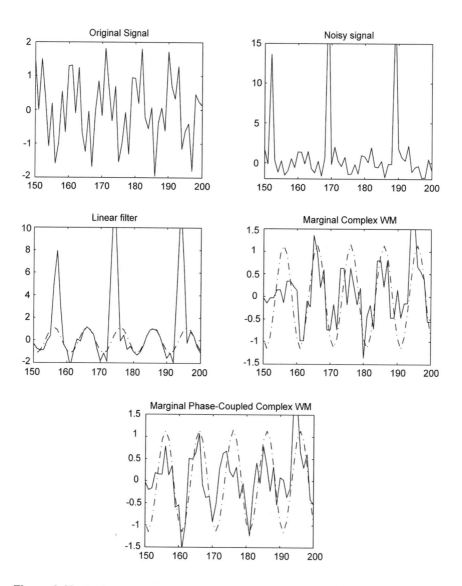

Figure 6.40 Real part of the output of the complex WM filters for the frequency response design problem with (α=1). (the real part of the ideal output is shown in dash-dot)

To show the theorem define $Y_i = e^{-j\theta_i} X_i = U_i + jV_i$.

$$\mathbf{E}\left\{\left|\mathrm{MEDIAN}(W_i \diamond X_i) - \sum h_i X_i\right|^2\right\} = \tag{6.152}$$

$$\mathbf{E}\left\{\left|\mathrm{MEDIAN}(|W_i| \diamond Y_i) - \sum q_i Y_i\right|^2\right\}$$

$$= \mathbf{E}\left\{ \left(\text{MEDIAN}(|W_i| \diamond U_i) - \sum b_i U_i + \sum c_i V_i \right)^2 \right\}$$

$$+ \mathbf{E}\left\{ \left(\text{MEDIAN}(|W_i| \diamond V_i) - \sum b_i V_i + \sum c_i U_i \right)^2 \right\}$$

where $q_i = e^{j\theta_i} h_i = b_i + j c_i$. Again from Mallows' theorem, (6.153) is minimized when $c = 0$ and $b_i = p_i$.

This characteristics permit the development of a design method for complex valued WM filters from spectral requirements using the algorithms described in Section 6.2 as follows

(1) Design a linear complex valued FIR filter $\mathbf{h} = (h_1, h_2, \ldots, h_N)$ given the impulse response and the other desired characteristics for the filter.

(2) Decouple the phases of the coefficients to form the vectors $|\mathbf{h}| = (|h_1|, |h_2|, \ldots, |h_N|)$ and $\boldsymbol{\Theta}(\mathbf{h}) = (\theta(h_1), \theta(h_2), \ldots, \theta(h_N))$, where $\theta(h_i)$ represents the phase of h_i.

(3) Normalize the vector $|\mathbf{h}|$ and find the closest WM filter to it using the algorithm based on the theory of sample selection probabilities, developed in Section 6.2, say $\mathbf{W}' = (W'_1, W'_2, \ldots, W'_N)$.

(4) The complex WM filter is given by $\mathbf{W} = \left[e^{j\theta(h_i)} W'_i \cdot |_{i=1}^N \right]$

EXAMPLE 6.16

Design 9-tap marginal phase-coupled complex weighted median filters with the characteristics indicated in Table 6.9.

Figure 6.41 shows that the frequency response characteristics of the complex WM are very close to the ones of their linear counterparts. The values of the weights for the linear and median filters are shown in Table 6.10 ∎

EXAMPLE 6.17

Repeat example 6.15 using the algorithm for the spectral design of complex valued weighted medians developed in Section 6.6.5.

(1) The complex valued linear filter to approximate is: $h = [-0.0123 - 0.0123i, -0.0420 - 0.1293i, -0.0108 + 0.0687i, -0.0541 + 0.0744i, 0.2693 - 0.1372i, 0.5998, 0.2693 + 0.1372i, -0.0541 - 0.0744i, -0.0108 - 0.0687i, -0.0420 + 0.1293i, -0.0123 + 0.0123i]^T$.

Table 6.9 Characteristics of the complex-weighted median filters to be designed

Filter	Cut off frequencies
Low-pass	-0.4, 0.7
Band-pass	-1 -0.5, 0.3 0.7
High-pass	-0.5, 0.8
Band-stop	-1 -0.4, 0.2 0.8

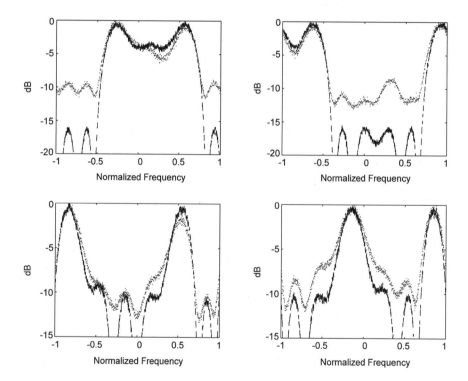

Figure 6.41 Approximated frequency response of the complex WM filters designed with the algorithm in Section 6.6.5 (*a*) low-pass, (*b*) high-pass, (*c*) band-pass, (*d*) band-stop (dotted: Marginal Phase Coupled Complex WM, dashed: Linear filter)

Table 6.10 Weights of the complex median filters designed using the algorithm in Section 6.6.5 and the linear filters used as reference.

Low-pass					Band-pass				
Linear		Median			Linear		Median		
-0.0324	- 0.0997i	-0.0215	-	0.0662i	0.0601	- 0.1844i	0.0299	-	0.0918i
-0.0172	+ 0.1080i	-0.0112	+	0.0707i	0.0212	+ 0.1530i	0.0112	+	0.0810i
-0.0764	+ 0.1050i	-0.0473	+	0.0651i	-0.1766	- 0.2270i	-0.0808	-	0.1038i
0.2421	- 0.1235i	0.1266	-	0.0646i	-0.1487	- 0.0647i	-0.0776	-	0.0338i
	0.5714		0.2723			0.5002		0.2109	
0.2421	+ 0.1235i	0.1266	+	0.0646i	-0.1487	+ 0.0647i	-0.0776	+	0.0338i
-0.0764	- 0.1050i	-0.0473	-	0.0651i	-0.1766	+ 0.2270i	-0.0808	+	0.1038i
-0.0172	- 0.1080i	-0.0112	-	0.0707i	0.0212	- 0.1530i	0.0112	-	0.0810i
-0.0324	+ 0.0997i	-0.0215	+	0.0662i	0.0601	+ 0.1844i	0.0299	+	0.0918i

High-pass					Band-stop				
Linear		Median			Linear		Median		
0.0324	+ 0.0997i	0.0189	+	0.0583i	-0.0601	+ 0.1844i	-0.0299	+	0.0918i
0.0172	- 0.1080i	0.0106	-	0.0666i	-0.0212	- 0.1530i	-0.0112	-	0.0810i
0.0764	- 0.1050i	0.0559	-	0.0769i	0.1766	+ 0.2270i	0.0808	+	0.1038i
-0.2421	+ 0.1235i	-0.1486	+	0.0758i	0.1487	+ 0.0647i	0.0776	+	0.0338i
	0.4286		0.2187			0.4998		0.2109	
-0.2421	- 0.1235i	-0.1486	-	0.0758i	0.1487	- 0.0647i	0.0776	-	0.0338i
0.0764	+ 0.1050i	0.0559	+	0.0769i	0.1766	- 0.2270i	0.0808	-	0.1038i
0.0172	+ 0.1080i	0.0106	+	0.0666i	-0.0212	+ 0.1530i	-0.0112	+	0.0810i
0.0324	- 0.0997i	0.0189	-	0.0583i	-0.0601	- 0.1844i	-0.0299	-	0.0918i

(2) Decouple the phases to form the vectors:

$$|\mathbf{h}| = [0.0174, 0.1359, 0.0696, 0.0920, 0.3022, 0.5998, 0.3022,$$
$$0.0920, 0.0696, 0.1359, 0.0174]^T$$
$$\mathbf{\Theta}(\mathbf{h}) = [-2.3548, -1.8850, 1.7272, 2.1991, -0.4712, 0, 0.4712,$$
$$-2.1991, -1.7272, 1.8850, 2.3548]^T$$

(3) After normalizing $|\mathbf{h}|$, the closest WM smoother to it is found as: $\mathbf{W}' = [0.0225, 0.0805, 0.0478, 0.0572, 0.1584, 0.2671, 0.1584, 0.0572, 0.0478, 0.0805, 0.0225]$.

(4) Coupling the phases of the linear weights to the median weights:

$$\mathbf{W} = \begin{bmatrix} -0.0159 - 0.0160i & -0.0249 - 0.0766i & -0.0074 + 0.0472i \\ -0.0336 + 0.0463i & 0.1412 - 0.0719i & 0.2671 \\ 0.1412 + 0.0719i & -0.0336 - 0.0463i & -0.0074 - 0.0472i \\ -0.0249 + 0.0766i & -0.0159 + 0.0160i & \end{bmatrix}$$

To compare the results obtained with the two algorithms, the frequency response of the median filters is approximated feeding the filters with independent complex

Gaussian noise and approximating the spectra of the outputs using the Welch method. The results are shown in Figure 6.42.

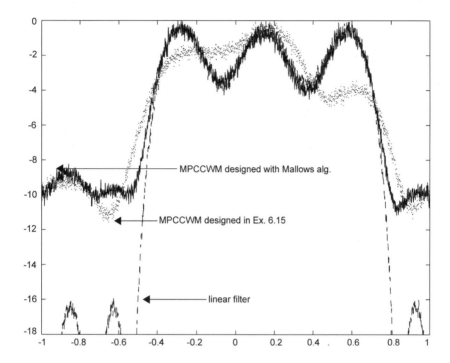

Figure 6.42 Approximated frequency response of the complex WM filters designed with and adaptive LMS algorithm (dotted), the algorithm for spectral design of complex valued weighted medians in Section 6.6.5(solid), and the linear filter used as a reference to design them (dashed).

6.7 WEIGHTED MEDIAN FILTERS FOR MULTICHANNEL SIGNALS

The extension of the weighted median for use with multidimensional (multichannel) signals is not straightforward. Sorting multicomponent (vector) values and selecting the middle value is not well defined as in the scalar case, see Barnett (1976) [27]. In consequence, the weighted median filtering operation of a multidimensional signal can be achieved in a number of ways among which the most well known are: marginal medians of orthogonal coordinates in Hayford (1902) [98], L_1-norm median, from Gini and Galvani (1929) [80] and Haldane (1948) [89] that minimizes the sum of distances to all samples, the halfplane median from Tukey (1975) [190] that minimizes the maximum number of samples on a halfplane, convex hull median from Barnett

(1976) [27] and Shamos (1976) [172] that is the result of continuous "peeling" off pairs of extreme samples, the simplex median from Oja (1983) [152] that minimizes the sum of the volumes of all simplexes[4] formed by the point and some samples, the simplex median of Liu (1990) [133] that maximizes the number of simplexes that contain it, and the hyperplane median from Rousseeuw (1999) [170] that maximizes the *hyperplane depth*. For historical reviews on multivariate medians, see Aloupis (2001) [1] and Small (1990) [178]. Other approaches can be found in Hardie and Arce (1991) [92], Koivunen (1996) [116], Pitas and Tsakalides (1991) [163], and Trahanias and Venestanopoulos (1996) [185].

A problem with many definitions of multivariate medians is that they have more conceptual meaning than practical use because of their high computational complexities. The algorithms used to compute the L_1 median often involve gradient techniques or iterations that can only provide numerical solutions as shown by Großand Strempel (1998) [88], and even the fastest algorithm up-to-date for the Oja median is about $O(n^3 \log n)$ in time, see Aloupis (2001) [1]. Moreover, they usually have difficulties with extension on more complex weighting structures. Many definitions are also difficult to analyze. Simple and mathematically tractable structures to perform multivariate weighted median filtering are described below.

6.7.1 Marginal WM filter

The simplest approach to WM filtering of a multidimensional signal is to process each component independently by a scalar WM filter. This operation is illustrated in Figure 6.43 where the green, blue, and red components of a color image are filtered independently and then combined to produce the filtered color image. A drawback associated with this method is that different components can be strongly correlated and, if each component is processed separately, this correlation is not exploited. The advantage of marginal processing is the computational simplicity. Marginal weighted median filters are, in general, very limited in most multichannel signal processing applications, as will be illustrated shortly.

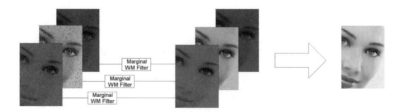

Figure 6.43 Center WM filter applied to each component independently. (Figure also appears in the Color Figure Insert)

---

[4]A simplex is a d-dimensional solid formed by $d+1$ points in R^d.

6.7.2   Vector WM filter

A more logical extension is found through the minimization of a weighted cost function which takes into account the multicomponent nature of the data. Here, the filtering operation processes all components jointly such that some of the cross-correlation between components is exploited. As it is shown in Figure 6.44 the three components are jointly filtered by a vector WM filter leading to a filtered color image. Vector WM filtering requires the extension of the original WM filter

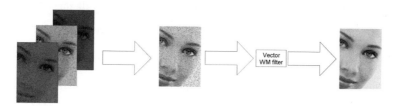

Figure 6.44 Center vector WM filter applied in the 3-dimensional space. (Figure also appears in the Color Figure Insert)

definition as follows (Astola 1990 [19]). The filter input vector is denoted as $\mathbf{X} = [\vec{X}_1 \ \vec{X}_2 \ \ldots \ \vec{X}_N]^T$, where $\vec{X}_i = [X_i^1 \ X_i^2 \ \ldots \ X_i^M]^T$ is the ith M-variate sample in the filter window. The filter output is $\vec{Y} = [Y^1 \ Y^2 \ \ldots \ Y^M]^T$. Recall that the weighted median of a set of 1-dimensional samples $X_i \ i = 1, \ldots, N$ is given by

$$Y = \arg\min_{\beta} \sum_{i=1}^{N} |W_i| |\mathrm{sgn}(W_i) X_i - \beta|. \tag{6.153}$$

Extending this definition to a set of M-dimensional vectors $\vec{X}_i$ for $i = 1, \ldots, N$ leads to

$$\vec{Y} = \arg\min_{\vec{\beta}} \sum_{i=1}^{N} |W_i| \|\vec{S}_i - \vec{\beta}\|, \tag{6.154}$$

where $\vec{Y} = [Y^1, Y^2, \ldots, Y^M]^T$, $\vec{S}_i = \mathrm{sgn}(W_i)\vec{X}_i$, and $\|\cdot\|$ is the L_2 norm defined as

$$\|\vec{\beta} - \vec{S}_i\| = \left((\beta^1 - S_i^1)^2 + (\beta^2 - S_i^2)^2 + \ldots + (\beta^M - S_i^M)^2 \right)^{\frac{1}{2}}. \tag{6.155}$$

The vector weighted median thus requires N scalar weights, with one scalar weight assigned per each input vector sample. Unlike the 1-dimensional case, $\vec{Y}$ is not generally equal in value to one of the $\vec{S}_i$. Indeed, there is no closed-form solution for $\vec{Y}$. Moreover, solving (6.154) involves a minimization problem in a M-dimensional space that can be computationally expensive. To overcome these difficulties, a suboptimal solution for (6.154) is found if $\vec{Y}$ is restricted to be one of the signed samples $\vec{S}_i$. This leads to the definition of the weighted vector median.

$$\vec{Y} = \arg\min_{\vec{\beta} \in \{\vec{S}_i\}} \sum_{i=1}^{N} |W_i| \|\vec{\beta} - \vec{S}_i\|. \tag{6.156}$$

That is, the vector WM filter output of $\vec{X}_1, \ldots, \vec{X}_N$ has the value of $\vec{Y}$, with $\vec{Y} \in \{\vec{S}_1, \ldots, \vec{S}_N\}$ such that

$$\sum_{i=1}^{N} W_i \|\vec{Y} - \vec{S}_i\| \leq \sum_{i=1}^{N} W_i \|\vec{S}_j - \vec{S}_i\| \qquad \text{for all } j = 1, \ldots, N. \quad (6.157)$$

This definition can be implemented as follows:

- For each signed sample $\vec{S}_j$, compute the distances to all the other signed samples $(\|\vec{S}_j - \vec{S}_i\|)$ for $i = 1, \ldots, N$ using (6.155).

- Compute the sum of the weighted distances given by the right side of (6.157).

- Choose as filter output the signed sample $\vec{S}_j$ that produces the minimum sum of the weighted distances.

In a more general case, the same procedure can be employed to calculate the vector weighted median of a set of input samples using other distance measures. The vector median in Astola (1990) [19] uses the norm, L_p defined as $\|\vec{X}\|_p = (\sum |X_i|^p)^{\frac{1}{p}}$ as a distance measure, transforming (6.154) into

$$Y = \arg\min_{\vec{X} \in \{\vec{S}_i\}} \sum_{i=1}^{N} W_i \|\vec{X} - \vec{S}_i\|_p. \qquad (6.158)$$

Several optimization algorithms for the design of the weights have been developed. One such method, proposed by Shen and Barner (2004) [173] is summarized below as an example.

By definition, the WVM filter is selection type and its output is one of the input samples as it is shown in (6.158). First it is necessary to find the closest sample to the desired output, say $\vec{X}_{e_{min}}$. The output of the filter is then calculated using the current weights. If the output of the filter is $\vec{X}_{e_{min}}$ the weights are considered optimal. Otherwise, the weights should be modified in order to obtain $\vec{X}_{e_{min}}$ as the output. The optimization process can be summarized as follows:

(1) Initialize the filter weights ($W_i = 1, i = 1, \ldots, N$).

(2) Calculate the distance of each input sample to the desired output as:

$$e_i = \|\vec{X}_i(n) - \vec{D}(n)\|, \quad i = 1, 2, \cdots, N. \qquad (6.159)$$

(3) Find the sample $\vec{X}_{e_{min}}$ such that the error e_i is minimum

$$\vec{X}_{e_{min}} = \arg\min_{\vec{X} \in \{\vec{S}_i\}} e_i = \arg\min_{\vec{X} \in \{\vec{S}_i\}} \|\vec{X}_i(n) - \vec{D}(n)\|. \qquad (6.160)$$

(4) If $\vec{X}_{e_{min}}$ is the current output of the WVM filter, set $\Delta W_i = 0$. Otherwise, compute the necessary weight changes so that $\vec{X}_{e_{min}}$ becomes the filter output using the set of weights $W_i(n) + \Delta W_i$. The ΔW_i are given by

$$\Delta W_i = \frac{d(\vec{X}_{e_{min}}) - d(\vec{X}_{j_0})}{\|\vec{X}_{j_0} - \vec{X}_i\| - \|\vec{X}_{e_{min}} - \vec{X}_i\|}, \quad i = 1, 2, \ldots, N. \quad (6.161)$$

where $d(\vec{X}_j) = \sum_{i=1}^{N} W_i \|\vec{X}_j - \vec{X}_i\|$ and $\vec{X}_{j_0}$ is the current filter output.

(5) Update the filter weights:

$$W_i(n+1) = W_i(n) + \mu \Delta W_i, \quad i = 1, 2, \ldots, N, \quad (6.162)$$

where μ is the iteration step size.

This algorithm is a greedy approach since it determines the weight changes based on local characteristics. Despite the existence of several optimization algorithms like the one just shown, weighted vector medians have not significantly spread beyond image smoothing applications. The limitations of the weighted vector median are deep and their formulation needs to be revisited from its roots.

With this goal in mind, a revision of the principles of parameter estimation reveals that the weighted vector median emerges from the location estimate of independent (but not identically distributed) vector valued samples, where only the scale of each input vector sample varies. The multichannel components of each sample are, however, still considered mutually independent. In consequence, the weighted vector median in (6.158) is cross-channel blind.

In the following, more general vector median filter structures are presented. These structures are capable of capturing and exploiting the spatial and cross-channel correlations embedded in the data. First, the vector location estimate of samples that are assumed to be mutually correlated across channels but independent (but not identical) in time is revisited. This model leads to a multichannel median structure that is computationally simple, yet it exploits cross-channel information. The structure can be adapted to admit positive and negative weights using sign coupling.

6.7.3 Weighted Multichannel Median Filtering Structures

As it was done in the scalar case, the multivariate filtering structure is derived from the Maximum Likelihood estimation of location, this time in a multivariate signal space. Consider a set of independent but not identically distributed vector valued samples, each obeying a joint Gaussian distribution with the same location parameter $\vec{\mu}$,

$$f(\vec{X}_i) = \frac{1}{(2\pi)^{\frac{M}{2}} |\mathbb{C}_i|^{\frac{1}{2}}} e^{-\frac{1}{2}(\vec{X}_i - \vec{\mu})^T \mathbb{C}_i^{-1} (\vec{X}_i - \vec{\mu})}, \quad (6.163)$$

where $\vec{X}_i$ and $\vec{\mu}$ are all M-variate column vectors, and $\mathbb{C}_i^{-1}$ is the $M \times M$ cross-channel correlation matrix of the sample $\vec{X}_i$. The Maximum Likelihood estimation of location $\vec{\mu}$ can be derived as

$$\vec{\mu} = \left(\sum_{i=1}^{N} \mathbb{C}_i \right) \left(\sum_{i=1}^{N} \mathbb{C}_i^{-1} \vec{X}_i \right). \tag{6.164}$$

As in the univariate case, a general multivariate filtering structure results from the maximum likelihood estimator as

$$Y = \sum_{i=1}^{N} \mathbb{W}_i^T X_i, \tag{6.165}$$

where $\mathbb{W}_i^T = \left(\sum_{i=1}^{N} \mathbb{C}_i \right) \mathbb{C}_i^{-1}$.

An example of an optimal filter design algorithm for this linear filtering structure is shown by Robinson (1983) [168]. It presents only one inconvenience: the overwhelming size of the weight matrix. For instance, to filter a 3-channel color image using a 5x5 window requires the optimization of 225 weights. Alternative filter structures requiring lesser weights are needed. The following approach, proposed by Li et al. (2004) [130] provides such implementation.

Weighted Multichannel Median (WMM) Filter I   In most multichannel applications, the signals from sub-channels are often correlated. Further, the correlation structure between subchannels may often be stationary or at least quasi-stationary for a period of time. In these cases, the assumption that the correlation matrices $\mathbb{C}_i^{-1}$ differ only by a scale factor is valid, that is

$$\mathbb{C}_i^{-1} = q_i \mathbb{C}^{-1}. \tag{6.166}$$

The corresponding MLE is then

$$\vec{\mu} = \left(\sum_{i=1}^{N} q_i \mathbb{C}^{-1} \right)^{-1} \left(\sum_{i=1}^{N} q_i \mathbb{C}^{-1} \vec{X}_i \right), \tag{6.167}$$

where $\left(\sum_{i=1}^{N} q_i \mathbb{C}^{-1} \right)^{-1}$ is a normalization constant and $\sum_{i=1}^{N} q_i \mathbb{C}^{-1} \vec{X}_i$ provides the filtering structure. Removing the normalization constant, the filtering structure can be formulated as

$$\vec{Y} = \sum_{i=1}^{N} V_i \mathbb{W}^T \vec{X}_i \tag{6.168}$$

$$= \sum_{i=1}^{N} V_i \begin{bmatrix} W^{11} & \cdots & W^{M1} \\ \vdots & \ddots & \vdots \\ W^{1M} & \cdots & W^{MM} \end{bmatrix} \begin{bmatrix} X_i^1 \\ \vdots \\ X_i^M \end{bmatrix}, \tag{6.169}$$

where V_i is the (time/spatial) weight applied to the ith vector sample in the observation window and W_{ij} is the cross-channel weight exploiting the correlation between the

ith and jth components of a sample. The filter thus consists of $M^2 + N$ weights. In the example of a RGB image with a 5×5 window, the number of weights would be reduced from 225 to $3^2 + 25 = 34$.

Even though it is mathematically intractable to derive a similar result as in (6.169) from a multivariate Laplacian distribution, it is still possible to define a nonlinear multivariate filter by direct analogy by replacing the summations in (6.169) with median operators. This filter is referred to as the Weighted Multichannel Median (WMM) and is defined as follows (Li et al. (2004) [130]).

$$\vec{Y} = \text{MEDIAN}(|V_i| \diamond \text{sgn}(V_i)\vec{Q}_i \mid_{i=1}^{N}), \tag{6.170}$$

where

$$\vec{Q}_i = \begin{bmatrix} \text{MEDIAN}(|W^{j1}| \diamond \text{sgn}(W^{j1})X_i^j \mid_{j=1}^{M}) \\ \text{MEDIAN}(|W^{j2}| \diamond \text{sgn}(W^{j2})X_i^j \mid_{j=1}^{M}) \\ \vdots \\ \text{MEDIAN}(|W^{jM}| \diamond \text{sgn}(W^{jM})X_i^j \mid_{j=1}^{M}) \end{bmatrix} \tag{6.171}$$

is an M-variate vector. As it was stated before, there is no unique way of defining even the simplest median over vectors, in consequence, the outer median in (6.170) can have several different implementations. Due to its simplicity and ease of mathematical analysis, a suboptimal implementation of (6.170) can be used, where the outer median in (6.170) is replaced by a vector of marginal medians. Thus, the Marginal Weighted Multichannel Median (Marginal WMM) is defined as in Li et al. (2004) [130].

$$\vec{Y} = \begin{bmatrix} \text{MED}(|V_i| \diamond \text{sgn}(V_i)Q_i^1 \mid_{i=1}^{N}) \\ \text{MED}(|V_i| \diamond \text{sgn}(V_i)Q_i^2 \mid_{i=1}^{N}) \\ \vdots \\ \text{MED}(|V_i| \diamond \text{sgn}(V_i)Q_i^M \mid_{i=1}^{N}) \end{bmatrix}, \tag{6.172}$$

where $Q_i^l = \text{MED}(|W^{jl}| \diamond \text{sgn}(W^{jl})X_i^j \mid_{j=1}^{M})$ for $l = 1, \ldots, M$.

Weighted Multichannel Median (WMM) Filter II There are some applications where the initial assumption about stationarity stated in (6.166) may not be appropriate. The need of a simpler filtering structure remains, and this is why a more general structure for median filtering of multivariate signals is presented as in Li et al. (2004) [130]. In such case replace (6.166) by

$$\mathbb{C}_i^{-1} = \text{diag}(q_i)\mathbb{C}^{-1} \tag{6.173}$$

$$= \begin{bmatrix} q_i^1 C^{11} & \cdots & q_i^1 C^{1M} \\ \vdots & \ddots & \vdots \\ q_i^M C^{M1} & \cdots & q_i^M C^{MM} \end{bmatrix}. \tag{6.174}$$

In this case, the cross-channel correlation is not stationary, and the q_i^j represent the correlation between components of different samples in the observation window. The linear filtering structure reduces to

$$Y = \sum \text{diag}(\vec{V}_i) \mathbb{W} \vec{X}_i \tag{6.175}$$

$$= \sum \begin{bmatrix} V_i^1 & \cdots & 0 \\ \vdots & \ddots & \vdots \\ 0 & \cdots & V_i^M \end{bmatrix} \begin{bmatrix} W^{11} & \cdots & W^{M1} \\ \vdots & \ddots & \vdots \\ W^{1M} & \cdots & W^{MM} \end{bmatrix} \begin{bmatrix} X_i^1 \\ \vdots \\ X_i^M \end{bmatrix} \tag{6.176}$$

$$= \sum \begin{bmatrix} V_i^1 \sum W^{j1} X_i^j \\ \vdots \\ V_i^M \sum W^{jM} X_i^j \end{bmatrix}, \tag{6.177}$$

where V_i^l is the weight reflecting the influence of the lth component of the ith sample in the lth component of the output. The weights W^{ij} have the same meaning as in the WMM filter I. Using the same analogy used in the previous case, a more general weighted multichannel median filter structure can be defined as

$$\vec{Y} = \begin{bmatrix} \text{MEDIAN}(|V_i^1| \diamond \text{sgn}(V_i^1)\text{MEDIAN}(|W^{j1}|) \diamond \text{sgn}(W^{j1})X_i^j|_{j=1}^M)|_{i=1}^N \\ \vdots \\ \text{MEDIAN}(|V_i^M| \diamond \text{sgn}(V_i^M)\text{MEDIAN}(|W^{jM}|) \diamond \text{sgn}(W^{jM})X_i^j|_{j=1}^M)|_{i=1}^N \end{bmatrix}. \tag{6.178}$$

This structure can be implemented directly, that is, it does not require suboptimal implementations like the previous one. The number of weights increases, but is still significantly smaller compared to the number of weights required by the complete version of the filter in (6.165). For the image filtering example, the number of weights will be $M \times (N + M) = 84$.

In the following section, optimal adaptive algorithms for the structures in (6.170) and (6.178) are defined.

6.7.4   Filter Optimization

Assume that the observed process $\vec{X}(n)$ is statistically related to a desired process $\vec{D}(n)$ of interest, typically considered a transformed or corrupted version of $\vec{D}(n)$. The filter input vector at time n is

$$\mathbf{X}(n) = [\vec{X}_1(n) \, \vec{X}_2(n) \, \ldots \, \vec{X}_N(n)]^T,$$

where $\vec{X}_i(n) = [X_i^1(n) \, X_i^2(n) \, \ldots \, X_i^M(n)]^T$. The desired signal is $\vec{D}(n) = [D^1(n) \, D^2(n) \, \ldots \, D^M(n)]^T$.

Optimization for the WMM Filter I   Assume that the time/spatial dependent weight vector is $\mathbf{V} = [V_1\ V_2\ \ldots\ V_N]^T$, and the cross-channel weight matrix is

$$
\mathbb{W} = \begin{bmatrix} W^{11} & \cdots & W^{1M} \\ \vdots & \ddots & \vdots \\ W^{M1} & \cdots & W^{MM} \end{bmatrix}.
$$

Denote $Q_i^l = \text{MED}(|W^{jl}| \diamond \text{sgn}(W^{jl})X_i^j\ |_{j=1}^M)$ for $l = 1, \ldots, M$, then the output of the marginal WMM can be defined as

$$
\hat{\vec{D}} = [\hat{D}^1\ \hat{D}^2\ \ldots\ \hat{D}^M]^T,
$$

where

$$
\hat{D}^l = \text{MED}(|V_i| \diamond \text{sgn}(V_i)Q_i^l\ |_{i=1}^N) \quad l = 1, \ldots, M. \tag{6.179}
$$

Applying the real-valued threshold decomposition technique as in Section 6.3.1, we can rewrite (6.179) to be analyzable as follows,

$$
\begin{aligned}
\hat{D}^l &= \frac{1}{2} \int \text{MED}(|V_i| \diamond \text{sgn}(\text{sgn}(V_i)Q_i^l - p^l)\ |_{i=1}^N)dp^l \\
&= \frac{1}{2} \int \text{sgn}(\mathbf{V}_a^T \mathbf{G}^{p^l})dp^l,
\end{aligned} \tag{6.180}
$$

where $\mathbf{V}_a = [|V_1|\ |V_2|\ \ldots\ |V_N|]^T$ and $\mathbf{G}^{p^l} = [\text{sgn}(\text{sgn}(V_1)Q_1^l - p^l)\ \ldots\ \text{sgn}(\text{sgn}(V_N)Q_N^l - p^l)]^T$. Similarly, by defining

$$
\vec{W}_a^l = [|W^{1l}|\ |W^{2l}|\ \ldots\ |W^{Ml}|]^T,
$$

$$
\vec{S}_i^{q_i^l} = [\text{sgn}(\text{sgn}(W^{1l})X_i^1 - q_i^l)\ \ldots\ \text{sgn}(\text{sgn}(W^{Ml})X_i^M - q_i^l)]^T,
$$

the inner weighted medians will have the following thresholded representation

$$
\begin{aligned}
Q_i^l &= \frac{1}{2} \int \text{MED}(|W^{jl}| \diamond \text{sgn}(\text{sgn}(W^{jl})X_i^j - q_i^l)\ |_{j=1}^M)dq_i^l \\
&= \frac{1}{2} \int \text{sgn}((W_a^l)^T S_i^{q_i^l})dq_i^l.
\end{aligned} \tag{6.181}
$$

Under the Least Mean Absolute Error (LMA) criterion, the cost function to minimize is

$$
J_1(\mathbf{V}, \mathbb{W}) = E\{\|\vec{D} - \hat{\vec{D}}\|_1\} \tag{6.182}
$$

$$
= E\left\{ \sum_{l=1}^M |D^l - \hat{D}^l| \right\}. \tag{6.183}
$$

Substitute (6.180) in (6.183) to obtain

$$J_1(\mathbf{V}, \mathbb{W}) = E\left\{\frac{1}{2}\sum_{l=1}^{M}\left|\int \mathrm{sgn}(D^l - p^l) - \mathrm{sgn}(\mathbf{V}_a^T \mathbf{G}^{p^l})dp^l\right|\right\}. \qquad (6.184)$$

Since the integrals in (6.184) act on strictly positive or strictly negative functions, the absolute value operators and the integral operators can thus be interchanged, leading to

$$J_1(\mathbf{V}, \mathbb{W}) = E\left\{\frac{1}{2}\sum_{l=1}^{M}\int \left|\mathrm{sgn}(D^l - p^l) - \mathrm{sgn}(\mathbf{V}_a^T \mathbf{G}^{p^l})\right|dp^l\right\}. \qquad (6.185)$$

Due to the linearity of the expectation, the summation, and the integration operations, (6.185) can then be rewritten as

$$J_1(\mathbf{V}, \mathbb{W}) = \frac{1}{2}\sum_{l=1}^{M}\int E\left\{\left|\mathrm{sgn}(D^l - p^l) - \mathrm{sgn}(\mathbf{V}_a^T \mathbf{G}^{p^l})\right|\right\}dp^l. \qquad (6.186)$$

Furthermore, since the absolute value operators inside the expectations in (6.186) can only take values in the set $\{0, 2\}$, they can be replaced by a properly scaled square operator resulting in

$$J_1(\mathbf{V}, \mathbb{W}) = \frac{1}{4}\sum_{l=1}^{M}\int E\left\{\left(\mathrm{sgn}(D^l - p^l) - \mathrm{sgn}(\mathbf{V}_a^T \mathbf{G}^{p^l})\right)^2\right\}dp^l. \qquad (6.187)$$

Taking the derivative of the above equation with respect to $\vec{V}$ results in

$$\frac{\partial}{\partial \mathbf{V}}J_1(\mathbf{V}, \mathbb{W}) = -\frac{1}{2}\sum_{l=1}^{M}\int E\left\{e^{p^l}\frac{\partial}{\partial \mathbf{V}}\mathrm{sgn}(\mathbf{V}_a^T \mathbf{G}^{p^l})\right\}dp^l, \qquad (6.188)$$

where $e^{p^l} = \mathrm{sgn}(D^l - p^l) - \mathrm{sgn}(\mathbf{V}_a^T \mathbf{G}^{p^l})$. For convenience, the non-differentiable sign function is approximated by the hyperbolic tangent function $\mathrm{sgn}(x) \approx \tanh(x) = \frac{e^x - e^{-x}}{e^x + e^{-x}}$. Since its derivative $\frac{d}{dx}\tanh(x) = \mathrm{sech}^2(x) = \frac{4}{(e^x + e^{-x})^2}$, it follows that

$$\frac{\partial}{\partial \mathbf{V}}\mathrm{sgn}(\mathbf{V}_a^T \mathbf{G}^{p^l}) \approx \mathrm{sech}^2(\mathbf{V}_a^T \mathbf{G}^{p^l})\begin{bmatrix} \mathrm{sgn}(V_1)G_1^{p^l} \\ \mathrm{sgn}(V_2)G_2^{p^l} \\ \vdots \\ \mathrm{sgn}(V_N)G_N^{p^l} \end{bmatrix}, \qquad (6.189)$$

where $G_i^{p^l} = \mathrm{sgn}(\mathrm{sgn}(V_i)Q_i^l - p^l)$ for $i = 1, \ldots, N$. Substituting (6.189) in (6.188) leads to the updates for the V_i

Table 6.11 Summary of the LMA Algorithm for the marginal WMM Filter I

Initialization
$$\vec{V} = \vec{1}; \mathbb{W} = \mathbb{I}$$

For Loop

$$\vec{\hat{D}}(n) = \begin{bmatrix} \mathrm{MED}(|V_i| \diamond \mathrm{sgn}(V_i)Q_i^1 \mid_{i=1}^N) \\ \vdots \\ \mathrm{MED}(|V_i| \diamond \mathrm{sgn}(V_i)Q_i^M \mid_{i=1}^N) \end{bmatrix}$$

$$\vec{e}(n) = \vec{D}(n) - \vec{\hat{D}}(n)$$

For $i = 1, \dots, N$

$$V_i(n+1) = V_i(n) + \mu_v \mathrm{sgn}(V_i(n))\vec{e}^T(n)\vec{G}_i^{\hat{D}}(n)$$

where $\vec{G}_i^{\hat{D}} = [G_i^{\hat{D}^1} \ \cdots \ G_i^{\hat{D}^M}]^T$ and $G_i^{\hat{D}^l} = \mathrm{sgn}(\mathrm{sgn}(V_i)Q_i^l - \hat{D}^l)$ for $l = 1, \dots, M$.

For $s, t = 1, \dots, M$.

$$W^{st}(n+1) = W^{st}(n) + \mu_w \mathrm{sgn}(W^{st}(n))e^t(n)(\vec{V}^T(n+1)\vec{A}^s(n))$$

where $\vec{A}^s = [A_1^s \ A_2^s \ \cdots \ A_N^s]^T$, and $A_i^s = \delta(\mathrm{sgn}(V_i)Q_i^l - \hat{D}^l)\mathrm{sgn}(\mathrm{sgn}(W^{st})X_i^s - Q_i^t)$ for $i = 1, \dots, N$.

$$V_i(n+1) = V_i(n) + 2\mu_v \left\{ -\frac{\partial}{\partial V_i} J_1(\mathbf{V}, \mathbb{W}) \right\} \tag{6.190}$$

$$= V_i(n) + \mu_v \left(\sum_{l=1}^M \int E\left\{ e^{p^l} \mathrm{sech}^2(\mathbf{V}_a^T \mathbf{G}^{p^l})\mathrm{sgn}(V_i)G_i^{p^l} \right\} dp^l \right).$$

Using the instantaneous estimate for the gradient, and applying an approximation similar to the one in Section 6.3.2, we obtain the adaptive algorithm for the time dependent weight vector $\vec{V}$ of the marginal WMM filter as follows,

$$V_i(n+1) = V_i(n) + \mu_v \mathrm{sgn}(V_i(n))\vec{e}^T(n)\vec{G}_i^{\hat{D}}(n), \tag{6.191}$$

where $\vec{G}_i^{\hat{D}} = [G_i^{\hat{D}^1} \ \cdots \ G_i^{\hat{D}^M}]^T$ and $G_i^{\hat{D}^l} = \mathrm{sgn}(\mathrm{sgn}(V_i)Q_i^l - \hat{D}^l)$ for $l = 1, \dots, M$.
To derive the updates for $\mathbf{W}$, it is easy to verify that

$$\frac{\partial}{\partial W^{st}} J_1(\mathbf{V}, \mathbb{W}) \approx -\frac{1}{2} \sum_{l=1}^M \int E\left\{ e^{p^l} \mathrm{sech}^2(\mathbf{V}_a^T \mathbf{G}^{p^l})\mathbf{V}_a^T \left(\frac{\partial}{\partial W^{st}} \mathbf{G}^{p^l} \right) \right\} dp^l, \tag{6.192}$$

$$\frac{\partial}{\partial W^{st}} \vec{G}^{p^l} \approx \begin{bmatrix} \mathrm{sech}^2(\mathrm{sgn}(V_1)Q_1^l - p^l)\mathrm{sgn}(V_1)\frac{\partial}{\partial W^{st}}Q_1^l \\ \vdots \\ \mathrm{sech}^2(\mathrm{sgn}(V_N)Q_N^l - p^l)\mathrm{sgn}(V_N)\frac{\partial}{\partial W^{st}}Q_N^l \end{bmatrix}, \tag{6.193}$$

$$\frac{\partial}{\partial W^{st}} Q_i^l \approx -\frac{1}{2} \int \text{sech}^2 \left(\vec{W}_a^{l\,T} \vec{S}_i^{q_i^l} \right) \frac{\partial}{\partial W^{st}} \left(\vec{W}_a^{l\,T} \vec{S}_i^{q_i^l} \right) dq_i^l$$

$$\approx \begin{cases} \text{sgn}(W^{st})\text{sgn}(\text{sgn}(W^{st})X_i^s - q_i^t) & l = t \\[2ex] 0 & \text{otherwise}. \end{cases} \quad (6.194)$$

Notice that in (6.194), the derivative that introduces one more sech ^{2} term is omitted since it is insignificant compared to the other one.

After some mathematical manipulations and similar arguments as in Section 6.3.2, the adaptive algorithm for the cross-channel weight matrix W can be simplified as follows

$$W^{st}(n+1) = W^{st}(n) + \mu_w \text{sgn}(W^{st}(n)) e^t(n)(\vec{V}^T(n)\vec{A}^s(n)), \quad (6.195)$$

where $\mathbf{A}^s = [A_1^s \, A_2^s \, \ldots \, A_N^s]^T$, and $A_i^s = \delta(\text{sgn}(V_i)Q_i^l - \hat{D}^l)\text{sgn}(\text{sgn}(W^{st})X_i^s - Q_i^t)$ for $i = 1, \ldots, N$, where $\delta(x) = 1$ for $x = 0$ and $\delta(x) = 0$ otherwise. Table 6.11 summarizes the LMA algorithm for the marginal WMM filter I.

EXAMPLE 6.18

A RGB color image contaminated with 10% correlated salt-and-pepper noise is processed by the WVM filter, and the marginal WMM filter separately. The observation window is set to 3×3 and 5×5. The optimal weights for the marginal WMM filter are obtained first by running the LMA algorithm derived above over a small part of the corrupted image. The same section of the noiseless image is used as a reference. A similar procedure is repeated to optimize the weights of the WVM filter. The adaptation parameters are chosen in a way such that the average absolute error obtained in the training process is close to its minimum for each filter. The resulting weights are then passed to the corresponding filters to denoise the whole image. The filter outputs are depicted in Figures 6.45 and 6.46.

As a measure of the effectiveness of the filters, the mean absolute error of the outputs was calculated for each filter, the results are summarized in Table 6.12. Peak signal-to-noise ratio (PSNR) was also used to evaluate the fidelity of the two filtered images.

The statistics in Table 6.12 show that the marginal WMM filter outperforms the WVM filter in this color image denoising simulation by a factor of 3 in terms of the mean absolute error, or 8–11dB in terms of PSNR. Moreover, the output of the marginal WMM filter is almost salt and pepper noise free. As a comparison, the output of the WVM filter is visually less pleasant with many unfiltered outliers. Notice that the output of the marginal WMM filter with the 3×3 observation window preserves more image details than that of the 5×5 realization, and has a better PSNR though the mean absolute errors in the two cases are roughly the same.

Figure 6.45 Multivariate medians for color images in salt-and-pepper noise, $\mu = 0.001$ for the WVM, $\mu_v, \mu_w = 0.05$ for the marginal WMM. From left to right and top to bottom: noiseless image, contaminated image, WVM with 3×3 window, marginal WMM with 3×3 window. (Figure also appears in the Color Figure Insert)

Figure 6.46   Multivariate medians for color images in salt-and-pepper noise, $\mu = 0.001$ for the WVM, $\mu_v, \mu_w = 0.05$ for the marginal WMM (continued). From left to right: WVM with 5×5 window, marginal WMM with 5×5 window. (Figure also appears in the Color Figure Insert)

Table 6.12   Average MAE and PSNR of the output images.

Filter	MAE		PSNR (dB)	
	3×3	5×5	3×3	5×5
Noisy signal	0.1506		14.66	
WVM	0.0748	0.0732	23.41	27.74
marginal WMM	0.0248	0.0247	32.26	32.09

Figure 6.47 shows the optimum weights obtained for all the filters used in this example.

The noise generated for this example was cross-channel correlated. As a result, Figures 6.47 (c) and (f), show that the optimum cross-channel weights for the 3×3 and 5×5 window are very similar, since they are based on the same statistics. Figures

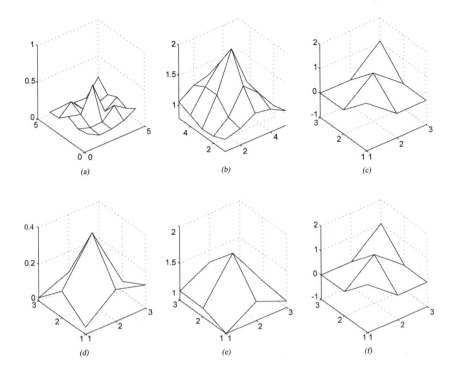

Figure 6.47 Optimum weights for the multivariate medians for color images in salt-and-pepper noise, (*a*) 5 × 5 WVM, (*b*) $\vec{V}$ in 5 × 5 marginal WMM I, (*c*) **W** in 5 × 5 marginal WMM I, (*d*) 3 × 3 WVM, (*e*) $\vec{V}$ in 3 × 3 marginal WMM I, (*f*) **W** in 3 × 3 marginal WMM I.

6.47 (*b*) and (*e*) show that, spatially, the marginal WMM filter I tries to emphasize the center sample of the window. This is an expected result since the noise samples are spatially independent. Finally, Figures 6.47 (*a*) and (*d*) show a distribution of the spatial weights that is not as smooth as the one shown in Figures 6.47 (*b*) and (*e*), this shows the negative effects that the cross channel correlation of the noise generates in the WVM filter.   ∎

Optimization of the WMM Filter II   The optimization process for the second WMM filtering structure is very similar to the one shown above (See Li et al. (2004) [130]). Assume that the time/spatial dependent weight matrix and the cross-channel weight matrix are:

$$\mathbf{V} = \begin{bmatrix} \vec{V}^1 \\ \vdots \\ \vec{V}^M \end{bmatrix} \quad \mathbb{W} = \begin{bmatrix} W^{11} & \cdots & W^{1M} \\ \vdots & \ddots & \vdots \\ W^{M1} & \cdots & W^{MM} \end{bmatrix}. \tag{6.196}$$

If Q_i^l and $\vec{S}_i^{q_i^l}$ are defined as in the previous case, the output of the filter can be written as

$$\hat{D} = [\hat{D}^1 \ \hat{D}^2 \ \dots \ \hat{D}^M]^T,$$

where

$$\hat{D}^l = \text{MED}(|V_i^l| \diamond \text{sgn}(V_i^l)Q_i^l \ |_{i=1}^N) \quad l = 1, \cdots, M \tag{6.197}$$

$$= \frac{1}{2} \int \text{MED}(|V_i^l| \diamond \text{sgn}(V_i^l)Q_i^l \ |_{i=1}^N)dp^l$$

$$= \frac{1}{2} \int \text{sgn}(\vec{V_a^l}^T \vec{G}^{p^l})dp^l, \tag{6.198}$$

where

$$\mathbf{V}_a^l = [|V_1^l| \ |V_2^l| \ \cdots \ |V_N^l|]^T \ ,$$

$$\mathbf{G}^{p^l} = [\text{sgn}(\text{sgn}(V_1^l)Q_1^l - p^l) \ \cdots \ \text{sgn}(\text{sgn}(V_N^l)Q_N^l - p^l)]^T \ .$$

Under the Least Mean Absolute (LMA) criterion, the cost function to minimize will be just like (6.187).

Taking the derivative of the above equation with respect to $\mathbf{V}$ and using similar approximations to the ones used on the previous case results in

$$\frac{\partial}{\partial V_s^t} J_1(\mathbf{V}, \mathbb{W}) = -\frac{1}{2} \int E\left\{ e^{p^t} \text{sech}^2((\mathbf{V}_a^l)^T \mathbf{G}^{p^t})\text{sgn}(V_s^t)G_s^{p^t} \right\} dp^t \tag{6.199}$$

where $e^{p^t} = \text{sgn}(D^t - p^t) - \text{sgn}((\mathbf{V}_a^l)^T \mathbf{G}^{p^t})$. Using instantaneous estimates for the expectation the updates for $\mathbf{V}$ result in

$$V_s^t(n+1) = V_s^t(n) + \mu_v e^t(n)\text{sgn}(V_s^t(n))\text{sgn}(\text{sgn}(V_s^t(n))Q_s^t(n) - \hat{D}^l(n))$$
$$\tag{6.200}$$

$$= V_s^t(n) + \mu_v e^t(n)\text{sgn}(V_s^t(n))G_s^{\hat{D}^t} \tag{6.201}$$

On the other hand, the updates for $\mathbb{W}$ are given by:

$$W^{st}(n+1) = W^{st}(n) + \mu_w \text{sgn}(W^{st}(n))e^t(n)((\mathbf{V}^l)^T(n)\mathbf{A}^{st}(n)), \tag{6.202}$$

that is basically the same as (6.195) with the difference that $\mathbf{V}$ is now a matrix and $\mathbf{A}^{st} = [\delta(\text{sgn}(V_i^t)Q_i^t - \hat{D}^t)\text{sgn}(\text{sgn}(W^{st})X_i^s - Q_i^t)|_{i=1}^N]$.

EXAMPLE 6.19 (ARRAY PROCESSING WITH THE WMM FILTER II)

To test the effectiveness of the WMM filter II, a simple array processing problem with real-valued signals is used. The system shown in Figure 6.48 is implemented.

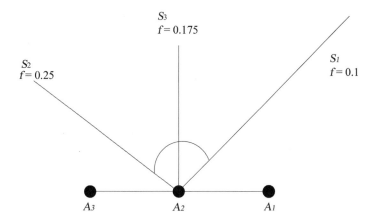

Figure 6.48 Array of Sensors

It consists of a 3 element array and 3 sources in the farfield of the array transmitting from different directions and at different frequencies as indicated in the figure. The goal is to separate the signals from all sources using the array in the presence of alpha stable noise.

In order to do so, a WVM filter, a marginal WMM filter and a WMM filter II all with a window size of 25 are used. The filters are optimized using the algorithms described earlier in this section, with a reference signal whose components are noiseless versions of the signals emitted by the sensors.

The results obtained are summarized in Figure 6.49 and Table 6.13. Figure 6.49

Table 6.13 Average MAE of the output signals.

Filter	MAE
Noisy signal	0.8248
WVM	0.6682
Marginal WMM I	0.5210
WMM II	0.3950

shows that the WMM filter II is able to extract the desired signals from the received signals at the sensors successfully. The WVM filter and the marginal WMM filter I are unable to do so. Linear filters were implemented with adaptive algorithms

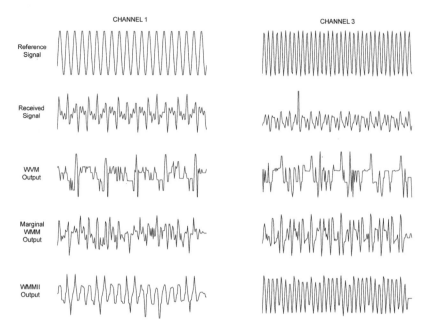

Figure 6.49 Input and output signals for array processing with multivariate medians. Each column corresponds to a channel (only channels one and three are shown) and the rows represent: the reference signal, the signal received at the sensors, the output of the WVM filter, the output of the marginal WMM filter and the output of the WMM filter II.

to optimize them for this problem but the impulsiveness of the noise made the optimization algorithms diverge. The final weights obtained with the optimization algorithms are shown in Figures 6.50 and 6.51.

Figure 6.50*a* shows why the WVM filter is not able to obtain a good result for this problem. The optimal weights are erratically distributed and in consequence, the output looks nothing like the desired signal. A similar conclusion can be reached for the weights of the marginal WMM filter in Fig. 6.50*b*. The outer weights of the WMM filter II are shown in in Figs. 6.50*c–e*, each one corresponding to a different channel of the signals. It can be seen how the weights show a certain periodicity with frequencies related to the ones of the signals we want to extract in each channel.

The inner weights for the marginal WMM filter and the WMM filter II are shown in Fig. 6.51*a* and *b* respectively. It can be seen that the extra time correlation included in this problem completely distorts the weights **W** of the marginal WMM filter. The weights **W** of the WMM filter II, on the other hand, reflect the cross-channel correlation of the signals. ∎

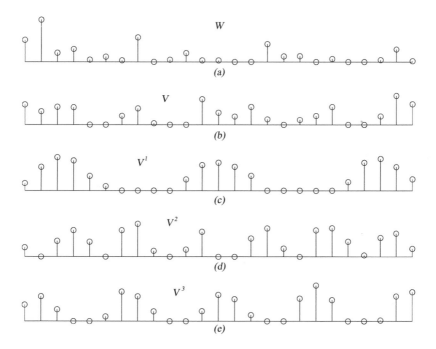

Figure 6.50 Optimized weights for the multivariate medians in the array processing example: (*a*) WVM, (*b*) Marginal WMM filter I $\vec{V}$, (*c*) first row of $\mathbf{V}$ for WMM filter II , (*d*) Second row of $\mathbf{V}$, (*e*) Third row of $\mathbf{V}$

Problems

6.1 Show that the ML estimate of location for samples observing a multivariate Gaussian distribution as in (6.1) reduces to $\hat{\beta} = \mathbf{W}^T \mathbf{X}$ as shown in (6.2).

6.2 Prove that the integral operation in (6.58) can be taken out of the median operation leading to (6.59).

6.3 Prove that the absolute value and integral operator in (6.70) can be interchanged leading to (6.71).

6.4 Show that $\frac{\partial}{\partial \mathbf{W}} \mathrm{sgn}\left(\mathbf{W}_a^T \mathbf{s}^q\right)$ in (6.75) is equivalent to the expression in (6.76).

6.5 Prove the BIBO stability of recursive WM filters stated in property (6.1)

6.6 Show that $\frac{\partial}{\partial W_i} \mathrm{sgn}(\mathbf{W}_a^T \mathbf{P}_R^q)$ in (6.146) reduces to (6.147).

6.7 Show (using sample selection probabilities) that a center weighted median filter with $W_c \geq N$ is an identity operator (i.e., the sample selection probability of the center sample is 1), where N is the number of taps, N odd.

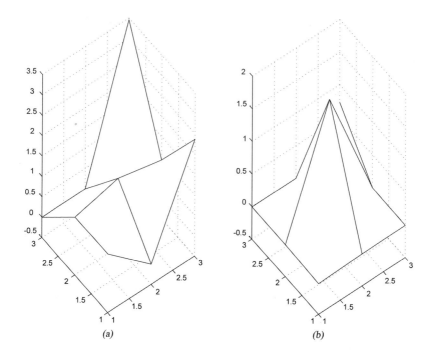

Figure 6.51 Optimized inner weights for the Multivariate medians in the array processing example. (*a*) Marginal WMM filter I, (*b*) WMM filter II.

6.8   Find the closest linear filter to the weighted median filter given by the weight vector: $\mathbf{W} = [1,\ 2,\ 3,\ 2,\ 1]$.

6.9   Show that

(a) The maximum likelihood estimator of location for the distribution in (6.163) equals

$$\vec{\mu} = \left(\sum_{i=1}^{N} \mathbb{C}_i \right) \left(\sum_{i=1}^{N} \mathbb{C}_i^{-1} \vec{X}_i \right). \tag{6.203}$$

(b) The same MLE reduces to

$$\vec{\mu} = \left(\sum_{i=1}^{N} q_i \mathbb{C}^{-1} \right)^{-1} \left(\sum_{i=1}^{N} q_i \mathbb{C}^{-1} \vec{X}_i \right), \tag{6.204}$$

under the condition in (6.166).

6.10   Given a vector weighted median filter defined by the weights $W_j|_{j=1}^{N}$. Show that ΔW_i as defined in (6.161) is the change required in the weight W_i to make the output of the vector weighted median change from the value $\vec{X}_{j_0}$ to the value $\vec{X}_{e_{min}}$.

7

Linear Combination of Order Statistics

Given the ordered set $X_{(1)}, X_{(2)}, \ldots, X_{(N)}$ corresponding to the N observation samples $X_1, X_2, \ldots, X_N$, an alternative approach to use the order statisticsorder statistics is to work with linear combinations of these. Simple linear combinations, of the form

$$Y = \sum_{i=1}^{N} W_i X_{(i)} \tag{7.1}$$

are known as L-statistics or L-estimates. Arnold et al. (1992) [16], David (1982) [58], and Hosking (1998) [102] describe their long history in statistics. L-statistics have a number of advantages for use in signal processing. If the random variable Z is a linear transformation of X, $Z = \alpha + \gamma X$, for $\gamma > 0$, then the order statistics of X and Z satisfy $Z_{(i)} = \alpha + \gamma X_{(i)}$, and L-statistics computed from them satisfy $Y^{(Z)} = \alpha \sum W_i + \gamma Y^{(X)}$. Thus, $Y^{(Z)} = \alpha + \gamma Y^{(X)}$ if $\sum W_i = 1$, a required condition to use L-statistics for location estimates. In addition, by appropriate choice of the weights W_i, it is possible to derive robust estimators whose properties are not excessively dependent on correct statistical assumptions. L-estimates are also a natural choice for censored estimation where the most extreme order statistics are ignored. As it is described later in this chapter, useful generalizations of L-statistics are obtained if the weights in (7.1) are made data dependent or if functions of order statistics are used in the linear combination. In particular, several hybrid filter classes are presented where L-filter attributes are complemented with properties of linear FIR filters.

7.1 *L*-ESTIMATES OF LOCATION

In the location estimation problem, the observation samples are of the form $X_i = \beta + Z_i$, where β is the constant location parameter to be estimated, and where Z_i is a zero mean sequence of independent and identically distributed noise samples with variance σ^2. For the sake of simplicity, we assume that the noise is symmetrically distributed. Given the observations $X_1, X_2, \ldots, X_N$ and the corresponding order statistics $X_{(i)}$, $i = 1, 2 \ldots, N$, the goal is to design an *L*-estimate of the location parameter β. Lloyd (1952) [134] showed how the location parameter can be more efficiently estimated with a linear combination of ordered samples than by the classical sample mean. The corresponding mean square error will always be smaller than, or equal to, that obtained with the sample mean or sample median.

Simplifying Lloyd's contribution that dealt with the simultaneous estimation of location and scale parameters, Bovik et al. (1983) [38] considered the restoration of a noisy constant signal, say β, with an *L*-estimate designed to minimize the mean square error. Using the fact that the unknown parameter β is constant, the simplified approach starts by relating the order statistics of the observations and the noise as $X_{(i)} = \beta + Z_{(i)}$. The *L*-estimate of location is then

$$\hat{\beta} = \sum_{i=1}^{N} W_i X_{(i)} \tag{7.2}$$

where the W_is form an N-dimensional vector of real coefficients. Further, the estimate in (7.2) is required to be unbiased such that $E\{\hat{\beta}\} = \beta$ leading to

$$
\begin{aligned}
E\{\hat{\beta}\} &= \sum_{i=1}^{N} W_i \, E\{X_{(i)}\} \\
&= \sum_{i=1}^{N} W_i \, E\{\beta + Z_{(i)}\} \\
&= \beta \sum_{i=1}^{N} W_i + \sum_{i=1}^{N} W_i \, E\{Z_{(i)}\}.
\end{aligned}
\tag{7.3}
$$

Assuming the noise samples $Z_{(i)}$ are independent, identically distributed, and zero mean with a symmetric probability density function $(f(z) = f(-z))$, then $E[Z_{(i)}] = -E[Z_{(N-i+1)}]$. Using this fact in (7.3), the estimate is unbiased if the weights are symmetric $(W_{N-i+1} = W_i)$ and if $\sum_{i=1}^{N} W_i = 1$. The above can be written in vector notation as

$$\hat{\beta} = \mathbf{W}^T(\beta \, \mathbf{e} + \mathbf{Z}_L) \tag{7.4}$$

where $\mathbf{e}$ is the N-long one-valued vector $\mathbf{e} = [1, 1, \ldots, 1]^T$, and where $\mathbf{Z}_L$ is the vector comprised of the noise component order statistics

$$\mathbf{Z}_L = [Z_{(1)}, Z_{(2)}, \ldots, Z_{(N)}]^T. \tag{7.5}$$

The mean-square estimation error can be written as

$$
\begin{aligned}
J(\mathbf{W}) &= E\left[\mid \beta - \hat{\beta} \mid^2\right] \\
&= E\left[\mid \mathbf{W}^T\mathbf{Z}_L \mid^2\right] \\
&= \mathbf{W}^T\mathbf{R}_L\mathbf{W},
\end{aligned} \tag{7.6}
$$

where the unbiasedness constraint $\mathbf{W}^T\mathbf{e} = 1$ was utilized, and where the correlation matrix $\mathbf{R}_L$ is given by

$$
\mathbf{R}_L = \begin{bmatrix}
EZ_{(1)}^2 & EZ_{(1)}Z_{(2)} & \cdots & EZ_{(1)}Z_{(N)} \\
EZ_{(2)}Z_{(1)} & EZ_{(2)}^2 & & EZ_{(2)}Z_{(N)} \\
\vdots & \vdots & \ddots & \vdots \\
EZ_{(N)}Z_{(1)} & EZ_{(N)}Z_{(2)} & \cdots & EZ_{(N)}^2
\end{bmatrix} \tag{7.7}
$$

where $EZ_{(i)}Z_{(j)}$ is the correlation moment of the ith and jth noise order statistics. The minimization of $J(\mathbf{W})$, subjected to the unbiasedness constraint $\mathbf{W}^T\mathbf{e} = 1$, is a quadratic optimization problem that can be solved by the method of Lagrange multipliers. The Lagrangian function of the constrained mean-square-error cost function $J(\mathbf{W})$ is written as

$$F(\lambda, \mathbf{W}) = \mathbf{W}^T\mathbf{R}_L\mathbf{W} + \lambda(\mathbf{W}^T\mathbf{e} - 1). \tag{7.8}$$

Taking the derivative of $F(\lambda, \mathbf{W})$ with respect to the weight vector $\mathbf{W}$ and setting it to zero leads to

$$2\mathbf{R}_L\mathbf{W} + \lambda\mathbf{e} = \mathbf{0}. \tag{7.9}$$

In order to solve for λ, the terms above are multiplied by $\mathbf{e}^T\mathbf{R}_L^{-1}$ resulting in

$$\lambda = \frac{-2}{\mathbf{e}^T\mathbf{R}_L^{-1}\mathbf{e}}, \tag{7.10}$$

which is then used in (7.9) to obtain the optimal L-estimate of location weights

$$\mathbf{W}_0 = \frac{\mathbf{R}_L^{-1}\mathbf{e}}{\mathbf{e}^T\mathbf{R}_L^{-1}\mathbf{e}}, \tag{7.11}$$

with the corresponding mean square error

$$J(\mathbf{W}_0) = J_{min} = \frac{1}{\mathbf{e}^T\mathbf{R}_L^{-1}\mathbf{e}}. \tag{7.12}$$

The estimate in (7.11) is optimal only among the restricted class of L-estimates. There is no assurance that other estimators, such as maximum likelihood estimators, may be more efficient. Since the sample mean is included in the class of *L*-estimates, optimal *L*-estimates will always do better than, or at least equal to, the linear estimate. The conditions that determine when an *L*-estimate will improve on the sample mean can, in fact, be determined [134].

Given that the correlation matrix $E\{\mathbf{Z}_L\mathbf{Z}_L^T\}$ is positive definite, by the use of a Cholesky decomposition it can be represented by a product of two triangular matrices

$$E\{\mathbf{Z}_L\mathbf{Z}_L^T\} = \mathbf{LL}^T. \tag{7.13}$$

Premultiplying and postmultiplying the matrix $\mathbf{R}_L$ by the vectors $\mathbf{e}^T$ and $\mathbf{e}$ leads to

$$
\begin{aligned}
\mathbf{e}^T E\{\mathbf{Z}_L\mathbf{Z}_L^T\}\mathbf{e} &= E\{\mathbf{e}^T\mathbf{Z}_L\mathbf{Z}_L^T\mathbf{e}\} \\
&= E\{\mathbf{e}^T\mathbf{Z}\,\mathbf{Z}^T\mathbf{e}\} \\
&= N\sigma^2
\end{aligned}
\tag{7.14}
$$

where $\mathbf{Z} = [Z_1, Z_2, \ldots, Z_N]^T$ is the vector of unordered noise samples, and where we have used the fact that the sum of the elements in $\mathbf{Z}_L$ is the same regardless of the order in which the elements are summed. From (7.13), it is also found that (7.14) is given by

$$
\begin{aligned}
\mathbf{e}^T\mathbf{LL}^T\mathbf{e} &= \mathbf{h}^T\mathbf{h} \\
&= \sum_{i=1}^N h_i^2
\end{aligned}
\tag{7.15}
$$

where $\mathbf{h} = \mathbf{L}^T\mathbf{e}$. Similarly, letting $\mathbf{k} = \mathbf{L}^{-1}\mathbf{e}$

$$
\begin{aligned}
\mathbf{e}^T E\{\mathbf{Z}_L\mathbf{Z}_L^T\}^{-1}\mathbf{e} &= \mathbf{e}^T \left(\mathbf{L}^{-1}\right)^T \left(\mathbf{L}^{-1}\right)\mathbf{e} \\
&= \mathbf{k}^T\mathbf{k} = \sum_{i=1}^N k_i^2 \\
&= J_{min}^{-1}.
\end{aligned}
\tag{7.16}
$$

Moreover, since

$$
\begin{aligned}
\sum_{i=1}^N h_i k_i &= \mathbf{h}^T\mathbf{k} \\
&= \mathbf{e}^T\mathbf{LL}^{-1}\mathbf{e} = N,
\end{aligned}
\tag{7.17}
$$

and invoking the Cauchy-Schwartz inequality

$$\left(\sum_{i=1}^N h_i k_i\right)^2 \le \sum_{i=1}^N h_i^2 \cdot \sum_{i=1}^N k_i^2 \tag{7.18}$$

Table 7.1 Optimal weights for the L-estimate of location $N = 9$ (adapted from [38])

Distribution	Weights								
	W_1	W_2	W_3	W_4	W_5	W_6	W_7	W_8	W_9
Uniform	0.5	0	0	0	0	0	0	0	0.5
Gaussian	0.11	0.11	0.11	0.11	0.11	0.11	0.11	0.11	0.11
Laplacian	-0.018	0.029	0.069	0.238	0.364	0.238	0.069	0.029	-0.018

equations (7.15) and (7.17) lead to

$$\sum_{i=1}^{N} k_i^2 = \mathbf{e}^T E\{\mathbf{Z}_L \mathbf{Z}_L^T\}^{-1} \mathbf{e} \geq \frac{N}{\sigma^2}. \tag{7.19}$$

Hence, the minimum L-estimate mean square error satisfies

$$J_{min} \leq \frac{\sigma^2}{N}, \tag{7.20}$$

with equality if and only if $\mathbf{h} = \mathbf{k}$ corresponding to the condition

$$E\{\mathbf{Z}_L \mathbf{Z}_L^T\}\mathbf{e} = \mathbf{e}. \tag{7.21}$$

The L-filter estimate will thus perform better than the sample mean whenever the row sums (or column sums) of the correlation matrix $E\{\mathbf{Z}_L \mathbf{Z}_L^T\}$ are not equal to one [134].

The optimal weights of the L-estimate will have markedly different characteristics depending on the parent distribution of the noise samples Z_i. Bovik et al. computed the optimal L-filter coefficients for various i.i.d. noise distributions. Table 7.1 shows the optimal coefficients for the L-location estimate when the additive noise ranges from uniformly distributed, to Gaussian distributed, to Laplacian distributed, for $N = 9$. The impulsive nature of the Laplacian noise is referred to as being heavy-tailed, while the bounded shape of the uniform distribution is referred to as being short-tailed.

Table 7.1 illustrates the L filter weights as the impulsive characteristics of the parent distribution are varied. For a Laplacian distribution, the center order statistics are emphasized since outliers disturbing the estimation will be placed in the outer order statistics. On the other hand, the L estimate in uniformly distributed noise reduces to the midrange, which relies on the first and last order statistic alone. Finally, as expected, in Gaussian noise, all order statistics are equally weighted leading the estimate to the sample mean. Tables for the optimal coefficients of L-filters for some common distributions are listed by Sarham and Greenberg (1962) [33].

Nearly Best L-Estimates   The disadvantage of Lloyd's and Bovik et al.'s approaches is that they require the tabulation of covariances of order statistics for every distribution and sample size for which it is used. Combinatorial formulas for their computation can lead to unfeasible complexity even for small sample sizes. Blom's (1962) [33] "unbiased nearly best linear estimate" overcomes this disadvantage by using asymptotic approximations of the covariances. A similar approach was used by Oten and Figueiredo (2003) [153] where a Taylor's expansion approximation is used. Suppose the L-estimate in (7.1) is location invariant and is to estimate a constant signal β embedded in zero mean noise with a symmetric density function $f(z)$ and distribution function $F(z)$.

The L-estimate in (7.1) can be rewritten as:

$$Y = \frac{\sum_{i=1}^{N} \omega_i X_{(i)}}{\sum_{i=1}^{N} \omega_i}. \tag{7.22}$$

The original weights W_i are related to the weights ω_i as follows:

$$W_i = \frac{\omega_i}{\sum_{i=1}^{N} \omega_i}. \tag{7.23}$$

The filter coefficients ω_i are symmetric about the median, that is $\omega_i = \omega_{(N-i+1)}$. Therefore, the filter can be rewritten using only half the coefficients as

$$\begin{aligned} Y &= \frac{\sum_{i=1}^{r} b_i \left[X_{(i)} + X_{(N-i+1)} \right]}{2 \sum_{i=1}^{r} b_i} \\ &= \frac{\sum_{i=1}^{r} b_i \left[2\beta + Z_{(i)} + Z_{(N-i+1)} \right]}{2 \sum_{i=1}^{r} b_i} \end{aligned} \tag{7.24}$$

where $r = \lceil \frac{N}{2} \rceil$ and $b_i = \omega_i$ except that for N odd $b_r = \frac{\omega_r}{2}$.

Following a procedure similar to that used in Section 7.1 the optimal coefficients $\mathbf{b} = [b_1, b_2, \ldots, b_r]^T$ are found as:

$$\mathbf{b}^T = \mathbf{C}^{-1}\mathbf{e} \tag{7.25}$$

where $\mathbf{C} = [c_{ij}]$ is the covariance matrix of the noise samples $Z_{(i)} + Z_{(N-i+1)}$. $\mathbf{C}$ is related to the covariance matrix $\mu = [\mu_{(ij):N}]$ of the order statistics of the noise samples $Z_{(i)}$ since

$$c_{ij} = \frac{\mu_{(ij):N} + \mu_{(i,N-j+1):N}}{2}. \tag{7.26}$$

A procedure to calculate the covariance matrix $\mathbf{C}$ is developed in the following. Let U be a uniformly distributed random variable in the range $(0, 1)$. The random variable obtained through the transformation $F^{-1}(U)$ obeys the distribution function $F(X)$, namely

$$X \overset{\mathrm{d}}{=} F^{-1}(U)$$

where $\stackrel{\mathrm{d}}{=}$ denotes equality in "distribution" and F^{-1} denotes the inverse of the distribution function. In general, if $X_1, X_2, \ldots, X_N$ are i.i.d. random variables taken from the distribution $F(X)$, and $U_1, U_2, \ldots, U_N$ are i.i.d. uniform $(0,1)$ random variables, then

$$(X_{(1)}, \ldots, X_{(N)}) \stackrel{\mathrm{d}}{=} (F^{-1}(U_{(1)}), \ldots, F^{-1}(U_{(N)})).$$

Applying a Taylor expansion to $F^{-1}(U_{(i)})$ around the point $\lambda_i = EU_{(i)} = \frac{i}{N+1}$ and neglecting the higher order terms the following representation is obtained

$$X_{(i)} \approx F^{-1}\left(\frac{i}{N+1}\right) + \left(U_{(i)} - \frac{i}{N+1}\right) F^{-1(1)}\left(\frac{i}{N+1}\right) \tag{7.27}$$

where $F^{-1(1)}\left(\frac{i}{N+1}\right)$ denotes the derivative of $F^{-1}(u)$ evaluated at $u = \frac{i}{N+1}$. Taking expectation on both sides of (7.27) leads to the approximate first order moment of the ith order statistic, such that

$$EX_{(i)} \approx F^{-1}\left(\frac{i}{N+1}\right) = \kappa_i. \tag{7.28}$$

Similarly, using (7.27) and (7.28) in $\mathrm{cov}(X_{(i)}, X_{(j)}) = EX_{(i)}X_{(j)} - EX_{(i)}EX_{(j)}$ leads to

$$\mu_{(i,j):N} = \mathrm{cov}(X_{(i)}, X_{(j)}) \approx \mathrm{cov}(U_{(i)}, U_{(j)}) F^{-1(1)}(\lambda_i) F^{-1(1)}(\lambda_j). \tag{7.29}$$

Since

$$\mathrm{cov}(U_{(i)}, U_{(j)}) = \frac{i(N+1-j)}{(N+1)^2(N+2)} = \frac{\lambda_i(1-\lambda_j)}{n+2} \tag{7.30}$$

and

$$F^{-1(1)}(u) = \frac{dx}{du} = \frac{1}{f(x)} = \frac{1}{f(F^{-1}(u))}, \tag{7.31}$$

the covariance approximation reduces to

$$\mu_{(i,j):N} = \mathrm{cov}(X_{(i)}, X_{(j)}) \approx \frac{i(N+1-j)}{(N+1)^2(N+2)} F^{-1(1)}\left(\frac{i}{N+1}\right) F^{-1(1)}\left(\frac{j}{N+1}\right)$$

$$\approx \frac{\lambda_i(1-\lambda_j)}{(N+2)f(\kappa_i)f(\kappa_j)}. \tag{7.32}$$

Here, its assumed that $F^{-1(1)}(u)$ exists at $u = \frac{i}{N+1}$. Consequently, $F(X)$ and $f(X)$ must be differentiable and continuous at $X = \frac{i}{N+1}$, respectively. Empirical results have shown that the covariance approximations in (7.32) are more precise if the term

$N + 2$ in the denominator is replaced by N. Using either representation does not make any difference since that term will be cancelled during the normalization of the coefficients. In the following, $N + 2$ will be replaced by N for simplicity. The c_{ij}s can be calculated now using (7.26) and (7.32) as:

$$
\begin{aligned}
c_{ij} &= \frac{\frac{\lambda_i(1-\lambda_j)}{Nf(\kappa_i)f(\kappa_j)} + \frac{\lambda_i(1-\lambda_{N-j+1})}{Nf(\kappa_i)f(\kappa_{N-j+1})}}{2} \\
&= \frac{\lambda_i}{2Nf(\kappa_i)}\left(\frac{1-\lambda_j}{f(\kappa_j)} + \frac{1-\lambda_{N-j+1}}{f(\kappa_{N-j+1})}\right) \\
&= \frac{\lambda_i}{Nf(\kappa_i)f(\kappa_j)}
\end{aligned}
\tag{7.33}
$$

where the fact that $\lambda_j = 1 - \lambda_{N-j+1}$ and $f(\kappa_j) = f(\kappa_{N-j+1})$ was used.

Once the matrix $\mathbf{C}$ has been approximated, its inverse $\mathbf{C}^{-1} = [\bar{c}_{ij}]$ can be calculated as well as:

$$
\begin{aligned}
\bar{c}_{ii} &\approx 4Nf^2(\kappa_i)(N+1) & 1 \le i \le r-1 \\
\bar{c}_{rr} &\approx 2Nf^2(\kappa_r)(N+1) & \\
\bar{c}_{ij} &\approx -2Nf(\kappa_i)f(\kappa_j)(N+1) & i = j-1 = 1, \ldots, r-1 \\
\bar{c}_{ij} &= 0 & |i-j| > 1.
\end{aligned}
\tag{7.34}
$$

These values can be substituted in (7.25) to calculate the coefficients b_i as

$$
b_i = \sum_{j=1}^{r} \bar{c}_{ij}.
\tag{7.35}
$$

In order to calculate the output of the filter, (7.24) must be evaluated. Since the coefficients b_i appear in the numerator and denominator of (7.24) the constants obtained in (7.35) can be eliminated leading to

$$
\begin{aligned}
b_1 &= f(\kappa_1)[-2f(\kappa_1) + f(\kappa_2)] \\
b_i &= f(\kappa_i)[f(\kappa_{i-1}) - 2f(\kappa_i) + f(\kappa_{i+1})], \quad i = 2, \ldots, r-1 \\
b_r &= f(\kappa_r)[f(\kappa_{r-1}) - f(\kappa_r)].
\end{aligned}
\tag{7.36}
$$

Having the coefficients $b_i, \ldots, b_r$, the actual filter coefficients can be calculated using (7.23). This method was used to recalculate the coefficients in Table 7.1 obtaining the values in Table 7.2.

7.2 L-SMOOTHERS

L-smoothers are obtained when *L*-estimates are computed at each location of a running window and, as such, they are more general than running median smoothers, as shown in Nodes and Gallagher (1982) [150], Bovik et al. (1983) [38], and Bednar and Watt

Table 7.2   Approximated weights for the *L*-estimate of location $N = 9$ (adapted from [153])

Distribution	Weights								
	W_1	W_2	W_3	W_4	W_5	W_6	W_7	W_8	W_9
Uniform	0.5	0	0	0	0	0	0	0	0.5
Gaussian	0.130	0.107	0.105	0.105	0.106	0.105	0.105	0.107	0.130
Laplacian	0.0	0.0	0.0	0.0	1	0.0	0.0	0.0	0.0

(1984) [30]. The application of *L*-smoothers to image denoising is illustrated in Pitas and Venestanopoulos (1992) [162], and Kotropoulos and Pitas (1996) [118]. Adaptive optimization of *L*-smoothers is discussed in Kotropoulos and Pitas (1992) [117], Pitas and Venestanopoulos (1991) [161], and Clarkson and Williamson (1992) [52].

DEFINITION 7.1 (RUNNING *L*-SMOOTHERS) *Given a set of N real valued weights* $W_1, W_2, \ldots, W_N$ *assigned to the order statistics* $X_{(1)}, X_{(2)}, \ldots, X_{(N)}$, *in the running window* $\mathbf{X}(n) = [X_1(n), X_2(n), \ldots, X_N(n)]^T$, *the L-smoother output is*

$$Y(n) = \sum_{i=1}^{N} W_i X_{(i)}. \tag{7.37}$$

Note that if the weights are chosen uniformly as $W_i = 1/N$, the *L*-smoother reduces to the running mean. In fact, the running mean is the only smoother that is both linear and an *L*-smoother. Another example of an *L*-smoother with a long history is the running range $Y(n) = X_{(1)} - X_{(N)}$ often used as an informal measure of the dispersion of the sample set [58].

EXAMPLE 7.1 (AM DEMODULATION)

Perhaps the simplest *L*-smoother is found by zero-weighting all order statistics except for one. That is $W_i = 0$ for $i = 1, \ldots, r - 1, r + 1, \ldots, N$ and $W_r = 1$ leading to the *rank*-smoother

$$Y(n) = r\text{th Largest Sample of } [X_1(n), X_2(n), \ldots, X_N(n)],$$

with the median smoother as a special case. Notably, the nonlinearity of even this simple smoother can produce useful results [150]. The demodulation of AM signals is one such example where the output of the rank-smoother is selected so as to tract the envelope function of the AM signal. Figure 7.1 depicts the AM demodulation of a

5-kHz tone signal on a 31-kHz carrier and sampled at 250kHz using an eighth-ranked-order operation with a running window of size 9. Figure 7.1*a* shows the envelope detection when no noise is present, whereas Figure 7.1*b* shows the envelope detection in an impulsive noise environment. Note that while impulsive noise is very disruptive with most envelope detectors, the output of the rank-order filter is hardly perturbed by the noise.

∎

Trimmed means are yet another subclass of *L*-smoothers that have received significant attention in statistics, see Hosking (1998) [102]. As described in Chapter 3, trimmed means are useful in situations where the signal under analysis may contain observations that are discordant with the rest of the signal. Because outlying values are mapped to the most extreme order statistics, trimmed smoothers give zero weight to these order statistics leading to

$$Y_r(n) = \frac{1}{N - 2r} \sum_{i=r+1}^{N-r} X_{(r)} \tag{7.38}$$

where $r = [N\alpha]$. Thus, the largest and smallest r observations, each representing a fraction α of the entire sample, are ignored when calculating the running mean. Appropriate choice of α, the amount of trimming, depends on the degree of robustness required — larger amounts of trimming provides higher protection against heavier-tailed noise. Crow and Siliqui (1967) [57] suggest a value of $\alpha = 1/5$ when possible distributions range from the Normal to the Laplacian, and values of α between $1/4$ and $1/3$, when noise distributions range from the Normal to the Cauchy.

As described by Bednar and Watt (1984) [30], trimmed means provide a connection between average smoothing and median smoothing. Consider a segment of the voiced waveform "a", shown at the bottom of Figure 7.2. This speech signal is placed at the input to several trimmed mean filters of size 9. The outputs of the trimmed means as we vary the trimming parameter r from zero to four are also shown in Figure 7.2, for $r = 0, 1, 2, 3, 4$. The vertical index denotes the trimming where the top signal is the median filtered output, the second signal from the top is the trimmed mean with $r = 3$ output signal, and successively the other trimmed means are displayed in Figure 7.2. The different characteristics of the filtered signals as we vary the trimming can be immediately seen. Notice that while the running mean results in a smooth blurring of the signal, the running median smoothes the signal with sharp discontinuities. This arises from the fact that the running median restricts the output value to be identical to the value of one of the input signals in the observation window. Depending on the amount of trimming, the alpha trimmed filter removes narrow impulses, but also does some edge smoothing.

These smoothing properties are even more pronounced if the filtering is performed repeatedly over the same input signal. Figure 7.3 shows the speech signal that has been repeatedly filtered for five consecutive times with the various trimmed mean filters. It is interesting to note that in the extreme trimming case of the median, after a few iterations, the resultant signal will not be modified by further median filtering.

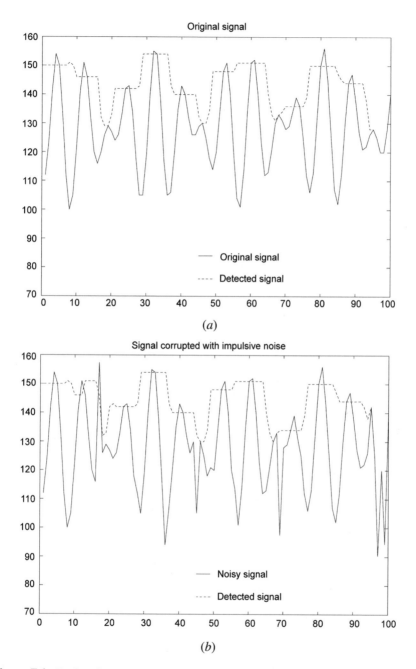

Figure 7.1   Rank-order AM demodulation. The window size is 9, and the output is the 8th largest in the window. Baseband signal is at 5kHz with a carrier of 31kHz. The sampling frequency is 250kHz. (*a*) noiseless reception. (*b*) noisy reception with impulsive noise.

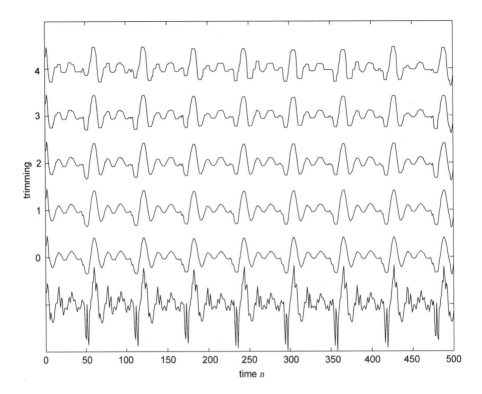

Figure 7.2 Trimmed mean filtering of a speech signal for various levels of trimming. The bottom waveform is the original speech signal

Thus, it can be seen that there exist fixed-point signals to the median filter that are not affected by the running median. This is only true for the median filter and not for a general trimmed mean, since the averaging operation in the trimmed mean will inevitably modify the output of the filter after every filter pass.

7.3 $L\ell$-FILTERS

Weighted medians only admitting positive weights were labeled "smoothers" in Chapter 5 since these lead to operations having low-pass characteristics. The low-pass characteristics of weighted median smoothers are well understood and are directly attributed to their nonnegative weights. L-smoothers, on the other hand, also exhibit low-pass characteristics even though positive and negative valued weights are allowed in their computation. In this case, the limitation does not arise as a consequence of the weights values, but from the fact that prior to the weighting operation, the observation samples are sorted, and as a result their temporal ordering is lost. Thus, the L-smoother weights cannot exploit the time ordering relationship of time-series that is

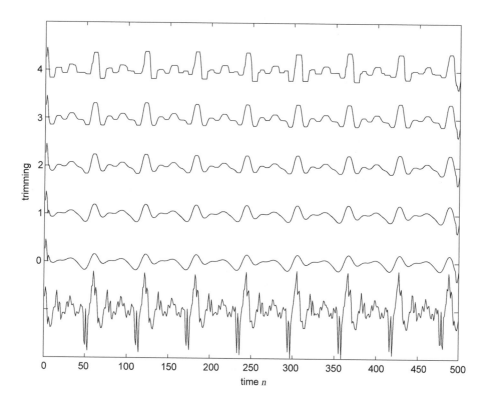

Figure 7.3 Five passes of trimmed mean filtering of a signal with various levels of trimming.

critical in signal processing applications. To overcome this limitation of L-smoothers, methods that combine rank-ordering and time-ordering of the observations samples have been developed. The class of $L\ell$-filters [1] is one approach introduced by Palmieri and Boncelet (1989) [155] and independently by Ghandi and Kassam (1991) [79]. Their approach is to modify the weight structure of L-smoothers such that the weights depend on both, the time-order and rank-order characteristics of the data.

DEFINITION 7.2 ($L\ell$-FILTERS) *Given the observation vector at time* n, $\mathbf{X}(n) = [X_1(n), X_2(n), \ldots, X_N(n)]^T$, *where* $X_i(n) = X(n + i - (K + 1))$ *with* $N = 2K + 1$, *and the ranks* R_i *for each of the samples* X_i, $i = 1, 2, \ldots, N$, *the* $L\ell$-*filter output is defined as*

$$Y(n) = \sum_{i=1}^{N} W_{i,R_i} X_i(n), \tag{7.39}$$

---

[1] Also referred to as *combination* filters.

where the weight given to the ith sample X_i, W_{i,R_i}, depends on the sample's rank R_i.

Since R_i can take on N possible weights, there are up to N different weights per sample. The $L\ell$-filter thus requires a total of N^2 weights. The name $L\ell$-filter was coined borrowing notation from order statistics, L referring to the rank-ordering and ℓ referring to the time-ordering of the observation samples.

While Definition 7.2 is succinct, it does not lead to a simple-to-optimize formulation since the linear combination in (7.39) involves data-dependent weights. An equivalent definition, using vectorial notation, will prove useful in this regard. Dropping the temporal index n, for notational simplicity, and expressing the temporal-order and rank-order observations as $\mathbf{X} = [X_1, X_2, \ldots, X_N]^T$ and $\mathbf{X}_L = [X_{(1)}, X_{(2)}, \ldots, X_{(N)}]^T$, the N^2-long vector $\mathbf{X}_{L\ell}$ that combines the rank and temporal ordering is defined as [79, 155]

$$\mathbf{X}_{L\ell}^T = [X_{1(1)},\ X_{1(2)}, \ldots,\ X_{1(N)} | \ldots, X_{i(j)}, \ldots | X_{N(1)}, X_{N(2)}, \ldots, X_{N(N)}],$$
(7.40)

where

$$X_{i(j)} = \begin{cases} X_i & \text{if } X_i \longleftrightarrow X_{(j)} \\ 0 & \text{else} \end{cases}$$
(7.41)

and where $X_i \longleftrightarrow X_{(j)}$ denotes the event that the ith element in $\mathbf{X}$ is the jth smallest in the sample set. Thus, the ith input sample is mapped into the bin of samples $X_{i(1)}, X_{i(2)}, \ldots, X_{i(N)}$, of which $N - 1$ are zero, and where only one is nonzero having the same value as X_i. The location of the nonzero sample, in turn, characterizes the ranking of X_i among the N input samples. Again, R_i represents the rank of X_i among the elements of $\mathbf{X}$. For example, the ranks of the elements in the observation vector $\mathbf{X} = [\,3,\ 5,\ 2\,]^T$ are $R_1 = 2$, $R_2 = 3$, and $R_3 = 1$ leading to the $L\ell$ vector

$$\mathbf{X}_{L\ell} = [\,0,\ 3,\ 0\,|\,0,\ 0,\ 5\,|\,2,\ 0,\ 0\,]^T.$$
(7.42)

In the case of rank ties among a subset of input samples, stable sorting is performed where a lower rank is assigned to the sample with the lower time indexing in the subset containing rank ties.

The decomposition $\mathbf{X} \in R^N \longleftrightarrow \mathbf{X}_{L\ell} \in R^{N^2}$ specified in (7.40) and (7.41) is a one-to-one nonlinear mapping where $\mathbf{X}$ can be reconstructed from $\mathbf{X}_{L\ell}$ as

$$\mathbf{X} = \left[\mathbf{I}_N \otimes \mathbf{e}_N^T\right] \mathbf{X}_{L\ell},$$
(7.43)

where $\mathbf{I}_N$ is an $N \times N$ identity matrix, $\mathbf{e}_N$ is an $N \times 1$ one-valued vector, and $\otimes$ is the matrix Kronecker product. Since the $\mathbf{X}_{L\ell}$ vector contains both, time and rank ordering information, it is not surprising that we can also obtain $\mathbf{X}_L$ from $\mathbf{X}_{L\ell}$ as

$$\mathbf{X}_L = \left[\mathbf{e}_N^T \otimes \mathbf{I}_N\right] \mathbf{X}_{L\ell}.$$
(7.44)

EXAMPLE 7.2

Consider again the length 3 observation vector $\mathbf{X} = [\,3,\,5,\,2\,]^T$ which is mapped to the $L\ell$ vector $\mathbf{X}_{L\ell} = [\,0,\,3,\,0\,|\,0,\,0,\,5\,|\,2,\,0,\,0\,]^T$. The vector $\mathbf{X}$ can be reconstructed from $\mathbf{X}_{L\ell}$ as

$$\mathbf{X} = \mathbf{e}_3^T \otimes \begin{bmatrix} 1 & 0 & 0 \\ 0 & 1 & 0 \\ 0 & 0 & 1 \end{bmatrix} [\,0,\,3,\,0\,|\,0,\,0,\,5\,|\,2,\,0,\,0\,]^T \qquad (7.45)$$

$$= \begin{bmatrix} \mathbf{e}_3^T & \mathbf{0}^T & \mathbf{0}^T \\ \mathbf{0}^T & \mathbf{e}_3^T & \mathbf{0}^T \\ \mathbf{0}^T & \mathbf{0}^T & \mathbf{e}_3^T \end{bmatrix} [\,0,\,3,\,0\,|\,0,\,0,\,5\,|\,2,\,0,\,0\,]^T, \qquad (7.46)$$

where $\mathbf{e}_3 = [1,1,1]^T$, and $\mathbf{0} = [0,0,0]^T$. Similarly, $\mathbf{X}_L$ is obtained from $\mathbf{X}_{L\ell}$ as

$$\mathbf{X}_L = [\,\mathbf{I}_3\,|\,\mathbf{I}_3\,|\,\mathbf{I}_3\,]\,[\,0,\,3,\,0\,|\,0,\,0,\,5\,|\,2,\,0,\,0\,]^T \qquad (7.47)$$

where $\mathbf{I}_3$ is the 3×3 identity matrix. ∎

The decomposition $\mathbf{X} \longrightarrow \mathbf{X}_{L\ell}$ thus maps a vector with time-ordered samples into a vector whose elements are both time- and rank-ordered. The rank of X_i, R_i, determines the value of all the elements $X_{i(1)}, X_{i(2)}, \ldots, X_{i(N)}$, regardless of how the other samples in the window are ranked. Having the $L\ell$ vector representation, the filter output definition follows naturally.

DEFINITION 7.3 ($L\ell$-FILTERS (VECTOR FORMULATION)) *The output of the $L\ell$-filter is given by the inner product*

$$Y(n) = \mathbf{W}^T \mathbf{X}_{L\ell}(n) \qquad (7.48)$$

where the weight vector is $\mathbf{W} = \left[(\mathbf{W}_1)^T | (\mathbf{W}_2)^T | \ldots | (\mathbf{W}_N)^T \right]^T$ *in which* $\mathbf{W}_i = [W_{i(1)}, W_{i(2)}, \ldots, W_{i(N)}]^T$ *is the N long tap weight vector associated with the ith input sample.*

7.3.1 Design and Optimization of $L\ell$-filters

The simplest method to design robust $L\ell$-filters is through the concept of trimming. Observations that are discordant with the main body of samples map to the most extreme order statistics, thus, $L\ell$-filters that give zero weight to these order statistics are robust in nature. Trimming in the $L\ell$ weight vector is accomplished easily as

$$\mathbf{W}^T = [0, W_{1(2)}, \ldots, W_{1(N-1)}, 0 | \ldots, W_{i(j)}, \ldots | 0, W_{N(2)}, \ldots, W_{N(N-1)}, 0],$$
$$(7.49)$$

where the first and last of the N weights associated with a sample are set to zero. The remaining $N - 2$ samples can be optimized, or they can all be assigned the corresponding weights values of an FIR filter so as to mimic the linear FIR spectral characteristics.

The optimization of $L\ell$-filters, in general, is straight forward. The inner product formulation of $L\ell$-filters is well suited for the development of optimization algorithms, since the filter's output is linear with respect to the observation samples. The goal is to minimize the error $e(n)$ between a desired signal $D(n)$ and the $L\ell$-filter estimate $Y(n)$. The optimization follows the well known Wiener filter solution of linear FIR filters, with the exception that the $N^2 \times N^2$ $L\ell$ correlation matrix of the $L\ell$ observation vector is used rather than the $N \times N$ correlation matrix associated with the observation vector $\mathbf{X}$ in the Wiener filter [99]. Under the mean-square-error criterion, the cost function is defined as $J(\mathbf{W}) = E\{e^2(n)\}$, where $e(n) = D(n) - Y(n)$. For each value assigned to the filter weight $\mathbf{W}$, a corresponding mean square estimation error occurs. The goal is then to obtain the weight vector values so that the error function is minimized. Using the vector definition of the $L\ell$-filter, the cost function can be expressed as

$$J(\mathbf{W}) = E[(D(n) - \mathbf{W}^T\mathbf{X}_{L\ell}(n))(D^T(n) - \mathbf{X}_{L\ell}^T(n)\mathbf{W})] \qquad (7.50)$$
$$= \sigma_d^2 - 2\mathbf{p}_{L\ell}^T\mathbf{W} + \mathbf{W}^T\mathbf{R}_{L\ell}\mathbf{W}, \qquad (7.51)$$

where $\mathbf{p}_{L\ell} = E\{D(n)\,\mathbf{X}_{L\ell}\}$ and where $\mathbf{R}_{L\ell}$ is the $N^2 \times N^2$ symmetric correlation matrix $\mathbf{R}_{L\ell} = E\{\mathbf{X}_{L\ell}\mathbf{X}_{L\ell}^T\}$. $J(\mathbf{W})$ is quadratic in $\mathbf{W}$; thus, for different values in $W_{1(1)}, W_{1(2)}, \ldots, W_{N(N)}$, $J(\mathbf{W})$ defines a bowl-shape surface in an $2N + 1$ dimensional space having a unique minimum. The global minimum is found through the gradient

$$\nabla = \frac{\partial(J(\mathbf{W}))}{\partial\mathbf{W}}$$
$$= \left[\frac{\partial(J(\mathbf{W}))}{\partial W_{1(1)}}, \frac{\partial(J(\mathbf{W}))}{\partial W_{1(2)}}, \ldots, \frac{\partial(J(\mathbf{W}))}{\partial W_{N(N)}}\right]^T$$
$$= 2\mathbf{R}_{L\ell}\mathbf{W} - 2\mathbf{p}_{L\ell}.$$

Setting the gradient to zero yields the optimal weights for the $L\ell$-filter

$$\mathbf{W}_o = \mathbf{R}_{L\ell}^{-1}\mathbf{p}_{L\ell}. \qquad (7.52)$$

The optimal solution, thus requires knowledge of the second order moments of the rank and time ordered vector $\mathbf{X}_{L\ell}$ and of the cross-correlation between this vector and the desired signal. In order to gain more intuition, the structure of the correlation matrix $\mathbf{R}_{L\ell} = E\{\mathbf{X}_{L\ell}\mathbf{X}_{L\ell}^T\}$, can be further analyzed. Partitioning the $L\ell$ vector as

$$\mathbf{X}_{L\ell} = [X_{1(1)}, X_{1(2)}, \ldots, X_{1(N)}|X_{2(1)}, \ldots, X_{2(N)}|\ldots|X_{N()}, \ldots, X_{N(N)}]$$
$$= [\mathbf{X}_1^T, \mathbf{X}_2^T, \ldots, \mathbf{X}_N^T]^T \qquad (7.53)$$

where $\mathbf{X}_i^T = [X_{i(1)}, X_{i(2)}, \ldots, X_{i(N)}]$, the $L\ell$ correlation matrix can be written as

$$\mathbf{R}_{L\ell} = E\{\mathbf{X}_{L\ell}\mathbf{X}_{L\ell}^T\} \tag{7.54}$$

$$= \begin{bmatrix} \mathbf{R}_{11} & \mathbf{R}_{12} & \cdots & \mathbf{R}_{1N} \\ \mathbf{R}_{21} & \mathbf{R}_{22} & \cdots & \mathbf{R}_{2N} \\ \vdots & \vdots & \ddots & \vdots \\ \mathbf{R}_{N1} & \mathbf{R}_{N2} & \ldots & \mathbf{R}_{NN} \end{bmatrix} \tag{7.55}$$

in which $\mathbf{R}_{uv} = E\{\mathbf{X}_u\mathbf{X}_v\}$ are the submatrices in (7.55). These submatrices are sparse as a result that many ordered sample combinations cannot occur, that is, two samples cannot have the same rank. In fact, all submatrices $\mathbf{R}_{uu}$, for $u = 1, 2, \ldots, N$, are diagonal having all their off-diagonal elements zero. Also, the sub matrices $\mathbf{R}_{uv}^j$ for $u \neq v$ form off-diagonal matrices whose diagonal elements are all zero. The correlation $\mathbf{R}_{L\ell}$ for $N = 3$, for example, is the expectation of the elements in

$$\left[\begin{array}{ccc|ccc|ccc} x_{(1)1}^2 & 0 & 0 & 0 & X_{(1)1}X_{(2)2} & X_{(1)1}X_{(2)3} & 0 & X_{(1)1}X_{(3)2} & X_{(1)1}X_{(3)3} \\ 0 & x_{(1)2}^2 & 0 & X_{(1)2}X_{(2)1} & 0 & X_{(1)2}X_{(2)3} & X_{(1)2}X_{(3)1} & 0 & X_{(1)2}X_{(3)3} \\ 0 & 0 & x_{(1)3}^2 & X_{(1)3}X_{(2)1} & X_{(1)3}X_{(2)2} & 0 & X_{(1)3}X_{(3)1} & X_{(1)3}X_{(3)2} & 0 \\ \hline 0 & X_{(2)1}X_{(1)2} & X_{(2)1}X_{(1)3} & x_{(2)1}^2 & 0 & 0 & 0 & X_{(2)1}X_{(3)2} & X_{(2)1}X_{(3)3} \\ X_{(2)2}X_{(1)1} & 0 & X_{(2)2}X_{(1)3} & 0 & x_{(2)2}^2 & 0 & X_{(2)2}X_{(3)1} & 0 & X_{(2)2}X_{(3)3} \\ X_{(2)3}X_{(1)1} & X_{(2)3}X_{(1)2} & 0 & 0 & 0 & x_{(2)3}^2 & X_{(2)3}X_{(3)1} & X_{(2)3}X_{(3)2} & 0 \\ \hline 0 & X_{(3)1}X_{(1)2} & X_{(3)3}X_{(1)3} & 0 & X_{(3)1}X_{(2)2} & X_{(3)1}X_{(2)3} & x_{(3)1}^2 & 0 & 0 \\ X_{(3)2}X_{(1)1} & 0 & X_{(3)2}X_{(1)3} & X_{(3)2}X_{(2)1} & 0 & X_{(3)2}X_{(2)3} & 0 & x_{(3)2}^2 & 0 \\ X_{(3)2}X_{(1)1} & X_{(3)3}X_{(1)2} & 0 & X_{(3)3}X_{(2)1} & X_{(3)3}X_{(2)2} & 0 & 0 & 0 & x_{(3)3}^2 \end{array}\right]$$

where the structure of the submatrices is readily seen. This example matrix provides us with a more intuitive understanding of the $L\ell$ correlation matrix. Note that for stationary observations, $\mathbf{R}_{L\ell}$ is not Toeplitz since, in general, $EX_{(i)j}^2 \neq EX_{(k)j}^2$ for $i \neq k$. On the other hand, in a stationary environment, $EX_{(i)j}X_{(k)l} = EX_{(i)m}X_{(k)n}$ for any j, m, l, n. Thus, the correlation matrix $\mathbf{R}_{L\ell}$ for all $L\ell$ estimators is block-Toeplitz.

EXAMPLE 7.3 (ROBUST WAVELET DENOISING)

Denoising by wavelet shrinkage has evolved into a popular application of wavelet analysis. The key concept in wavelet denoising lies in the fact that most signals encountered in practice have a relatively small number of wavelet coefficients with significant energy. Wavelet shrinkage, developed by Johnstone and Donoho (1995) [64] decomposes a signal by a discrete wavelet transform and then selects the significant coefficients based on thresholding. Coefficients that fall below a threshold are removed and those that exceed the threshold are either kept in their original magnitude (hard thresholding) or are preserved with their magnitude reduced by the threshold level (soft thresholding). By an appropriate choice of the threshold value, signals corrupted by Gaussian noise are effectively denoised. The noise is assumed white and its power is assumed to be much smaller than the signal power. Under these conditions, wavelet shrinkage has been shown superior to traditional linear filtering methods. Figure 7.4 depicts the wavelet shrinkage of the noisy blocks and doppler signals with soft thresholding. The noise is Gaussian and a 3-level wavelet decomposition is used with Daubechies wavelet filter coefficients of length 6. The wavelet shrinkage removes some of the signal's power, but is very effective at removing the white noise power that was uniformly distributed throughout the various levels of decomposition. A particular strength of wavelet shrinkage is its ability to preserve sharp features.

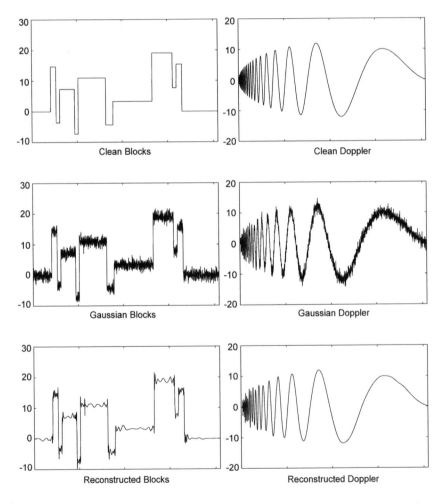

Figure 7.4 Wavelet denoising of block and doppler signals corrupted by Gaussian noise.

When the noise power is no longer small in comparison to the signal power, or when the noise has heavier-than-Gaussian tails, traditional wavelet denoising is not as effective since the noise components can lead to coefficients with significant magnitude in the wavelet decomposition and consequently cannot be removed by simple thresholding. Figure 7.5 depicts the block and doppler signals corrupted by contaminated Gaussian noise and the signal reconstructions after wavelet shrinkage. The "Gaussian" component of the noise in the reconstruction attained by wavelet shrinkage is effectively filtered out, but the majority of the outlier samples remain.

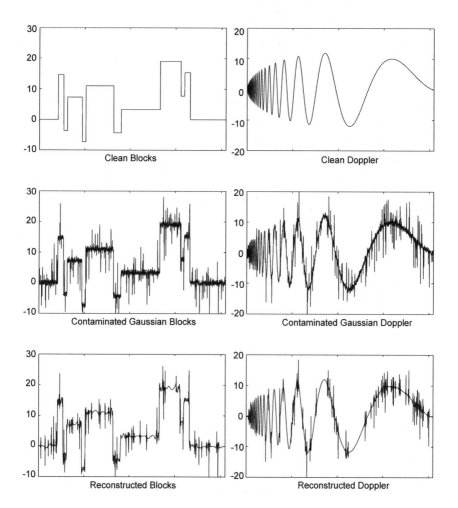

Figure 7.5 Wavelet denoising of block and doppler signals corrupted by contaminated Gaussian noise.

One approach to overcome the effects of outliers is to prefilter the input signal by a median or similar filter prior to applying wavelet shrinkage. Prefiltering will remove the impulsive noise but it will also affect the underlying signal structure as the characteristics of wavelet shrinkage are not used at this initial stage. Lau et al. (1996) [123] introduced a different approach to robust signal denoising. In their approach, the low-pass and high-pass filters used in the first stage of the wavelet decomposition are replaced by $L\ell$ filters designed to mimic the finite impulse response of the wavelet FIR filters. The $L\ell$ filter weights are selected such that the smallest and largest samples in the observation vector are weighted by zero. The $L\ell$ weights associated with the ith input sample in the running filter are thus $\mathbf{W}_i = [0, W_{i(2)}, \ldots, W_{i(N-1)}, 0]^T$. The remaining coefficients in $\mathbf{W}_i$ are optimized so that the $L\ell$ filter response mimics that of the corresponding wavelet filter when no noise is present. Since outliers are for the most part removed in the first level of the decomposition, the filters used in the remaining levels of the decomposition, and in the reconstruction after shrinkage, are simply the traditional wavelet FIR filters.

Figure 7.6 depicts the denoising of the block and doppler signals corrupted by contaminated Gaussian noise. The outputs of the optimal $L\ell$ filter with 12 data-dependent coefficients are shown in the figure. The output is free of outliers but the "Gaussian" component of the noise is not removed adequately. The output of the robust wavelet shrinkage, on the other hand, eliminates both the Gaussian and nonGaussian noise components. The robust wavelet shrinkage in this example uses three levels of decomposition and length 6 Daubechies filter coefficients at all levels with the exception of the first level of decomposition where $L\ell$ filters that mimic Daubechies filter's response are designed.

■

7.4 $L^J\ell$ PERMUTATION FILTERS

In $L\ell$-filtering, the weight given to a sample X_i is determined by the sample's rank R_i. Thus, X_i is weighted by one of N possible weights stored in memory. Kim and Arce (1994) [114] showed that in some cases, it is useful to extend this concept such that the weight given to X_i is not only dependent on its rank R_i, but also on the rank of some other sample, X_j, in the observation window. In this case, X_i is given one of $N(N-1)$ weights stored in memory[2]. A total of $N^2(N-1)$ weights would be required for the weighting of the N observation samples.

This concept can be generalized progressively such that at the end, the weight given to a sample X_i depends not only on its rank R_i, but also on the ranks of the remaining $N-1$ samples in the window. In the most general case, a sample in the so called *permutation filter* would be assigned one of $N!$ weights [26, 155].

DEFINITION 7.4 (PERMUTATION FILTERS) *Given the observation vector at time* n, $\mathbf{X}(n) = [X_1, X_2, \ldots, X_N]^T$, *and the ranks* R_i *for each of the samples* X_i, $i = 1, 2, \ldots, N$, *the permutation filter output is defined as the linear combination*

---

[2]For each value of R_i, there are an additional $N-1$ possible values that R_j can take.

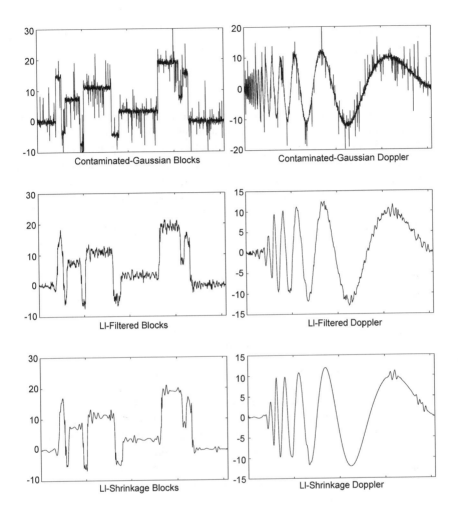

Figure 7.6 $L\ell$ filtering and Robust wavelet denoising of block and doppler signals corrupted by contaminated Gaussian noise.

$$Y(n) = \sum_{i=1}^{N} W_{i,p_{\mathbf{X}}} X_i(n), \tag{7.56}$$

where the weight given to the ith sample X_i, $W_{i,p_{\mathbf{X}}}$, is indexed by the permutation $p_{\mathbf{X}}$ of the index elements in $\mathbf{X}$ into the corresponding ranks R_i

$$p_{\mathbf{X}} = \begin{pmatrix} 1 & 2 & \dots & N \\ R_1 & R_2 & \dots & R_N \end{pmatrix}. \tag{7.57}$$

As an example, the index permutation associated with the observation $\mathbf{X} = [3, 5, 2]^T$ is

$$p\mathbf{x} = \begin{pmatrix} 1 & 2 & 3 \\ 2 & 3 & 1 \end{pmatrix}. \tag{7.58}$$

A permutation filter, in this example, would allow 3! weights per sample. The sample permutation shown above would point to one weight in this set. Permutation filters in their most general form are not practical as their memory requirements grow factorially with N. Several approaches to simplify their structure have been proposed by Barner and Arce (1986) [26], Kim and Arce (1994) [114], and Palmieri and Boncelet (1989) [155]. The formulation in [114] is particularly useful as it provides a progressive coloring of the permutation space.

Consider the observation vector $\mathbf{X} = [X_1, X_2, \ldots, X_N]^T$ and its corresponding sorted vector $\mathbf{X}_L = [X_{(1)}, X_{(2)}, \ldots, X_{(N)}]^T$. Define the rank indicator vector

$$\boldsymbol{\mathcal{R}}_i = [\mathcal{R}_{i1}, \mathcal{R}_{i2}, \ldots, \mathcal{R}_{iN}]^T, \tag{7.59}$$

where

$$\mathcal{R}_{ik} = \begin{cases} 1 & \text{if } X_i \leftrightarrow X_{(k)} \\ 0 & \text{else} \end{cases} \tag{7.60}$$

where $X_i \leftrightarrow X_{(k)}$ denotes that the ith temporal sample occupies the kth-order statistic. The variable R_i is then defined as the rank of X_i; hence, $\mathcal{R}_{iR_i} = 1$ by definition. Assuming the rank indicator vector $\boldsymbol{\mathcal{R}}_i$ is specified, if we would like to jointly characterize the ranking characteristics of X_i and its adjacent sample, X_{i+1}, contained in $\mathbf{X}$, then an additional indicator vector is needed which does not contain the information provided by $\boldsymbol{\mathcal{R}}_i$. Hence, we define the reduced rank indicator of X_{i+1}, as $\boldsymbol{\mathcal{R}}_i^1$, where we have removed the R_ith element from the rank indicator vector $\boldsymbol{\mathcal{R}}_{i+1}$. The two indicators $\boldsymbol{\mathcal{R}}_i$ and $\boldsymbol{\mathcal{R}}_i^1$ fully specify the rank permutation characteristics of X_i and X_{i+1}.

We can generalize this concept by characterizing the rank permutation characteristics of a set of j samples. Here, a reduced rank indicator of $X_{i\oplus a}$, $\boldsymbol{\mathcal{R}}_i^a$, is formed by removing the R_ith, $R_{i\oplus 1}$st, $\ldots$, $R_{i\oplus(a-1)}$th elements from the rank indicator vector $\boldsymbol{\mathcal{R}}_{i\oplus a}$ where $\oplus$ denotes the Modulo N addition $i \oplus a = (i + a)$ Mod N.[3] The parameter a specifies the sample, $X_{i\oplus a}$, whose rank information is being considered in addition to the rank information of the samples $(X_i, X_{i\oplus 1}, \ldots, X_{i\oplus(a-1)})$, such that, $a = 1$ considers the rank information of the sample $X_{i\oplus 1}$ when the rank information of the sample X_i is known, and $a = 2$ considers the rank information of the sample $X_{i\oplus 2}$ when the rank information of the samples $(X_i, X_{i\oplus 1})$ is known. For example, if $\mathbf{X} = [6, 3, 10, 1]^T$ and $\mathbf{X}_L = [1, 3, 6, 10]^T$, then the rank indicator vectors and their respective rank parameters are

---

[3]The Modulo N operation defined here is in the group $\{1, 2, \ldots, N\}$, such that (N Mod $N = N$) and ($N + 1$ Mod $N = 1$).

$$\mathcal{R}_1 = [0,0,1,0]^T, \ R_1 = 3, \quad \mathcal{R}_2 = [0,1,0,0]^T, \ R_2 = 2,$$
$$\mathcal{R}_3 = [0,0,0,1]^T, \ R_3 = 4, \quad \mathcal{R}_4 = [1,0,0,0]^T, \ R_4 = 1. \tag{7.61}$$

The reduced rank indicator vectors $\mathcal{R}_3^1$ and $\mathcal{R}_3^2$ are, for instance,

$$\mathcal{R}_3^1 = [\underbrace{1,0,0, \cancel{0}^{R_3}}_{\mathcal{R}_4}]^T = [1,0,0]^T$$
$$\mathcal{R}_3^2 = [\underbrace{\cancel{0}^{R_4},0,1, \cancel{0}^{R_3}}_{\mathcal{R}_1}]^T = [0,1]^T, \tag{7.62}$$

where the R_3th sample was removed from $\mathcal{R}_{3\oplus1} = \mathcal{R}_4$ to obtain $\mathcal{R}_3^1$ and where the R_3th and R_4th samples were deleted from $\mathcal{R}_{3\oplus2} = \mathcal{R}_1$ to get $\mathcal{R}_3^2$. Note that the notation $\cancel{0}^{R_i}$ used in (7.62) represents the deletion of the sample "0" which is the R_ith element of the rank indicator vector. The $\mathcal{R}_3^1$ indicates that the sample X_4 is the first ranked sample among (X_1, X_2, X_4), and similarly $\mathcal{R}_3^2$ indicates that X_1 is the second ranked sample among (X_1, X_2). The general idea behind the reduced rank indicator $\mathcal{R}_i^a$ is that it characterizes the rank information of the sample $X_{i\oplus a}$ under the situation in which the rank information of the samples $(X_i, X_{i\oplus1}, \ldots, X_{i\oplus(a-1)})$ is known.

The rank indicator vector and the reduced rank indicator vectors are next used to define the rank permutation indicator $\mathbf{P}_i^j$ as

$$\mathbf{P}_i^j = \mathcal{R}_i \otimes \mathcal{R}_i^1 \otimes \ldots \otimes \mathcal{R}_i^{j-1} \tag{7.63}$$

for $1 \leq j \leq N$, where $\otimes$ denotes the matrix Kronecker product. Note that while the vector $\mathcal{R}_i$ is of length N, the vector $\mathbf{P}_i^j$ in (7.63) has length of $P_N^j \overset{def}{=} N(N-1)\cdots(N-j+1)$ which represents the number of permutations choosing j samples from N distinct samples. The vector $\mathbf{P}_i^j$ effectively characterizes the relative ranking of the samples $(X_i, X_{i\oplus1}, \ldots, X_{i\oplus(j-1)})$, that is, the rank permutation of $(X_i, X_{i\oplus1}, \ldots, X_{i\oplus(j-1)})$. Hence, $\mathbf{P}_i^0$ does not unveil any rank information, whereas $\mathbf{P}_i^1$ provides the rank information of X_i but ignores the rank of the other $N-1$ input samples. Similarly, $\mathbf{P}_i^2$ provides the rank information of X_i and $X_{i\oplus1}$ but eludes the ranking information of the other $N-2$ input samples. Clearly, $\mathbf{P}_i^N$ accounts for the ranks of all input samples X_1 through X_N. In order to illustrate the formulation of the vector $\mathbf{P}_i^j$, again let the observation vector take on the values $\mathbf{X} = [6,3,10,1]^T$ and $\mathbf{X}_L = [1,3,6,10]^T$. The rank indicator vectors, $\mathcal{R}_i$, for this example vector $\mathbf{X}$ were listed in (7.61), then the rank permutation indicators for $j=2$ are found as

$$\mathbf{P}_1^2 = \mathcal{R}_1 \otimes \mathcal{R}_1^1 = [0,0,1,0]^T \otimes [0,1,0]^T$$
$$\mathbf{P}_2^2 = \mathcal{R}_2 \otimes \mathcal{R}_2^1 = [0,1,0,0]^T \otimes [0,0,1]^T$$
$$\mathbf{P}_3^2 = \mathcal{R}_3 \otimes \mathcal{R}_3^1 = [0,0,0,1]^T \otimes [1,0,0]^T \tag{7.64}$$
$$\mathbf{P}_4^2 = \mathcal{R}_4 \otimes \mathcal{R}_4^1 = [1,0,0,0]^T \otimes [0,1,0]^T.$$

To see how the $\mathbf{P}_i^j$ characterizes the rank permutation, let us carry out the matrix Kronecker product in the first equation in (7.64), that is,

$$\mathbf{P}_1^2 = [(0,0,0),(0,0,0),(0,1,0),(0,0,0)]^T \tag{7.65}$$

where parentheses are put for ease of reference. Note that the 1 located in the second position in the third parentheses in $\mathbf{P}_1^2$ implies that the rank of X_1 is 3 and the rank of X_2 among (X_2, X_3, X_4) is 2. Thus, $\mathbf{P}_1^2$ obtained in this example clearly specifies the rank permutation of X_1 and X_2 as $(R_1, R_2) = (3,2)$. Notice that the vectors $\mathbf{P}_i^3$ can be found recursively from (7.64) as $\mathbf{P}_i^3 = \mathbf{P}_i^2 \otimes \mathcal{R}_i^2$. In general, it can be easily seen from (7.63) that this recursion is given by $\mathbf{P}_i^j = \mathbf{P}_i^{j-1} \otimes \mathcal{R}_i^{j-1}$.

The rank permutation indicator forms the basis for the rank permutation vectors $\mathbf{X}_{L^j\ell}$ defined as the NP_N^j long vector

$$\mathbf{X}_{L^j\ell} = \left[X_1\,(\mathbf{P}_1^j)^T \mid X_2\,(\mathbf{P}_2^j)^T \mid \cdots \mid X_N\,(\mathbf{P}_N^j)^T \right]^T. \tag{7.66}$$

Note that $\mathbf{X}_{L^j\ell}$ places each X_i based on the rank of j time-ordered samples $(X_i, X_{i\oplus 1}, \ldots, X_{i\oplus(j-1)})$. Consequently, we refer to it as the $L^j\ell$ vector, where the j superscript stands for the j sample ranks used to determine the weight given to each observation sample. It should be mentioned here that there are other ways of defining rank permutation indicators. For instance, we could let $\mathbf{P}_i^j$ characterize the rank permutation of the samples $(X_{i\oplus 1}, X_{i\oplus 3}, \ldots, X_{i\oplus(2j+1)})$, or it can characterize the rank permutation of $(X_1, X_2, \ldots, X_j)$ regardless of the index i. Here, we use the definition of $\mathbf{P}_i^j$ in (7.63) since it provides a systematic approach to the design.

DEFINITION 7.5 (PERMUTATION $L^j\ell$ FILTERS) *Given the observation sample* $\mathbf{X} = [X_1, X_2, \ldots, X_N]^T$ *and the corresponding $L^j\ell$ vector of length NP_N^j given in (7.66)), the $L^j\ell$ estimate is defined as*

$$Y_j(n) = \mathbf{W}_{L^j\ell}^T\,\mathbf{X}_{L^j\ell} \tag{7.67}$$

where the weight vector is

$$\mathbf{W}_{L^j\ell} = \left[(\mathbf{W}_1^j)^T | (\mathbf{W}_2^j)^T | \ldots | (\mathbf{W}_N^j)^T \right]^T \tag{7.68}$$

in which $\mathbf{W}_i^j$ is the P_N^j long tap weight vector, and where $Y_j(n)$ is the $L^j\ell$ estimate.

Notice that for $j = 0$, the permutation filter $L^0\ell$ reduces to a linear FIR filter. For $j = 1$, the permutation filter is identical to the $L\ell$ filter introduced earlier.

Optimization   Given the $L^j\ell$ filtering framework, the goal is to minimize the error $e(n)$ between the desired signal $D(n)$ and the permutation filter estimate. Under the MSE criterion, the optimization is straightforward, since the output of the $L^j\ell$ filter is linear with respect to the samples in $\mathbf{X}_j$. Hence, it is simple to show that the optimal $L^j\ell$ filter is found as

$$\mathbf{W}_{L^j\ell}^{opt} = \mathbf{R}_{L^j\ell}^{-1}\mathbf{p}_{L^j\ell} \tag{7.69}$$

where $\mathbf{p}_{L^j\ell} = \{D(n)\,\mathbf{X}_{L^j\ell}\}$ and $\mathbf{R}_{L^j\ell}$ is the $P_N^j \times P_N^j$ correlation matrix

$$\mathbf{R}_{L^j\ell} = E\{\mathbf{X}_{L^j\ell}\mathbf{X}_{L^j\ell}^T\} \tag{7.70}$$

$$= \begin{bmatrix} \mathbf{R}_{11}^j & \mathbf{R}_{12}^j & \cdots & \mathbf{R}_{1N}^j \\ \mathbf{R}_{21}^j & \mathbf{R}_{22}^j & \cdots & \mathbf{R}_{2N}^j \\ \vdots & \vdots & \ddots & \vdots \\ \mathbf{R}_{N1}^j & \mathbf{R}_{N2}^j & \cdots & \mathbf{R}_{NN}^j \end{bmatrix} \tag{7.71}$$

in which

$$\mathbf{R}_{uv}^{L^j\ell} = E\{x_u x_v \mathbf{P}_u^j (\mathbf{P}_v^j)^T\} \tag{7.72}$$

From (7.72), it can be seen that the diagonal submatrix of $\mathbf{R}_{L^j\ell}$, $\mathbf{R}_{uu}^{L^j\ell}$, constitutes a diagonal matrix whose off-diagonal elements are all zeros. Also, the off-diagonal submatrix $\mathbf{R}_{uv}^{L^j\ell}$ for $u \neq v$ forms an off-diagonal matrix whose diagonal elements are all zeros. The solution of the optimal filter in (7.69) will be unique only when the correlation matrix in (7.71) is nonsingular. Certainly if the correlation matrix is singular, a solution could be found by use of the singular value decomposition [99]. Although in most of the applications encountered in practice, where broadband noise is present, the $L^j\ell$ autocorrelation matrices will be nonsingular.

7.5 HYBRID MEDIAN/LINEAR FIR FILTERS

$L\ell$ filters and $L^j\ell$ filters exploit the rank and temporal ordering characteristics of the data. The major drawback of these filters, however, is the large number of filter weights needed. Their complexity increases very rapidly with the window size N and with the parameter j, limiting practical implementations to relatively small sample sizes. Alternative filtering approaches which combine the rank-ordering and temporal-ordering of the underlying signals have been developed as described next.

7.5.1 Median and FIR Affinity Trimming

An alternative approach to combining the attributes of linear FIR filters and L-filters is through modified sample-trimming strategies that exploit the temporal and rank characteristics of the data. Recall that trimmed means discard a fraction α of all observation samples, regardless of their dispersion characteristics. When outliers are not present, trimmed means rapidly loose efficiency with increasing α. Several location estimators have been proposed to overcome this limitation of trimmed means. Lee and Kassam (1985) [125] introduced the *modified trimmed mean (MTM)*, where samples that differ considerably from the sample median are discarded prior to averaging. The MTM location estimate is formed as

$$\hat{\beta}_q = \frac{\sum_{i=1}^N a_i X_i}{\sum_{i=1}^N a_i} \tag{7.73}$$

where

$$a_i = \begin{cases} 1 & \text{if } |X_i - \text{MEDIAN}(X_j|_{j=1}^N)| \le q \\ 0 & \text{else} \end{cases} \tag{7.74}$$

where q is a user-defined parameter that determines the amount of trimming. A similar approach to trimming using distances to a reference point was proposed by Pomalaza and McGillem (1984)[164]. Much like trimmed means, modified trimmed means are robust estimators that can be used as running smoothers. A variation of (7.73), referred to as the *double window modified trimmed mean (DW MTM)*, employs two overlapping smoother windows [125]. In this case, the sample median of the smallest running window is used as a reference point.

Another related class of smoothers takes on the form (7.73) with the difference that the trimming reference point is not the sample median, but the center sample in the running window. The *K-nearest neighbor* smoother [59] and the *sigma* smoother [124] are two smoothers with this trimming structure where the coefficients a_i take on the values

$$a_i = \begin{cases} 1 & \text{if } |X_i - X_c| \le q \\ 0 & \text{else} \end{cases} \tag{7.75}$$

where X_c is the center sample in the observation window. Thus, those samples with values close enough to the central sample X_c are averaged. The parameter q is used to trim out the samples whose values differ significantly from the value of X_c. This structure has remarkable detail-preserving characteristics but is very fragile to outliers and impulsive noise.

Having the modified trimmed mean and the K-nearest neighbor estimators at hand, it is natural to extend these smoothers to a filtering structure where samples are not only trimmed and averaged, but are temporally weighted as well. In this manner, the rank and temporal ordering characteristics are captured at once.

Weighted Median Affine Filters   Introduced by Flaig et al. (1998) [73], weighted median affine filters use a weighted median as the trimming reference point, the trimming is *soft* rather than *hard*, and the samples are weighted averaged according to their temporal ordering.

DEFINITION 7.6 *Given the set of N observations $\{X_1, X_2, \ldots, X_N\}$ in an observation window, a set of N real-valued affinity weights $\{C_1, C_2, \ldots, C_N\}$, and a set of N filter weights $\{W_1, W_2, \ldots, W_N\}$, the trimming reference $\mu(n)$ is defined as the weighted median*

$$\mu(n) = \text{MEDIAN}(|C_1| \diamond \text{sgn}(C_1)X_1, \ldots, |C_N| \diamond \text{sgn}(C_N)X_N) \tag{7.76}$$

where $|C| \diamond X = \underbrace{X, X, \ldots, X}_{|C| \; times}$. *The (normalized) WM affine FIR filter is defined as:*

$$Y_\gamma(n) = K(n) \sum_{i=1}^N g\left(\frac{X_i - \mu(n)}{\gamma}\right) W_i X_i \tag{7.77}$$

where $K(n)$ is the normalization constant $K(n) = \left[\sum_{i=1}^{N} |W_i| g \left(\frac{X_i - \mu(n)}{\gamma} \right) \right]^{-1}$.
The function $g(\cdot)$ measures the affinity of the ith observation sample with respect to
the weighted median reference $\mu(n)$. The dispersion parameter γ is user defined.

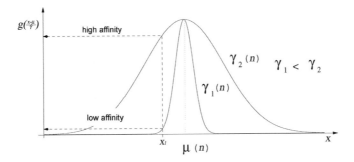

Figure 7.7 The affinity function assigns a low or high affinity to the sample X_i depending
on the location and dispersion parameters $\mu(n)$ and $\gamma(n)$.

The filter structure in (7.77) weights each observation twice: first, according to its
reliability through $g(\cdot)$, and second, according to its natural order through the W_is.
WM affine estimates are therefore based on observations that are both, reliable and
favorable due to their natural order. Observations that fail to meet either, or both,
criteria have only a limited influence on the estimate.

The affinity function can take on many forms. The exponential distance

$$g \left(\frac{X_i - \mu(n)}{\gamma} \right) = \exp \left(-\frac{(X_i - \mu(n))^2}{\gamma^2} \right) \qquad (7.78)$$

is commonly used. Figure 7.7 depicts the affinity measure assigned by (7.78) as
samples deviate from the point of reference $\mu(n)$. When the context is clear, WM
affine FIR filters are referred to as WM affine filters.

Note that the values of $g(\cdot)$ depend on the distance of the observation X_i to the
weighted median $\mu(n)$. While the affinity weighting leaves samples located close to
$\mu(n)$ unaltered, the magnitude of samples distant from $\mu(n)$ is, in general, reduced.
Note, that the total weight ascribed to the observations in (7.77) varies. Thus, the
normalization $K(n)$ is needed to guarantee unbiasedness of the filter as a location
estimator or low-pass filter, for instance.

The flexibility provided by the tunable affinity function translates to the filter characteristics of the estimator. By varying the dispersion parameter γ certain properties
of the WM affine filter can be stressed:

> *Large values of γ emphasize the linear properties of the filter whereas small
> values of γ put more weight on its order statistics properties.*

Of special interest are the limiting cases. For $\gamma \to \infty$, the affinity function is
constant on its entire domain. The estimator, therefore, weights all observations
merely according to their natural order, that is

$$\lim_{\gamma \to \infty} Y_\gamma(n) = \frac{\sum_{i=1}^{N} W_i X_i}{\sum_{i=1}^{N} |W_i|} \tag{7.79}$$

and the WM affine estimator reduces to a normalized *linear FIR filter*. For $\gamma \to 0$, on the other hand, the affinity function shrinks to a δ–impulse at $\mu(n)$. Thus, the weights W_i are disregarded and the estimate is equal to the weighted median $\mu(n)$, such that

$$\lim_{\gamma \to 0} Y_\gamma(n) = \mu(n). \tag{7.80}$$

The WM affine filter assumes a particularly simple form when the reference point is equal to the sample median, which has proven useful as a reference point in the MTM filter. Accordingly, this estimator is referred to as the *median affine filter*.

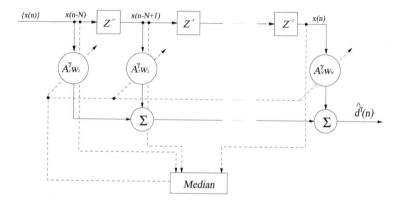

Figure 7.8   Structure of the median affine filter, with $A_i^\gamma = g\left(\frac{X_i - \mu(n)}{\gamma}\right)$.

Figure 7.8 shows a schematic diagram of the (unnormalized) median affine filter. The similarity to a linear transversal FIR filter is obvious. It is interesting to note that the median affine estimator subsumes the MTM filter [125]. The latter is obtained by using an affinity function with rectangular shape and uniform weights.

Medianization of Linear FIR Filters   In order to apply median affine filters, design procedures to determine the dispersion parameter γ and the values of the weights C_i and W_i, for $i = 1, 2, \ldots, N$ are needed. A gradient optimization approach is derived in [73]. This adaptive method, similar in structure and complexity to the LMS algorithm, requires a desired training signal. The set of weights designed by this algorithm minimize the mean square error criterion, assuming stationary observations. Table 7.3 summarizes the adaptive algorithm that can be used to optimize the affinity function parameters and filter weights.

An alternate design approach that is simple and intuitive can be derived from the fact that the median affine filter behaves like a linear FIR filter for $\gamma \to \infty$. Setting γ to a large initial value, one can take advantage of the multitude of linear filter design

Table 7.3 Summary of the Median Affine Adaptive Optimization Algorithm

Parameters:	N = number of taps ν_W = positive weight adaptation constant ν_γ = positive dispersion adaptation constant
Initial Conditions:	$W_i(n) = 0$, $i = 1, 2, \ldots, N$; γ set to a large value
Data (a) Given:	The N observation samples X_i at location n and the desired response at time n, $D(n)$
(b) Compute:	$W_i(n+1) =$ estimate of tap weight at time $n+1$, $i = 1, \ldots, N$
Computation:	$n = 0, 1, \ldots$ $e(n) = D(n) - Y_\gamma(n)$ $W_i(n+1) = W_i(n) + \nu_W e(n)\left(g_i \sum_{k=1}^{N} W_k g_k(\mathrm{sgn}(W_k)X_i - \tanh(W_i)X_k)\right)$ $\gamma(n+1) = \gamma(n) + \nu_\gamma e(n)\left(\sum_{i=1}^{N}(X_i - Y_\gamma(n))W_i \frac{(X_i - \mu(n))^2}{\gamma^2} e^{-(X_i - \mu(n))^2/\gamma}\right)$ where g_i stands for the abbreviated affinity function in (7.78).

methods to find the W_i coefficients of the median affine filter. Holding the W_is constant, the filter performance can, in general, be improved by gradually reducing the value of γ until a desired level of robustness is achieved. During the actual filtering process γ is fixed. Since this process strengthens the median-like properties while weakening the influence of the FIR filter weights, this design approach is referred to as the *medianization* of a linear FIR filter.

FIR Affine L-Filters The trimming affinity function used in (7.74) can be used within the L-filter framework to define the class of *FIR affine L-filters*, which are dual to WM affine FIR filters. In this case, the affinity function is utilized to measure the distance of the order statistics to the reference point that is defined as an FIR estimate. Therefore, the mode of the affinity function is positioned on the FIR estimate. Order-statistics that are distant from the FIR filter output reference are discarded from the L-estimate. The *sigma* and *K-nearest neighbor* estimates are special cases of this filter class. The FIR affine L-filter is defined as follows:

DEFINITION 7.7 *Consider the observations $X_1, X_2, \ldots, X_N$ and their corresponding order statistics $X_{(1)}, X_{(2)}, \ldots, X_{(N)}$. Given a set of N affinity weights $\{C_1, C_2, \ldots, C_N\}$, and a set of N filter weights $\{W_1, W_2, \ldots, W_N\}$, the trimming reference $\mu(n)$ is the FIR filter output $\mu(n) = \sum_{i=1}^{N} C_i X_i$. The (normalized) FIR affine L-filter is then defined as:*

$$Y_\gamma(n) = K(n) \sum_{i=1}^{N} g\left(\frac{X_{(i)} - \mu(n)}{\gamma}\right) W_i X_{(i)} \qquad (7.81)$$

where $K(n)$ is the normalization constant $K(n) = [\sum_{i=1}^{N} |W_i| g\left(\frac{X_{(i)} - \mu(n)}{\gamma}\right)]^{-1}$.
The function $g(\cdot)$ measures the affinity of the ith order-statistic $X_{(i)}$ with respect to the FIR filter output reference $\mu(n)$. The dispersion parameter γ is user defined.

When the context is clear, FIR affine L-filters are referred to as FIR affine filters. FIR affine filters weight the order statistics first according to their affinity to the FIR estimate $\mu(n)$, and second according to their rank. The estimate, therefore, is based mainly on those order statistics that are simultaneously close to the FIR estimate and preferable due to their rank-order. The affinity function can take on many forms. The exponential distance

$$g\left(\frac{X_{(i)} - \mu(n)}{\gamma}\right) = \exp\left(-\frac{(X_{(i)} - \mu(n))^2}{\gamma}\right) \qquad (7.82)$$

is commonly used. Like the WM affine filter, FIR affine filters reduce to their basic structure at the limits of the dispersion parameter γ:

$$\lim_{\gamma \to 0} Y_\gamma(n) = \sum_{i=1}^{N} C_i X_i \qquad (7.83)$$

and

$$\lim_{\gamma \to \infty} Y_\gamma(n) = \frac{\sum_{i=1}^{N} W_i X_{(i)}}{\sum_{i=1}^{N} |W_i|}. \qquad (7.84)$$

Thus, the FIR affine L-filter reduces to an FIR filter with coefficients C_i and to a (normalized) L-filter with coefficients W_i for $\gamma \to 0$ and $\gamma \to \infty$, respectively.

A special case of the FIR affine filter emerges when the coefficients C_i are chosen such that the FIR filter reduces to an identity operation, such that the order statistics are related to the center observation sample. The obtained estimator is referred to as the *center affine filter*. Clearly, the center affine filter reduces to an identity operation, that is, it is *linearized*, for $\gamma \to 0$.

EXAMPLE 7.4 (ELIMINATION OF INTERFERENCE OF THE DWD)

A powerful tool for the analysis of signals are their time-frequency representations (TFRs). The Wigner distribution (WD)

$$WD_x(t, f) = \int_\tau x\left(t + \frac{\tau}{2}\right) x^*\left(t - \frac{\tau}{2}\right) e^{-j2\pi f\tau} d\tau, \qquad (7.85)$$

in particular satisfies a number of desirable mathematical properties and features optimal time-frequency concentration, see Cohen (1995) [53]. Its use in practical applications, however, has been limited by the presence of cross terms. The Wiener distribution of the sum of two signals $x(t) + y(t)$

$$WD_{x+y}(t, f) = WD_x(t, f) + 2\mathbf{Re}\left(WD_{x,y}(t, f)\right) + WD_y(t, f) \qquad (7.86)$$

includes the cross term $2\mathbf{Re}\left(WD_{x,y}(t, f)\right)$ where $WD_{x,y}$ is defined as:

$$WD_{x,y}(t, f) = \int_{-\infty}^{\infty} x\left(t + \frac{\tau}{2}\right) y^*\left(t - \frac{\tau}{2}\right) e^{-j2\pi f\tau} d\tau. \qquad (7.87)$$

Cross terms are problematic, specially if the WD is to be studied by a human analyst. Consequently, they have been studied extensively. It is known that cross terms lie between two auto components and are oscillatory with a frequency that increases with the time-frequency distance between the auto components, leading to oscillations of relatively high frequency (See Fig. 7.9a). These oscillations can be attenuated by a smoothing operation that usually produces: (a) A (desired) partial attenuation of the interference terms, (b) a (undesired) broadening of signal terms, that is, a loss of time-frequency concentration, (c) a (sometimes undesired) loss of some of the mathematical properties of the WD (See Fig. 7.9b).

A filtering scheme that achieves effect (a) while avoiding effect (b) and also effect (c) if required is needed. Beginning with a time-frequency distribution, in this case the Discrete WD (DWD), a filtering process should be carried out over the whole TFD plane. A *center affine filter*, that is, an affine filter whose reference point is the center sample in the observation window, provides a solution to this problem. See Arce and Hasan (2000) [11]. The requirements of the problem are to obtain a response equal to zero if the observation window is in the cross-term region and equal to the center sample of the window if the observation window is in the auto-term region. To achieve this, some specifications about the filtering structure need to be done:

(1) The Gaussian affinity function is used to calculate the affinity of the samples.

(2) The absolute value of the samples is used instead of their actual value to calculate the affinities. The reference point is set as the absolute value of the center sample.

(3) The tuning parameter γ is made proportional to the local variance at which the window is centered. Thus, a higher value of γ is obtained for observations in the cross-term region and a smaller one for observations in the auto-term region.

(4) The TFR is obtained by filtering the original WD samples using the affinity functions of the corresponding absolute valued samples.

To estimate the tuning parameter γ, the variance of some selected samples of the observation window (a subwindow of size $l \times l$ around the center sample) is calculated. In the cross-term region, the samples chosen to calculate the variance are positive and negative and have similar magnitudes. The corresponding variance will be high and so will be γ. On the other hand, the samples in the auto-term region are all positive and have similar magnitudes leading to a much lower variance and a low value of γ.

The test signal $X(n)$ is a 128-point computer generated signal made up of a Gaussian pulse and a parabolic frequency modulated signal.

$$X(n) = r_{1,128}(n)e^{-j2\pi(-0.154n+0.0048n^2)} + r_{15,40}(n)e^{-\frac{(n-27)^2}{100}}e^{\frac{j\pi n}{2}} \qquad (7.88)$$

where $r_{a,b}(n)$ is the gating function:

$$r_{a,b}(n) = \begin{cases} 1, & a \le n \le b \\ 0, & \text{otherwise.} \end{cases} \qquad (7.89)$$

Figure 7.9a shows the DWD of the signal having auto components well localized but numerous high amplitude oscillating cross-terms. Figure 7.9b shows the pseudo-smoothed DWD (using a 13-point Gaussian time smoothing window and 31-point Gaussian frequency smoothing window) of the test signal, reducing the cross terms by both frequency and time direction smoothing. Its interpretation is much easier, but the signal component localization becomes coarser. Figure 7.9c shows the Choi-Williams distribution (1989) [45] of the given signal with the kernel width $\sigma = 1$. Again, most of the cross terms are gone at the cost of reduced localization. Obvious problems, whenever the signal components overlap in time and frequency, are also visible. Figure 7.10a shows the representation given in Jones and Baraniuk (1993) [108]. A signal-dependent kernel is designed for the test signal using a radial Gaussian kernel. This representation is referred to as the Baraniuk-Jones method-1. This method fails to track the smoothly varying parabolic chirp as result that component looks like two connected linear chirps. Figure 7.10b shows the representation given in Jones and Baraniuk (1995) [109]. Here a time-adaptive radially Gaussian kernel is used. The results are adequate, although some loss in auto component localization occurs. In addition, the computation cost is high as the kernel is computed in a local window sliding over the signal. Figure 7.10c shows the Affine Center filtered TFR. An almost complete reduction in cross terms is attained without losing the resolution and localization provided by the Wigner distribution.   ∎

EXAMPLE 7.5 (ISAR IMAGE DENOISING)

To further illustrate the attributes of the center affine filter, consider the denoising of the ISAR image shown in Figure 7.11a[4]. ISAR images emerge from the mapping of the reflectivity density function of the target onto the range-Doppler plane.

---

[4]Data provided by Victor C. Chen, Airborne Radar, Radar Division Naval Research Laboratory, Washington DC.

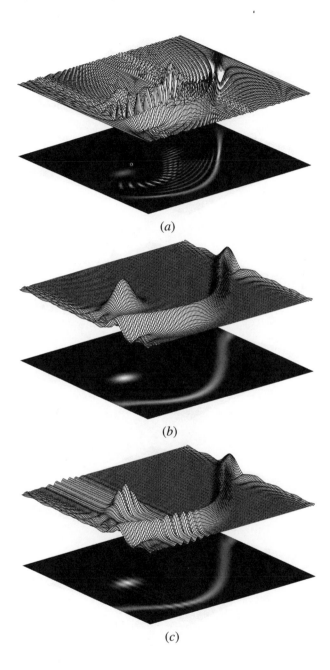

(a)

(b)

(c)

Figure 7.9 Time-frequency representation of the signal, $x_b(n)$, using (a) Wigner distribution, (b) pseudo-smoothed Wigner distribution, (c) Choi-Williams distribution with spread factor $\sigma = 1$.

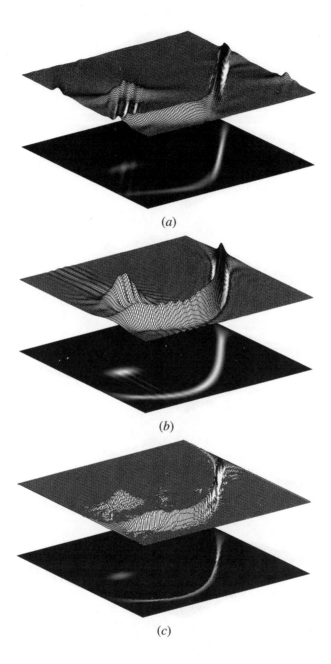

(a)

(b)

(c)

Figure 7.10 Time-frequency representation of the signal, $x_b(n)$ (continued). (a) Baraniuk-Jones distribution (Method 1), (b) Baraniuk-Jones (method 2), and (c) Center Affine filtered WD, $L = 11$.

Difficulties in target identification arise from the fact that radar backscatters from the target are typically embedded in heavy clutter noise. It is desirable to remove the noise without altering the target backscatters, which exhibit pulse-like features in nonGaussian noise. We compare the performance of a center affine filter to that of a weighted median filter and an $L\ell$-filter. The ISAR image is a 128×128, 8 bits/pixel intensity image of a B-727. The various filters were adaptively optimized to extract signal features embedded in background ISAR noise. The $L\ell$-filter was found by standard minimization of the MSE. All filters utilize a 5×5 observation window and were trained in a single run over the entire training images.

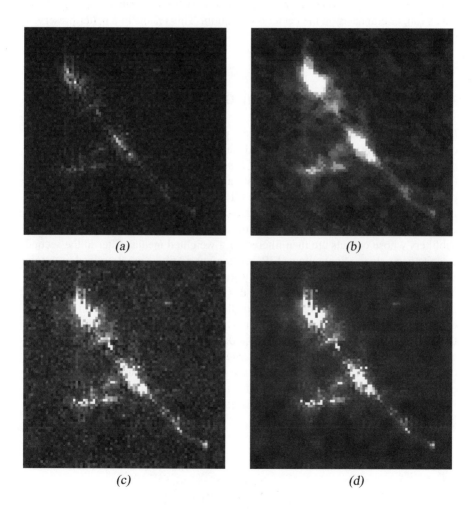

(a)   *(b)*

(c)   *(d)*

Figure 7.11 ISAR feature enhancing: (*a*) unprocessed ISAR image, (*b*) WM filter output, (*c*) $L\ell$-filter output, (*d*) center affine filter output.

Figure 7.11 shows the $L\ell$, WM, and center affine filtering outputs, respectively. The WM filter is effective at reducing the noise level but the signal details are blurred. The $L\ell$-filter output shown in Figure 7.11c preserves the plane details better but is not effective in removing the clutter noise. The center affine filter removes the background noise to a large extent while preserving the plane in all its details. While operating in the background noise, all observations are close to the center sample. Thus, the center affine filter behaves like an L-filter smoothing the clutter noise. When encountering backscatters from the plane, the center affine filter considers only those pixels with similar intensity as the center sample, and thus preserves the plane details. Both the WM filter and the $L\ell$-filter suffer from the inflexibility of their structure. The WM filter assigns a high weight to the center sample. This weight, however, is not large enough to preserve single outlying observations. The $L\ell$-filter puts stronger emphasis on the center observation. This results in a better preserved plane at the cost of poor noise suppression. ∎

7.6 LINEAR COMBINATION OF WEIGHTED MEDIANS

$L\ell$-filters and Median/Linear hybrid filters provide two distinct approaches to the combination of the attributes of linear and median filters. A third approach, proposed by Choi et al.(2001) [46], is the class of Linear Combination of Weighted Medians (LCWM). This class of filters represents a simple alternative to the ones shown previously. The LCWM is the dual of the FIR-Median Hybrid (FMH) filter described in Nieminen and Neuvo (1987) [147] depicted in Figure 7.12a. FMH filters are composed of two stages. In the first, the input signal is filtered by several FIR subfilters whose outputs are then filtered by a weighted median filter in the second stage to generate the final output. FIR-Median hybrid filters are described in Astola et al. (1989) [20], have been extensively studied in Nieminen et al. (1987) [147], Yin and Neuvo (1993) [200], and Yin et al. (1996) [201].

Since in the LCWM, the first stage is dedicated to the weighted median filters, the outlier rejection capabilities of these filters is greater compared to that of FMH filters that perform linear operations on the input data, including the outliers. The LCWM also leads itself to simple analysis and design. The structure of the LCWM is based on the structure used in the design of linear-phase FIR high-pass filters. These are easily obtained by changing the sign of the filter coefficients of a FIR low-pass filter in the odd positions. In consequence, this filter can be represented as the difference between two low-pass subfilters.

A general N-tap FIR filter is given by:

$$\begin{aligned} Y(n) &= \sum_{k=0}^{N-1} h(k)X(n-k) \\ &= \mathbf{h}^{\mathrm{T}}\mathbf{X} \end{aligned} \tag{7.90}$$

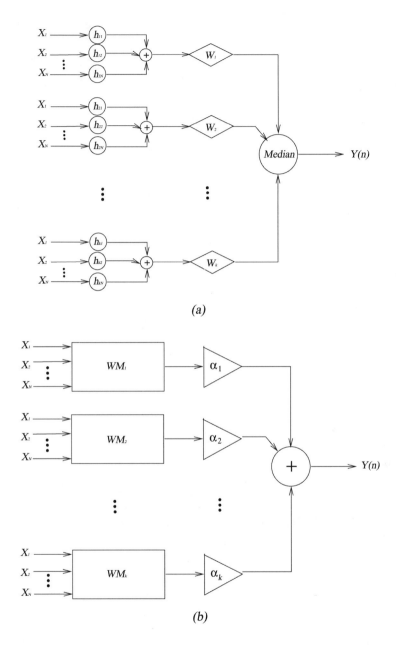

(a)

(b)

Figure 7.12 FIR-median hybrid filter and linear combination of weighted median filter structures.

where $\mathbf{h} = [h(0),\ h(1), \ldots,\ h(N-1)]^T$ represents the filter coefficients and $\mathbf{X} = [X(n),\ X(n-1), \ldots,\ X(n-N+1)]^T$ is the input vector. A low pass FIR filter can be obtained by restricting the coefficients to positive values. Let these positive coefficients be denoted by $h_L(k)$. To obtain the high pass FIR, change the sign of the coefficients in the odd positions after reversing the filter window in the time domain. This results in a filter given by:

$$
\begin{aligned}
Y_H(n) &= \sum_{k=0}^{N-1} (-1)^k h_L(N-k-1)X(n-k) \\
&= \sum_{k=0}^{N-1} h_H(k)X(n-k).
\end{aligned}
\tag{7.91}
$$

This filter can be divided into two subfilters as follows:

$$
\begin{aligned}
Y_H(n) &= \sum_{k=0}^{N-1} h_H(k)X(n-k) \\
&= \sum_{k=0}^{N-1} b_1(k)X(n-k) - \sum_{k=0}^{N-1} b_2(k)X(n-k) \\
&= Y_1(n) - Y_2(n),
\end{aligned}
\tag{7.92}
$$

where $b_1(k)$ and $b_2(k)$ are given by:

$$
b_1(k) = \begin{cases} h_H(k) & \text{if } h_H(k) \geq 0 \\ 0 & \text{otherwise} \end{cases}
$$

$$
b_2(k) = \begin{cases} |h_H(k)| & \text{if } h_H(k) < 0 \\ 0 & \text{otherwise.} \end{cases}
\tag{7.93}
$$

The two filters in (7.93) have only non-negative coefficients. They can be normalized so the final output can be written as:

$$
Y(n) = \alpha_1 Y_1^N(n) - \alpha_2 Y_2^N(n)
\tag{7.94}
$$

where

$$
Y_1^N(n) = \frac{\sum_{k=0}^{N-1} b_1(k)X(n-k)}{\sum_{k=0}^{N-1} b_1(k)} \quad Y_2^N(n) = \frac{\sum_{k=0}^{N-1} b_2(k)X(n-k)}{\sum_{k=0}^{N-1} b_2(k)},
\tag{7.95}
$$

where $\alpha_1 = \sum_{k=0}^{N-1} b_1(k)$ and $\alpha_2 = \sum_{k=0}^{N-1} b_2(k)$. In essence, (7.94) synthesizes a general high-pass or band-pass filter, based on linear combination of low pass FIR filters. This concept is next applied to the construction of linear combinations of weighted median filters.

7.6.1 LCWM Filters

The nonlinear counterpart of the linear combination of FIR filters in (7.94) is given by:

$$\tilde{Y}(n) = \alpha_1 Y_1^{WM}(n) - \alpha_2 Y_2^{WM}(n) \tag{7.96}$$

where Y_1^{WM} and Y_2^{WM} are WM smoothers, that is, they admit only positive weights. These smoothers are designed based on their spectral response using the algorithms derived in Section 6.2.

EXAMPLE 7.6

Consider the linear FIR high pass filter $h = [\frac{1}{6}, -\frac{1}{6}, \frac{1}{6}, -\frac{1}{6}, \frac{1}{6}, -\frac{1}{6},]^T$. The coefficients can be rewritten in the form of (7.94) as:

$$\alpha_1 = \frac{1}{2}, \qquad \alpha_2 = \frac{1}{2}$$

$$b_1 = [\frac{1}{6}, 0, \frac{1}{6}, 0, \frac{1}{6}, 0]$$

$$b_2 = [0, \frac{1}{6}, 0, \frac{1}{6}, 0, \frac{1}{6}].$$

According to Mallows' theory and using the spectral design of weighted medians shown in Section 6.2, it can be shown that the median equivalents to the filters b_1 and b_2 are $W_1 = \langle 1, 0, 1, 0, 1, 0 \rangle$ and $W_2 = \langle 0, 1, 0, 1, 0, 1 \rangle$. The frequency response of both, linear and median filters were approximated [5] and the results are shown in Figure 7.13. Notice that both filters produce very similar results.

∎

The procedure described above can be generalized to include more than two WM filters with overlapping or non-overlapping windows. The general form of Equation (7.94) is:

$$Y(n) = \sum_{i=1}^{K} \alpha_i \sum_{j=0}^{N-1} b_i(j) X(n-j) \tag{7.97}$$

where K is the number of subfilters, and $b_i(j)$ is the $j + 1$st coefficient of the ith subfilter. A matrix containing all the subfilters can be defined as:

$$\mathbf{b}_i = [b_i(0), b_i(1), \dots, b_i(N-1)] \tag{7.98}$$

---

[5]To approximate the frequency response of the LCWM filters, 10,000 samples of standard Gaussian i.i.d. samples were inputted and the spectra of the outputs was calculated with the Welch method [192]. The procedure was repeated 50 times to get an ensemble average.

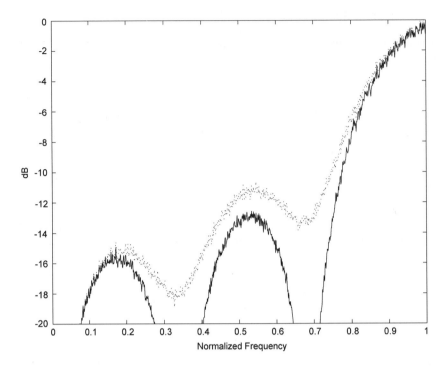

Figure 7.13 Estimated frequency response of a LCWM filter (dotted) and the linear FIR filter used as reference for its design (solid).

$$\mathbf{B} = \begin{bmatrix} \mathbf{b}_1 \\ \vdots \\ \mathbf{b}_K \end{bmatrix}, \tag{7.99}$$

so that (7.97) can be represented as:

$$y(n) = \vec{\alpha}^{\mathbf{T}}\mathbf{B}\mathbf{X} = \mathbf{h}^{\mathbf{T}}\mathbf{X} \tag{7.100}$$

where $\vec{\alpha} = [\alpha_1, \ldots, \alpha_K]^T$ is the weighting factor vector and $\mathbf{h} = [h_1, h_2, , \ldots, h_N]^T$. Solving the previous for $\vec{\alpha}$:

$$\vec{\alpha}^{\mathbf{T}} = \mathbf{h}^{\mathbf{T}}\mathbf{B}^{-1}. \tag{7.101}$$

Using the theory developed in Section 6.2, the nonlinear equivalents of $\mathbf{b}_i$ and $\mathbf{B}$ can be found as $\mathbf{w}_i$ and $\mathbf{W}$. The LCWM filter emerges from (7.97) as:

$$\begin{aligned} Y_{LCWM}(n) &= \sum_{i=1}^{K} \alpha_i \langle \mathbf{w}_i \rangle \\ &= \vec{\alpha}^{\mathbf{T}} \langle \mathbf{W} \rangle \end{aligned} \tag{7.102}$$

where $\langle \mathbf{w}_i \rangle = \text{MEDIAN}[\mathbf{w} \diamond \mathbf{X}]$ and

$$\langle \mathbf{W} \rangle = \begin{bmatrix} \langle \mathbf{w}_1 \rangle \\ \vdots \\ \langle \mathbf{w}_K \rangle \end{bmatrix}. \tag{7.103}$$

The structure of the filter is depicted in Figure 7.12*b*.

7.6.2 Design of LCWM filters

Having defined the structure of LCWM filters, a procedure to design them to follow a desired spectral profile is of interest. Choi et al. (2001) [46] provide a systematic approach. The first step is to design a prototype FIR using any of the standard FIR design tools. Next a LCWM is obtained by using a transformation of the prototype.

Before proceeding to the design of the LCWM filters, a brief review of some basic concepts of linear algebra is necessary. A real N-dimensional vector space $\mathbf{R}^N$ is spanned by N linearly independent vectors $\mathbf{b}_1, \ldots, \mathbf{b}_N$, each one with N components. This set of vectors is a basis of the space. That is, each vector in the space can be represented as a linear combination of elements of the basis as:

$$\begin{aligned} \mathbf{h}^T &= \alpha_1 \mathbf{b}_1 + \ldots + \alpha_N \mathbf{b}_N \\ &= \vec{\alpha}^{\mathbf{T}} \mathbf{B}. \end{aligned} \tag{7.104}$$

Suppose $\mathbf{h}$ represents a FIR filter. Then the filter can be represented as a linear combination of subfilters with coefficients given by the vectors $\mathbf{b}_1, \ldots, \mathbf{b}_N$. Equation (7.104) is identical to (7.101). The central issue is now how to determine the basis $\mathbf{B}$.

The first step is to define the vector space. In order to do that, the size $M < N$ of the subfilters $\mathbf{b}_i$ in (7.104) must be predefined by the user. The number of ways a subvector of M elements can be chosen from a vector of N elements is $\binom{N}{M}$. These subvectors are represented in a $\binom{N}{M} \times N$ matrix called $\mathbf{B}_{N,M}$ and they constitute the vector space. Each row of the matrix represents a different subvector and each one of its elements will be a one if the corresponding element belongs to the subvector or a zero if it does not.

Once $\mathbf{B}_{N,M}$ has been built, a basis for the space of subvectors of M elements of an N element vector can be built by choosing N linearly independent rows of $\mathbf{B}_{N,M}$ using a recursive row search algorithm. These vectors are stored in another matrix $(\mathbf{B}_{\mathbf{P}(N,M)})$. The row search algorithm can be summarized as:

$$\mathbf{B}_{\mathbf{P}(N,M)} = \begin{bmatrix} \mathbf{1}_{N-1 \times 1} & \mathbf{B}_{\mathbf{P}(N-1,M-1)} \\ 0 & \mathbf{B}_{N-1,M}(1, 1:N-1) \end{bmatrix} \tag{7.105}$$

where $\mathbf{B}_{N-1,M}(1, 1 : N - 1)$ represents the first row of the matrix $\mathbf{B}_{N-1,M}$ and is a row vector composed by $N - 1$ ones followed by all zeros. $\mathbf{B}_{\mathbf{P}(N,1)} = \mathbf{I}_N$, where $\mathbf{I}_N$ is the $N \times N$ identity matrix, and $\mathbf{B}_{\mathbf{P}(N,N)} = \mathbf{1}_{1 \times N}$.

EXAMPLE 7.7 (ROW SEARCH ALGORITHM)

Use the row search algorithm to calculate $\mathbf{B}_{\mathbf{P}(6,3)}$.

According to (7.105), in order to calculate $\mathbf{B}_{\mathbf{P}(6,3)}$, $\mathbf{B}_{\mathbf{P}(5,2)}$ has to be calculated first, and the calculation of $\mathbf{B}_{\mathbf{P}(5,2)}$ requires the calculation of $\mathbf{B}_{\mathbf{P}(4,1)}$, that is, the 4×4 identity matrix. The process is summarized as:

$$
\mathbf{B}_{\mathbf{P}(4,1)} = \begin{bmatrix} 1 & 0 & 0 & 0 \\ 0 & 1 & 0 & 0 \\ 0 & 0 & 1 & 0 \\ 0 & 0 & 0 & 1 \end{bmatrix} \Rightarrow \mathbf{B}_{\mathbf{P}(5,2)} = \begin{bmatrix} 1 & 1 & 0 & 0 & 0 \\ 1 & 0 & 1 & 0 & 0 \\ 1 & 0 & 0 & 1 & 0 \\ 1 & 0 & 0 & 0 & 1 \\ 0 & 1 & 1 & 0 & 0 \end{bmatrix}
$$

$$
\Rightarrow \mathbf{B}_{\mathbf{P}(6,3)} = \begin{bmatrix} 1 & 1 & 1 & 0 & 0 & 0 \\ 1 & 1 & 0 & 1 & 0 & 0 \\ 1 & 1 & 0 & 0 & 1 & 0 \\ 1 & 1 & 0 & 0 & 0 & 1 \\ 1 & 0 & 1 & 1 & 0 & 0 \\ 0 & 1 & 1 & 1 & 0 & 0 \end{bmatrix} \tag{7.106}
$$

∎

In order to obtain the matrix $\mathbf{B}$ of linear filters in (7.104), the "1"s in $\mathbf{B}_{\mathbf{P}(N,M)}$ should be replaced by their corresponding filter coefficients. For reasons that will be clear shortly, each "1" in $\mathbf{B}_{\mathbf{P}(N,M)}$ will be replaced by $\frac{1}{M}$ to obtain $\mathbf{B}$. In this way all the linear subfilters will perform standard mean operations. Once $\mathbf{B}$ is found, $\vec{\alpha}$ can be calculated using (7.101).

To design the LCWM filter, the median equivalent to each one of the subfilters in $\mathbf{B}$ should be found. According to the theory of Mallows and given that all the coefficients in the linear filters are the same, their median equivalents will be standard median operators, that is, all their nonzero weights will be set to one. In consequence $\mathbf{W} = \mathbf{B_p}$.

EXAMPLE 7.8

Design a LCWM with 3-tap subfilters from the 6-tap linear filter $\mathbf{h} = [-0.3327, 0.8069, -0.4599, -0.1350, 0.0854, 0.0352]^T$.

In this case $\mathbf{B}_{6,3}$, $\mathbf{B}_{\mathbf{P}(6,3)}$ and $\mathbf{B}$ will be:

$$\mathbf{B}_{6,3} = \begin{bmatrix} 1\ 1\ 1\ 0\ 0\ 0 \\ 1\ 1\ 0\ 1\ 0\ 0 \\ 1\ 1\ 0\ 0\ 1\ 0 \\ 1\ 1\ 0\ 0\ 0\ 1 \\ 1\ 0\ 1\ 1\ 0\ 0 \\ 1\ 0\ 1\ 0\ 1\ 0 \\ 1\ 0\ 1\ 0\ 0\ 1 \\ 1\ 0\ 0\ 1\ 1\ 0 \\ 1\ 0\ 0\ 1\ 0\ 1 \\ 1\ 0\ 0\ 0\ 1\ 1 \\ 0\ 1\ 1\ 1\ 0\ 0 \\ 0\ 1\ 1\ 0\ 1\ 0 \\ 0\ 1\ 1\ 0\ 0\ 1 \\ 0\ 1\ 0\ 1\ 1\ 0 \\ 0\ 1\ 0\ 1\ 0\ 1 \\ 0\ 1\ 0\ 0\ 1\ 1 \\ 0\ 0\ 1\ 1\ 1\ 0 \\ 0\ 0\ 1\ 1\ 0\ 1 \\ 0\ 0\ 1\ 0\ 1\ 1 \\ 0\ 0\ 0\ 1\ 1\ 1 \end{bmatrix} \qquad \mathbf{B}_{\mathbf{P}(6,3)} = \begin{bmatrix} 1\ 1\ 1\ 0\ 0\ 0 \\ 1\ 1\ 0\ 1\ 0\ 0 \\ 1\ 1\ 0\ 0\ 1\ 0 \\ 1\ 1\ 0\ 0\ 0\ 1 \\ 1\ 0\ 1\ 1\ 0\ 0 \\ 0\ 1\ 1\ 1\ 0\ 0 \end{bmatrix} = \mathbf{W}$$

$$\mathbf{B} = \begin{bmatrix} \frac{1}{3}\ \frac{1}{3}\ \frac{1}{3}\ 0\ 0\ 0 \\ \frac{1}{3}\ \frac{1}{3}\ 0\ \frac{1}{3}\ 0\ 0 \\ \frac{1}{3}\ \frac{1}{3}\ 0\ 0\ \frac{1}{3}\ 0 \\ \frac{1}{3}\ \frac{1}{3}\ 0\ 0\ 0\ \frac{1}{3} \\ \frac{1}{3}\ 0\ \frac{1}{3}\ \frac{1}{3}\ 0\ 0 \\ 0\ \frac{1}{3}\ \frac{1}{3}\ \frac{1}{3}\ 0\ 0 \end{bmatrix}$$

(7.107)

$\vec{\alpha}$ can be calculated now as:

$$\vec{\alpha}^T = \mathbf{h}^T \mathbf{B}^{-1} = [0.0430,\ 1.0176,\ 0.2563,\ 0.1057,\ -2.4207,\ 0.9980] \quad (7.108)$$

The frequency responses of the filters obtained are shown in Figure 7.14

∎

7.6.3 Symmetric LCWM Filters

If the FIR filter used as a reference to design the LCWM filter is linear phase, the number of subfilters can be reduced since a $(2N + 1)$-tap linear filter has only $N + 1$ independent coefficients (a $2N$-tap linear filter has N).

Denote by $\mathbf{h}' = [h'_1,\ h'_2, \ldots,\ h'_{N+1}]^T$ with $h'_i = h(N + i - 1)$ the vector consisting of the independent coefficients of the linear FIR $\mathbf{h}$ (the second half of $\mathbf{h}$). The reduced number of coefficients leads to a reduced $(N + 1) \times (N + 1)$ basis matrix $\mathbf{B}_{\mathbf{p}}'$. If the LCWM to be designed is made up of $(2M + 1)$-tap subfilters, $\mathbf{B}_{\mathbf{p}}'$ will consist of $N + 1$ independent rows of $\mathbf{B}_{N+1,M+1}$. The corresponding $\mathbf{B}_{\mathbf{p}}$ is found by left unfolding $\mathbf{B}_{\mathbf{p}}'$ with respect to its first column, given the relationship between $\mathbf{h}$ and $\mathbf{h}'$. The matrix $\mathbf{B}$ contains the median subfilters of the LCWM. To calculate the linear combination coefficients $\vec{\alpha}$, the linear equivalents of the filters in $\mathbf{B}_p$ have to be calculated using sample selection probabilities to obtain the matrix $\mathbf{B}$.

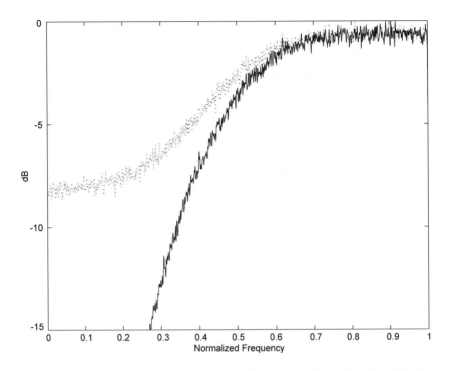

Figure 7.14   Frequency response of a LCWM with 3-tap medians (dotted) and the linear FIR taken as a reference for its design (solid)

Given the symmetry properties that have been used during the process, the matrix **B** will also be symmetric. The last $N + 1$ columns of **B** are taken to build the matrix **B'** from which $\vec{\alpha}$ is calculated as[6]:

$$\vec{\alpha}^T = \mathbf{h'^T}\mathbf{B'}^{-1} \tag{7.109}$$

EXAMPLE 7.9

Design a LCWM with subfilters of 5 and 6 taps based on a 7-tap bandpass filter with cutoff frequencies $[0.3\ 0.7]$. The symmetric linear filter is $h = [0,\ -0.1597,\ 0,\ 0.6806,\ 0,\ -0.1597,\ 0]^T$. For this case $N = 3$ and $M = 2$, resulting in:

---

[6]If $\mathbf{h'}$ is built as the first half of $\mathbf{h}$ instead of the second, $\mathbf{B_p'}$ should be right unfolded to create $\mathbf{B_p}$ and $\mathbf{B'}$ will be constituted by the first $N + 1$ columns of $\mathbf{B}$

$$\mathbf{B'_p} = \mathbf{B}_{4,3} = \begin{bmatrix} 1 & 1 & 1 & 0 \\ 1 & 1 & 0 & 1 \\ 1 & 0 & 1 & 1 \\ 0 & 1 & 1 & 1 \end{bmatrix} \Rightarrow \mathbf{B_p} = \begin{bmatrix} 0 & 1 & 1 & 1 & 1 & 0 \\ 1 & 0 & 1 & 1 & 0 & 1 \\ 1 & 1 & 0 & 1 & 0 & 1 & 1 \\ 1 & 1 & 1 & 0 & 1 & 1 & 1 \end{bmatrix}. \tag{7.110}$$

The first three rows of $\mathbf{B}_p$ represent 5-tap weighted medians so the SSPs for this vectors are $\frac{1}{5}$. The last one represents a 6-tap weighted median and the SSPs for this vector are $\frac{1}{6}$.

$$\mathbf{B} = \begin{bmatrix} 0 & \frac{1}{5} & \frac{1}{5} & \frac{1}{5} & \frac{1}{5} & \frac{1}{5} & 0 \\ \frac{1}{5} & 0 & \frac{1}{5} & \frac{1}{5} & \frac{1}{5} & 0 & \frac{1}{5} \\ \frac{1}{5} & \frac{1}{5} & 0 & \frac{1}{5} & 0 & \frac{1}{5} & \frac{1}{5} \\ \frac{1}{6} & \frac{1}{6} & \frac{1}{6} & 0 & \frac{1}{6} & \frac{1}{6} & \frac{1}{6} \end{bmatrix} \Rightarrow \mathbf{B'} = \begin{bmatrix} \frac{1}{5} & \frac{1}{5} & \frac{1}{5} & 0 \\ \frac{1}{5} & \frac{1}{5} & 0 & \frac{1}{5} \\ \frac{1}{5} & 0 & \frac{1}{5} & \frac{1}{5} \\ 0 & \frac{1}{6} & \frac{1}{6} & \frac{1}{6} \end{bmatrix}. \tag{7.111}$$

Having $\mathbf{B'}$, $\vec{\alpha}$ can be calculated as:

$$\vec{\alpha}^{\mathbf{T}} = [0.6806, \ 0, \ -0.1597, \ 0] \times \begin{bmatrix} \frac{1}{5} & \frac{1}{5} & \frac{1}{5} & 0 \\ \frac{1}{5} & \frac{1}{5} & 0 & \frac{1}{5} \\ \frac{1}{5} & 0 & \frac{1}{5} & \frac{1}{5} \\ 0 & \frac{1}{6} & \frac{1}{6} & \frac{1}{6} \end{bmatrix}^{-1}$$

$$= [0.8682, \ 1.6667, \ 0.8682, \ -3.0418]. \tag{7.112}$$

The frequency response of the resultant LCWM and the original linear FIR are shown in Figure 7.15

■

EXAMPLE 7.10 (ROBUST FREQUENCY-SELECTIVE FILTERING)

Filters that are jointly robust and frequency–selective are of interest. Here, the filter coefficients of a linear FIR bandpass filter are used to design a $L\ell$, a median affine filter, and a LCWM filter. Let $h_1, h_2, \ldots, h_N$ be the coefficients of an FIR bandpass filter of size $N = 41$ (designed with a Hamming window and cutoff frequencies 0.2ω and 0.4ω, where ω denotes the Nyquist frequency) and output $\hat{D}_{FIR}$. The $L\ell$-bandpass-filter is designed by setting the weights associated with observation X_i equal to C_i whenever X_i is not an extreme order statistic. When X_i corresponds to the two smallest or largest observations, its weight is set to zero. Thus, the weight vector of the $L\ell$-filter is given by

$$\mathbf{W}_{L\ell}^T = [0, 0, h_1, \ldots, h_1, 0, 0 | 0, 0, h_2, \ldots, h_2, 0, 0 | \ldots | 0, 0, h_N, \ldots, h_N, 0, 0].$$

This weighting pattern rejects outlying observations and weights the remaining samples by the corresponding FIR filter coefficients.

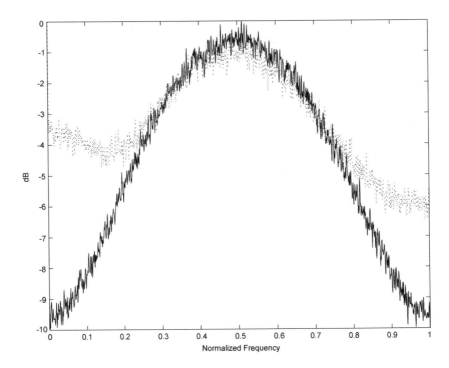

Figure 7.15   Frequency response of a bandpass LCWM (dotted) and the symmetric linear filter (solid) used as a reference.

The median affine bandpass filter is designed according to the "FIR medianization" method, that is, the coefficients are chosen equal to those of the linear bandpass and γ is successively decreased until a desired level of robustness is achieved, in this case $\gamma = 8$. This value is fixed during the filtering process. To improve the signal tracking ability of the median affine filter, the sample median of a subobservation-window of size $M = 5$ serves as a reference point. The LCWM filter was designed using standard medians of window size 5 and the method developed in Section 7.6.2 resulting in

$$
\begin{aligned}
\vec{\alpha} = [\ & 0.0415, & 0.0159, & -0.0470, & -0.0621, & -0.0191, & 0.0245, & 0., \\
& -0.0391, & 0.0491, & 0.2609, & 0.3294, & 0., & -0.5561, & -0.7722, \\
& -0.2814, & 0.5725, & 0.9958, & 0.5725, & -0.2814, & -0.7722, & -0.5561, \\
& 0., & 0.3294, & 0.2609, & 0.0491, & -0.0391, & 0., & 0.0245, \\
& -0.0191, & -0.0621, & -0.0470, & 0., & 0.0256, & 0.0187, & 0.0033, \\
& -0.0026, & 0., & -0.0178, & -0.0024, & 0.0035, & 0.0009]
\end{aligned}
$$

The performance of the above filters on a 2048-sample quad-chirp (sinusoidal waveform with quadratically increasing frequency) clean and additively contaminated by α-stable noise ($\alpha = 1.2$) to simulate an impulsive environment is compared in Figures 7.16 and 7.17. The corresponding mean square and absolute errors for the noisy case are given in Table 7.4. As expected, the impulsive noise is detrimental for the linear filter. The $L\ell$-filter is robust, but suffers from the blind rejection of the extreme order statistics, which results in artifacts over most of the frequency spectrum. This effect is even stronger for a higher signal-to-noise ratio (not shown). The median affine filter preserves the desired frequency band well, while strongly attenuating the impulsive noise, which is reflected in a MSE that is roughly half of that achieved by the $L\ell$-filter. By comparison of the linear estimate and that of the median affine filter, it can be seen that the latter behaves similar to the linear filter whenever no impulses are present. Finally, note that the $L\ell$-filter uses N^2 coefficients, whereas the median affine filter requires only $N + 1$ coefficients.

Table 7.4 Comparison of filter performance on noise-corrupted chirp.

Filter	MSE	MAE
(a) None	1.07×10^6	23.72
(b) FIR	6.0863	0.5169
(c) LCWM	0.1115	0.1665
(d) $L\ell$	0.0261	0.1280
(e) Median affine	0.0198	0.1036

■

Problems

7.1 Let $X_1, X_2, \ldots, X_N$ be N observations of a constant parameter β in noise such that $X_i = \beta + Z_i$ where Z_i is a zero mean sequence of i.i.d. noise samples, with a parent density function $f(z) = \frac{1}{2}\delta(1 - z) + \frac{1}{2}\delta(1 + z)$.

(a) Find the probability density function of $Z_{(i)}$.

(b) Find the joint probability density function of $Z_{(i)}$ and $Z_{(j)}$ for $i < j$.

(c) For $N = 2$, find the best unbiased L-estimate of the constant parameter.

7.2 Derive the equations for the LMS algorithm for the design of the Median Affine filter shown in Table 7.3, following the next steps:

(a) Define the cost function $J(\gamma, \mathbf{W}) = E\{(d - \hat{d})^2\}$, where $\hat{d}$ represents the output of the median affine filter and d the desired signal.

(b) Find the derivative of the cost function with respect to γ as a function of $\frac{\partial \hat{d}}{\partial \gamma}$. Then show that:

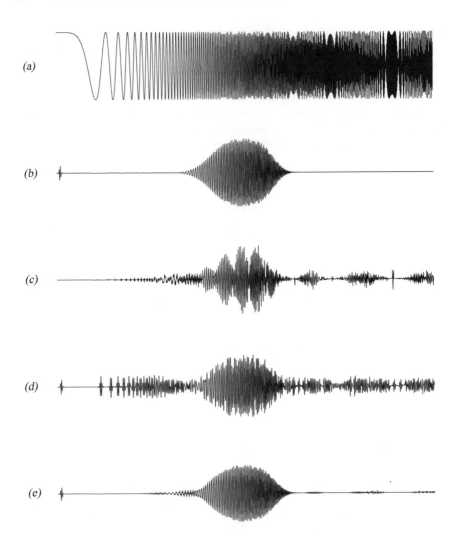

Figure 7.16 (*a*) The chirp signal and the outputs of the (*b*) linear FIR bandpass, (*c*) the LCWM, (*d*) the $L\ell$, (*e*) the median affine filter.

$$\frac{\partial \hat{d}}{\partial \gamma} = \frac{\sum_{i=1}^{N} W_i g(X_i - \hat{d})(X_i - \mu(n))^2}{\gamma^2 \sum_{i=1}^{N} W_i g(X_i - \mu(n))}. \tag{7.113}$$

(c) Calculate $\frac{\partial J}{\partial \gamma}$ and replace it in the gradient based algorithm

$$\gamma(n+1) = \gamma(n) - \nu_\gamma \frac{\partial J}{\partial \gamma} \tag{7.114}$$

to obtain the expression in Table 7.3.

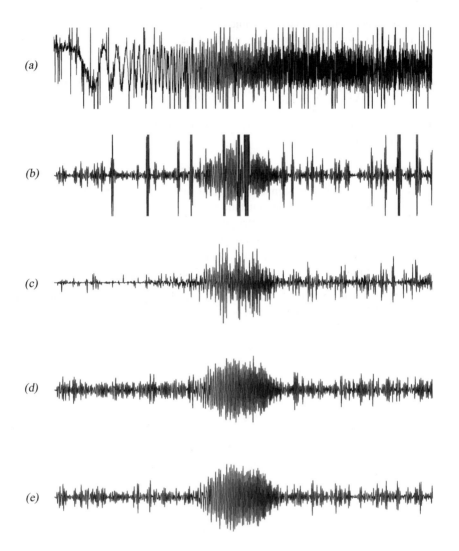

Figure 7.17 (*a*) The noise-corrupted chirp, the outputs of the (*b*) linear FIR bandpass, (*c*) the LCWM, (*d*) the $L\ell$, (*e*) the median affine filter.

(d) Repeat (b) and (c) derivating with respect to the weight W_i and using the gradient based algorithm

$$W_i(n+1) = W_i(n) - \nu_W \frac{\partial J}{\partial W_i} \qquad (7.115)$$

obtain the expression in Table 7.3.   .

7.3 Consider a binary i.i.d. sequence $\{X(n)\}$ with a parent density function $f(X) = p\delta(X - 1) + (1 - p)\delta(X)$, where $X = 1$ with probability p and $X = 0$ with probability $1 - p$.

(a) Find the correlation matrix of the observation vector $\mathbf{X}(n) = [X(n), X(n - 1), \ldots, X(n - N + 1)]^T$, for $N = 2$. Is this correlation matrix singular?

(b) Find the correlation matrix of the $L^1\ell$ observation vector $\mathbf{X}_1$ for $N = 3$ (use stable sorting if needed).

7.4 Consider the filtering problem where the observations X_i follow the model $X_i = S + Z_i$, where the signal is constant and where the noise is i.i.d. The L estimate of location (i.e., the constant signal S), for an N long observation is $\hat{S} = \sum_{j=1}^{N} W_j X_{(j)}$ where $X_{(j)}$ is the jth sample order statistic. Constraining the estimation to be location invariant (i.e., $\mathbf{W}^T \mathbf{e} = 1$ where $\mathbf{e}^T = [1, 1, \ldots 1]^T$), find the expression for the optimal set of L filter coefficients which will minimize the MSE. Assume the noise is symmetrically distributed about zero.

7.5 Design two LCWM filters based on an 11-tap high-pass FIR filter with a cutoff frequency of 0.5 and medians of size 5. Use the algorithms in Sections 7.6.2 and 7.6.3 and compare the results.

Part III

Signal Processing with the Stable Model

8

Myriad Smoothers

The motivations for using stable models, as described in Chapter 2, are simple yet profound. Firstly, good empirical fits are often found through the use of stable distributions on data exhibiting skewness and heavy tails. Secondly, there is solid theoretical justification that nonGaussian stable processes emerge in practice. The third argument for modeling with stable distributions is perhaps the most significant and compelling. Stable distributions satisfy an important generalization of the central limit theorem, which states that the only possible limit of normalized sums of independent and identically distributed terms is stable. A wide variety of impulsive processes found in many applications arise as the superposition of many small independent effects. While Gaussian models are clearly inappropriate, stable distributions thus have the theoretical underpinnings to accurately model these type of impulsive processes [149, 207]. Stable models are thus appealing since the generalization of the central limit theorem explains the apparent contradictions of its ordinary version, which could not naturally explain the presence of heavy-tailed signals.

Having the rich modeling characteristics of stable distributions at hand, several signal processing approaches have been developed which are suitable for processing and analyzing stable processes. One approach which has received considerable attention is based on the concept of *Fractional Lower Order Moments* (FLOMs) [149]. FLOM based signal processing has been studied extensively, as detailed in Nikias and Shao (1995). Section 8.1 in this chapter provides an introduction to FLOM smoothers and its applications. A second approach to signal processing for stable models is described, based on the so called *Weighted Myriad* filters derived from the theory of M-estimation. Much like the Gaussian assumption has motivated the development of linear filtering theory, *weighted myriad filters* are motivated by the need of a flexible filter class with increased efficiency in nonGaussian impulsive environments. The remainder sections of this chapter focus on this second approach.

The foundation for these algorithms lies in the definition of the *sample myriad* as a maximum likelihood estimate of location derived from the stable model, leading to cost functions of the form

$$\rho(x) = \log[K^2 + x^2], \tag{8.1}$$

where K is a tunable parameter. Along the range of tuning values of K, the sample myriad enjoys optimality properties in several practical impulsive models. The possibility of tuning the parameter K, provides the myriad filter with a rich variety of modes of operation that range from *highly resistant* mode-type estimators to the very efficient class of linear FIR filters. Since the myriad filter class subsumes that of linear FIR filters, weighted myriad filters can also be tuned to operate efficiently under the Gaussian model. Much like the mean and median have had a profound impact on signal processing, the sample myriad and its related myriad filtering framework lead to a powerful theory upon which *efficient* signal processing algorithms have been developed for applications exhibiting impulsive processes.

8.1   FLOM SMOOTHERS

Under the framework of Gaussian processes, the sample mean of a random variable can be described as the parameter $\bar{\beta}$ that minimizes the second moment of the shifted variable $X - \beta$ over all possible shifts β. This is

$$\mathrm{E}(X) = \bar{\beta} = \arg\min_{\beta} \mathrm{E}(X - \beta)^2. \tag{8.2}$$

Given that second-order moments do not exist with stable processes, but fractional-order moments do, the second moment in (8.2) can be replaced by fractional lower-order moments (FLOMs) to obtain the following measure of location

$$\beta_p = \arg\min_{\beta} \mathrm{E}(|X - \beta|^p), \quad p < 2. \tag{8.3}$$

FLOM smoothers follow from (8.3) where FLOM estimates are computed in the running window $\mathbf{X}(n) = [X_1(n), X_2(n), \ldots, X_N(n)]^T$ as

$$Y(n) = \arg\min_{\beta} \sum_{i=1}^{N} |X_i(n) - \beta|^p \tag{8.4}$$

with $p < 2$. The behavior of FLOM smoothers is markedly dependant on the choice of p. As $p \to 2$, FLOM smoothers resemble the running mean. As p is reduced in value, FLOM smoothers become more robust and its output can tract discontinuities more effectively. Figure 8.1 depicts the effect of varying the value of p in the smoothing of a speech signal. As described in Chapter 4, FLOM smoothers also arise from the location estimation problem under the generalized Gaussian distribution. A plot of the FLOM smoother cost function, found in Figure 4.2 for different values of p, reveals that for $p < 1$, FLOM smoothers are selection type where the output value is equal to that of one of the input samples. The cost function exhibits several local minima, located in the values of the input samples. For $p > 1$, the cost function is convex and the output is not necessarily equal in value to one of the input samples. FLOM smoothers thus represent a family of smoothers indexed by the value assigned to the parameter p.

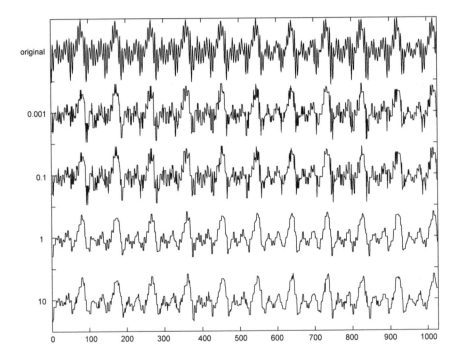

Figure 8.1   FLOM smoothing of a speech signal for different values of p and window size 5.

A drawback of the FLOM smoother class, however, is that their computation in (8.4) is in general nontrivial. A method to overcome this limitation is to force the output of the smoother to be identical in value to one of the input samples. Selection type FLOM smoothers are suboptimal in the sense that the output does not achieve the minimum of the cost function. Selection type FLOM smoothers have been studied by Astola (1999) [23] and are referred to as gamma filters. The gamma filter for

$p = 2$ is particularly interesting as its output can be shown to be the input sample that is closest in value to the sample mean [86].

EXAMPLE 8.1 (IMAGE DENOISING)

Consider the denoising of the image shown in Figure 8.2a which is the contaminated version of the image shown in Figure 8.4a. The contaminating noise is salt and pepper with a probability of 5%. FLOM and gamma smoothers are used with window sizes of 3×3 and 5×5.

Figure 8.2 shows the output of a FLOM smoother for different values of the parameter p and window size 3×3. When $p < 1$ as in Figure 8.2b and c, the smoother tends to choose as the output one of the most repeated samples. This explains the clusters of positive or negative impulses that are still visible in the output of the smoother, specially for the smallest value of p. Figure 8.2d shows the output for $p = 1$. In this case, the FLOM smoother is equivalent to the median operator. When $p = 2$ the smoother is equivalent to a mean operator. In this case (Fig. 8.3a), the blurriness in the output is caused by the averaging of the samples in the observation window. Figure 8.3b shows the output of the smoother for $p = 10$. As the parameter p grows, the FLOM operator gets closer to a midrange smoother.

Figure 8.4 shows the denoising of the image in Figure 8.2a, where the gamma smoother is used. Since the FLOM and gamma smoother are equivalent for $p \leq 1$, only the results for greater values of p are shown. Figure 8.4b shows the output for $p = 2$. The result is similar to the one shown in Figure 8.3a, except for certain details that reveal the selection-type characteristics of the gamma smoother. The image in Figure 8.3a is smoother while Figure 8.4b has more artifacts. These effects are more visible in Figures 8.4c and 8.5b where the images seem to be composed of squares giving them the look of a tiled floor. Figures 8.4d and 8.5a and b show the output of a 5×5 gamma smoother. The smoothing of these images is greater than that of their 3×3 counterparts. The artifacts are more severe for the larger window size when p is too large or too small.

■

8.2 RUNNING MYRIAD SMOOTHERS

The sample myriad emerges as the maximum likelihood estimate of location under a set of distributions within the family of α-stable distributions, including the well known Cauchy distribution. Since their introduction by Fisher in 1922 [70], myriad type estimators have been studied and applied under very different contexts as an efficient alternative to cope with the presence of impulsive noise [2, 3, 39, 90, 167, 181]. The most general form of the myriad, where the potential of tuning the so-called *linearity parameter* in order to control its behavior is fully exploited, was first introduced by Gonzalez and Arce in 1996 [82]. Depending on the value of this

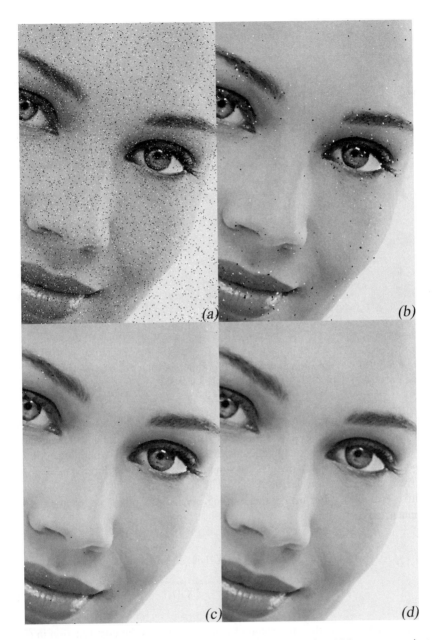

Figure 8.2 FLOM smoothing of an image for different values of p. (a) Image contaminated with salt-and-pepper noise (PSNR=17.75dB) and outputs of the FLOM smoother for: (b) $p = 0.01$ (PSNR=26.12dB), (c) $p = 0.1$ (PSNR=31.86dB), (d) $p = 1$ (median smoother, PSNR=37.49dB).

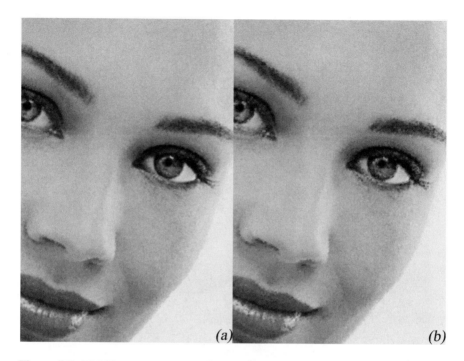

Figure 8.3   FLOM smoothing of an image for different values of p (continued). (a) $p = 2$ (mean smoother, PSNR=33.53dB), (b) $p = 10$ (PSNR=31.15).

free parameter, the sample myriad can present drastically different behaviors, ranging from *highly resistant* mode-type estimators to the familiar (Gaussian-efficient) sample average. This rich variety of operation modes is the key concept explaining optimality properties of the myriad in the class of symmetric α-stable distributions.

Given an observation vector $\mathbf{X}(n) = [X_1(n), X_2(n), \ldots, X_N(n)]$ and a fixed positive (tunable) value of K, the running myriad smoother output at time n is computed as

$$Y_K(n) = \mathrm{MYRIAD}[K; X_1(n), X_2(n), \ldots, X_N(n)]$$
$$= \arg\min_\beta \prod_{i=1}^{N} \left[K^2 + (X_i(n) - \beta)^2 \right]. \tag{8.5}$$

The myriad $Y_K(n)$ is thus the value of β that minimizes the cost function in (8.5). Unlike the sample mean or median, the definition of the sample myriad in (8.5) involves the free-tunable parameter K. This parameter will be shown to play a critical role in characterizing the behavior of the myriad. For reasons that will become apparent shortly, the parameter K is referred to as the *linearity parameter*. Since the $\log$ function is monotonic, the myriad is also defined by the equivalent expression

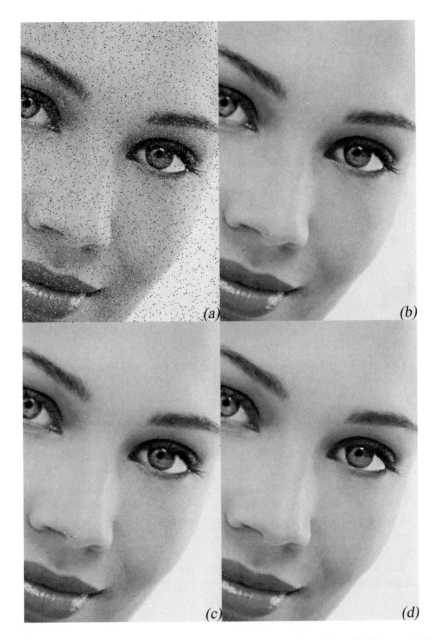

Figure 8.4 Gamma smoothing of an image for different values of p and different window sizes. (*a*) Original image and output of the 3×3 gamma smoother for (*b*) $p = 2$ (sample closest to the mean, PSNR=32.84dB), (*c*) $p = 10$ (PSNR=32.32dB), and the 5×5 gamma smoother for (*d*) $p = 0.1$ (PSNR=28.84dB).

Figure 8.5 Gamma smoothing of an image for different values of p and different window sizes (continued). (a) $p = 1$ (PSNR=29.91dB), (b) $p = 10$ (PSNR=28.13dB).

$$Y_K(n) = \arg\min_{\beta} \sum_{i=1}^{N} \log\left[K^2 + (X_i(n) - \beta)^2\right]. \tag{8.6}$$

In general, for a fixed value of K, the minima of the cost functions in (8.5) and (8.6) leads to a unique value. It is possible, however, to find sample sets for which the myriad is not unique. The event of getting more than one myriad is not of critical importance, as its associated probability is either negligible or zero for most cases of interest.

To illustrate the calculation of the sample myriad and the effect of the linearity parameter, consider the sample myriad, $\hat{\beta}_K$, of the set $\{-3, 10, 1, -1, 6\}$:

$$\hat{\beta}_K = \text{MYRIAD}(K; -3, 10, 1, -1, 6) \tag{8.7}$$

for $K = 20, 2, 0.2$. The myriad cost functions in (8.5), for these three values of K, are plotted in Figure 8.6. The corresponding minima are attained at $\hat{\beta}_{20} = 1.8$, $\hat{\beta}_2 = 0.1$, and $\hat{\beta}_{0.2} = 1$, respectively. The different values taken on by the myriad as the parameter K is varied is best understood by the results provided in the following properties.

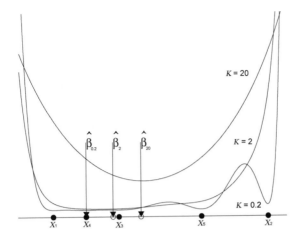

Figure 8.6 Myriad cost functions for different values of k

PROPERTY 8.1 (LINEAR PROPERTY) *Given a set of samples, $X_1, X_2, \ldots, X_N$, the sample myriad $\hat{\beta}_K$ converges to the sample average as $K \to \infty$. This is,*

$$\lim_{K \to \infty} \hat{\beta}_K = \lim_{K \to \infty} \text{MYRIAD}(K; X_1, \ldots, X_N)$$

$$= \frac{1}{N} \sum_{i=1}^{N} X_i. \tag{8.8}$$

To prove this property, first note that[1] $\hat{\beta}_K \leq X_{(N)}$ by checking that for any i, and for $\beta > X_{(N)}, K^2 + (X_i - \beta)^2 > K^2 + (X_i - X_{(N)})^2$. In the same way, $\hat{\beta}_K \geq X_{(1)}$. Hence,

$$\hat{\beta}_K = \arg \min_{X_{(1)} \leq \beta \leq X_{(N)}} \prod_{i=1}^{N} [K^2 + (X_i - \beta)^2] \tag{8.9}$$

$$= \arg \min_{X_{(1)} \leq \beta \leq X_{(N)}} \left\{ K^{2N} + K^{2N-2} \sum_{i=1}^{N} (X_i - \beta)^2 + f(K) \right\}, \tag{8.10}$$

where $f(K) = O(K^{2N-4})$ and O denotes the *asymptotic order* as $K \to \infty$ [2]. Since adding or multiplying by constants does not affect the arg min operator, Equation (8.10) can be rewritten as

---

[1]Here, $X_{(i)}$ denotes the ith-order statistic of the sample set.
[2]Given nonnegative functions f and g, we write $f = O(g)$ if and only if there is a positive constant C and an integer N such that $f(x) \leq Cg(x) \: \forall \: x > N$.

$$\hat{\beta}_K = \arg\min_{X_{(1)} \le \beta \le X_{(N)}} \left\{ \sum_{i=1}^{N} (X_i - \beta)^2 + \frac{O(K^{2N-4})}{K^{2N-2}} \right\}. \tag{8.11}$$

Letting $K \to \infty$, the term $O(K^{2N-4})/K^{2N-2}$ becomes negligible, and

$$\hat{\beta}_K \to \arg\min_{X_{(1)} \le \beta \le X_{(N)}} \left\{ \sum_{i=1}^{N} (X_i - \beta)^2 \right\} = \frac{1}{N} \sum_{i=1}^{N} X_i.$$

■

Plainly, an infinite value of K converts the myriad into the sample average. This behavior explains the name *linearity* given to this parameter: the larger the value of K, the closer the behavior of the myriad to a linear estimator. As the myriad moves away from the linear region (large values of K) to lower linearity values, the estimator becomes more resistant to the presence of impulsive noise. In the limit, when K tends to zero, the myriad leads to a location estimator with particularly good performance in the presence of *very* impulsive noise. In this case, the estimator treats every observation as a possible outlier, assigning more credibility to the most repeated values in the sample. This "mode-type" characteristic is reflected in the name *mode-myriad* given to this estimator.

DEFINITION 8.1 (SAMPLE MODE-MYRIAD) *Given a set of samples* $X_1, X_2, \ldots,$ X_N*, the mode-myriad estimator,* $\hat{\beta}_0$*, is defined as*

$$\hat{\beta}_0 = \lim_{K \to 0} \hat{\beta}_K, \tag{8.12}$$

where $\hat{\beta}_K = \text{MYRIAD}(K; X_1, X_2, \ldots, X_N)$.

The following property explains the behavior of the mode-myriad as a kind of generalized sample mode, and provides a simple method for determining the mode-myriad without recurring to the definition in (8.5).

PROPERTY 8.2 (MODE PROPERTY) *The mode-myriad* $\hat{\beta}_0$ *is always equal to one of the most repeated values in the sample. Furthermore,*

$$\hat{\beta}_0 = \arg\min_{X_j \in \mathcal{M}} \prod_{i=1, X_i \ne X_j}^{N} |X_i - X_j|, \tag{8.13}$$

where $\mathcal{M}$ *is the set of most repeated values.*

Proof : Since K is a positive constant, the definition of the sample myriad in (8.5) can be reformulated as $\hat{\beta}_K = \arg\min_\beta P_K(\beta)$, where

$$P_K(\beta) = \prod_{i=1}^{N} \left[1 + \frac{(X_i - \beta)^2}{K^2} \right]. \tag{8.14}$$

When K is very small, it is easy to check that

$$P_K(\beta) = O\left(\frac{1}{K^2}\right)^{N-r(\beta)},$$

where $r(\beta)$ is the number of times the value β is repeated in the sample set, and O denotes the asymptotic order as $K \to 0$. In the limit, the exponent $N - r(\beta)$ must be minimized in order for $P_K(\beta)$ to be minimum. Therefore, the mode-myriad $\hat{\beta}_0$ will lie on a maximum of $r(\beta)$, or in other words, $\hat{\beta}_0$ will be one of the most repeated values in the sample.

Now, let $r = \max_j r(X_j)$. Then, for $X_j \in \mathcal{M}$, expanding the product in (8.14) gives

$$P_K(X_j) = \left\{ \prod_{i, X_i \neq X_j} \frac{(X_i - X_j)^2}{K^2} \right\} + O\left(\frac{1}{K^2}\right)^{N-r-1}. \qquad (8.15)$$

Since the first term in (8.15) is $O(\frac{1}{K^2})^{N-r}$, the second term is negligible for small values of K, and $\hat{\beta}_0$ can be calculated as

$$
\begin{aligned}
\hat{\beta}_0 &= \arg\min_{X_j \in \mathcal{M}} P_K(X_j) \\
&= \arg\min_{X_j \in \mathcal{M}} \prod_{i, X_i \neq X_j} \frac{(X_i - X_j)^2}{K^2} \\
&= \arg\min_{X_j \in \mathcal{M}} \prod_{i, X_i \neq X_j} |X_i - X_j|
\end{aligned}
$$

∎

An immediate consequence of the mode property is the fact that running-window smoothers based on the mode-myriad are *selection-type*, in the sense that their output is always, by definition, one of the samples in the input window. The mode myriad output will always be one of the most repeated values in the sample, resembling the behavior of a sample mode. This *mode property*, indicates the high effectiveness of the estimator in locating heavy impulsive processes. Also, being a sample mode, the mode myriad is evidently a selection-type estimator, in the sense that it is always equal, by definition, to one of the sample values. This selection property, shared also by the median, makes mode-myriad smoother a suitable framework for image processing, where the application of selection-type smoothers has been shown convenient.

EXAMPLE 8.2 (BEHAVIOR OF THE MODE MYRIAD)

Consider the sample set $\vec{X} = [1, 4, 2.3, S, 2.5, 2, 5, 4.25, 6]$. The mode Myriad of $\vec{X}$ is calculated as the sample S varies from 0 to 7. The results are shown in Figure

8.7. It can be seen how the myriad with $K \to 0$ favors clusters of samples. The closest samples are 2, 2.3 and 2.5 and, when the value of S is not close to any other sample, the output of the mode-myriad is the center sample of this cluster. It can also be seen that, when two samples have the same value, the output of the myriad takes that value. Look for example at the spikes in $S = 1$ or $S = 6$. There is also another cluster of samples: 4 and 4.25. The plot shows how, when S gets closer to these values, this becomes the most clustered set of samples and the myriad takes on one of the values of this set.

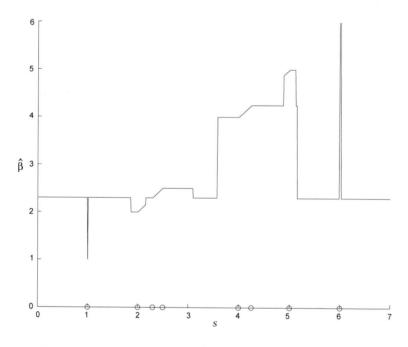

Figure 8.7 Mode myriad of a sample set with one variable sample. The constant samples are indicated with "◯"

EXAMPLE 8.3 (MODE-MYRIAD PERFORMANCE IN α-STABLE NOISE)

To illustrate the performance of the mode myriad, in comparison to the sample mean and sample median, this example considers the location estimation problem in i.i.d. α-stable noise for a wide range of values of the tail parameter α. Figure 8.8 shows the estimated mean absolute errors (MAE) of the sample mean, the sample median, and the mode-myriad when used to locate the center of an i.i.d. symmetric α-stable sample of size $N = 5$. The result comes from a Monte Carlo simulation with

$200,000$ repetitions. The values of the tail parameter range from $\alpha = 2$ (Gaussian case) down to $\alpha = 0.3$ (very impulsive). Values of α slightly smaller than 2 indicate a distribution close to the Gaussian, in which case the sample mean outperforms both the median and the mode-myriad estimator. As α is decreased, the noise becomes more impulsive and the sample mean rapidly loses efficiency, being outperformed by the sample median for values of α less than 1.7. As α approaches 1, the estimated MAE of the sample mean explodes. In fact, it is known that for $\alpha < 1$, it is more efficient to use any of the sample values than the sample mean itself. As α continues to decrease, the sample median loses progressively more efficiency with respect to the mode-myriad estimator, and at $\alpha \approx 0.87$, the mode-myriad begins to outperform the sample median. This is an expected result given the optimality of the mode-myriad estimator for small values of α. For the last value in the plot, $\alpha = 0.3$, the mode-myriad estimator has an estimated efficiency ten times larger than the median. This increase in relative efficiency is expected to grow without bounds as α approaches 0 (recall that $\alpha = 0$ is the optimality point of the mode-myriad estimator). ■

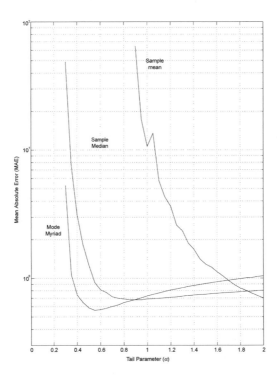

Figure 8.8 Estimated Mean Absolute Error of the sample mean, sample median and mode-myriad location estimator in α-stable noise ($N = 5$).

EXAMPLE 8.4 (DENOISING OF A VERY IMPULSIVE SIGNAL.)

This example illustrates the denoising of a signal which has been corrupted with very impulsive noise. The performance of the mode-myriad location estimator with that of the FLOM estimator is compared . The observation is a corrupted version of the "blocks" signal shown in Figure 8.9(a). The signal corrupted with additive stable noise with $\alpha = 0.2$ is shown in Figure 8.9(b) where a different scale is used to illustrate the very impulsive noise environment. The value of α is not known a priori. The mean square error between the original signal and the noisy observation is 8.3×10^{88}. The following running smoothers (location estimators) are applied, all using a window of size $N = 121$: the sample mean (MSE = 6.9×10^{86}) shown in Figure 8.9(c), the sample median (MSE = 3.2×10^4) shown in Figure 8.9(d), the FLOM with $p = 0.8$ (MSE = 77.5) in Figure 8.9(e), and the mode-myriad location estimator (MSE = 4.1) shown in Figure 8.9(f).

As shown in the figure, at this level of impulsiveness, the sample median and mean break down. The FLOM does not perform as well as the mode-myriad due to the mismatch of p and α. The performance of the FLOM estimator would certainly improve, but the parameter p would have to be matched closely to the stable noise index, a task that can be difficult. The mode-myriad, on the other hand performs well without the need of parameter tuning.

■

Geometrical Interpretation of the Myriad   Myriad estimation, defined in (8.5), can be interpreted in a more intuitive manner. As depicted in Figure 8.10(a), the sample myriad, $\hat{\beta}_K$, is the value that minimizes the product of distances from point A to the sample points $X_1, X_2, \ldots, X_6$. Any other value, such as $X = \beta'$, produces a higher product of distances. This can be shown as follows: Let D_i be the distance between the point A and the sample X_i. The points A, $\hat{\beta}$ and X_i form a right triangle with hypotenuse D_i. In consequence D_i is calculated as:

$$D_i^2 = K^2 + (X_i - \hat{\beta})^2. \tag{8.16}$$

Taking the product of the square distances to all the samples and searching for the minimum over β results in:

$$\arg\min_{\beta} \prod_{i=1}^{N} D_i^2 = \arg\min_{\beta} \prod_{i=1}^{N} \left[K^2 + (X_i(n) - \beta)^2 \right]. \tag{8.17}$$

matching the definition in (8.5).

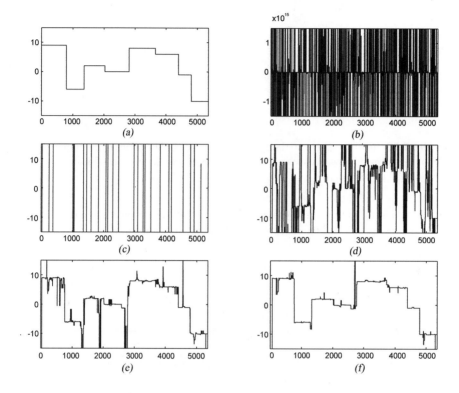

Figure 8.9 Running smoothers in stable noise ($\alpha = 0.2$). All smoothers of size 121; (*a*) original blocks signal, (*b*) corrupted signal with stable noise, (*c*) the output of the running mean, (*d*) the running median, (*e*) the running FLOM smoother, and (*f*) the running mode-myriad smoother.

As K is reduced, the myriad searches clusters as shown in Figure 8.10(*b*). If K is made large, all distances become close and it can be shown that the myriad tends to the sample mean.

EXAMPLE 8.5
___

Given the sample set $X = \{1, 1, 2, 10\}$ compute the sample myriad for values of K of 0.01, 5, and 100.

The outputs of the sample myriad are: 1.0001, 2.1012, and 3.4898. In the first case, since the value of K is small, the output goes close to the mode of the sample set, that is 1. Intermediate values of K give an output that is close to the most clustered set of samples. The largest value of K outputs a value that is very close to the mean of the set, such as 3.5. Figure 8.11 shows the sample set and the location

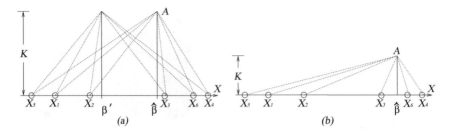

Figure 8.10 (*a*) The sample myriad, $\hat{\beta}$, minimizes the product of distances from point A to all samples. Any other value, such as $x = \beta'$, produces a higher product of distances; (*b*) the myriad as K is reduced.

of the myriad for the different values of K. It is noticeable how raising the value of K displaces the value of the myriad from the mode to the mean of the set.

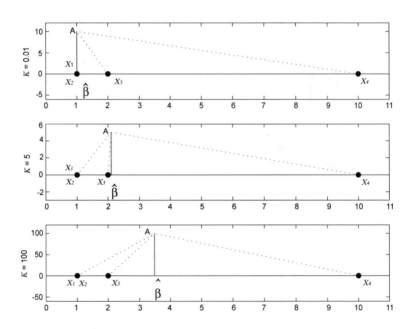

Figure 8.11 Sample myriad of the sample set $\{1, 1, 2, 10\}$ for (*a*) $K = 0.01$, (*b*) $K = 5$, (*c*) $K = 100$.

The Tuning of K The linear and mode properties indicate the behavior of the myriad estimator for large and small values of K. From a practical point of view, it is important to determine if a given value of K is large (or small) enough for the linear (or mode) property to hold approximately. With this in mind, it is instructive to look at the myriad as the maximum likelihood location estimator generated by a Cauchy distribution with dispersion K (geometrically, K is equivalent to half the interquartile range). Given a fixed set of samples, the ML method locates the generating distribution in a position where the probability of the specific sample set to occur is maximum.

When K is large, the generating distribution is highly dispersed, and its density function looks flat (see the density function corresponding to K_2 in Fig. 8.12). If K is large enough, *all* the samples can be accommodated inside the interquartile range of the distribution, and the ML estimator visualizes them as well-behaved (no outliers). In this case, a desirable estimator would be the sample average, in complete agreement with the linear property. From this consideration, it should be clear that a fair approximation to the linear property can be obtained if K is large enough so that *all* the samples can be seen as well-behaved under the generating Cauchy distribution. It has been observed experimentally that values of K on the order of the data range, $K \sim X_{(N)} - X_{(1)}$, often make the myriad an acceptable approximation to the sample average.

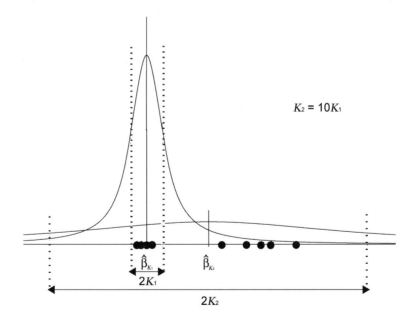

Figure 8.12 The role of the linearity parameter when the myriad is looked as a maximum likelihood estimator. When K is large, the generating density function is spread and the data are visualized as well-behaved (the optimal estimator is the sample average). For small values of K, the generating density becomes highly localized, and the data are visualized as very impulsive (the optimal estimator is a cluster locator).

Intermediate values of K assume a sample set with some outliers and some well behaved samples. For example, when $K = \frac{1}{2}[X_{(\frac{3}{4}N)} - X_{(\frac{1}{4}N)}]$ half the samples will be outside an interval around the myriad of length $2K = X_{(\frac{3}{4}N)} - X_{(\frac{1}{4}N)}$ and will be considered as outliers. On the other side, when K is small, the generating Cauchy distribution is highly localized, and its density function looks similar to a positive impulse. The effect of such a localized distribution is conceptually equivalent to observing the samples through a magnifying lens. In this case, most of the data look like possible outliers, and the ML estimator has trouble locating a large number of observations inside the interquartile range of the density (see the density function corresponding to K_1 in Fig. 8.12). Putting in doubt most of the data at hand, a desirable estimator would tend to maximize the number of samples inside the interquartile range, attempting to position the density function in the vicinity of a data cluster. In the limit case, when $K \to 0$, the density function gets infinitely localized, and the only visible clusters will be made of repeated value sets. In this case, one of the most crowded clusters (i.e., one of the most repeated values in the sample) will be located by the estimator, in accordance with the mode property. From this consideration, it should be clear that a fair approximation to the mode property can be obtained if K is made significantly smaller than the distances between sample elements. Empirical observations show that K on the order of

$$K \sim \min_{i,j} |X_i - X_j|, \tag{8.18}$$

is often enough for the myriad to be considered approximately a mode-myriad.

The myriad estimator thus offers a rich class of modes of operation that can be easily controlled by tuning the linearity parameter K. When the noise is Gaussian, for example, large values of the linearity can provide the optimal performance associated with the sample mean, whereas for highly impulsive noise statistics, the resistance of mode-type estimators can be achieved by using myriads with low linearity. The tradeoff between efficiency at the Gaussian model and resistance to impulsive noise can be managed by designing appropriate values for K (see Fig. 8.13).

Figure 8.13 Functionality of the myriad as K is varied. Tuning the linearity parameter K adapts the behavior of the myriad from impulse-resistant mode-type estimators (small K) to the Gaussian-efficient sample mean (large K).

To illustrate the above, it is instructive to look at the behavior of the sample myriad shown in Figure 8.14. The solid line shows the values of the myriad as a function of K for the data set $\{0, 1, 3, 6, 7, 8, 9\}$. It can be observed that, as K increases, the myriad tends asymptotically to the sample average. On the other hand, as K is decreased, the myriad favors the value 7, which indicates the location of the cluster formed by the samples $6, 7, 8, 9$. This is a typical behavior of the myriad for small

K: it tends to favor values where samples are more likely to occur or cluster. The term *myriad* is coined as a result of this characteristic.

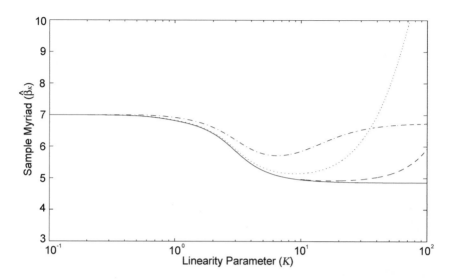

Figure 8.14 Values of the myriad as a function of K for the following data sets: (solid) original data set $= 0, 1, 3, 6, 7, 8, 9$; (dash-dot) original set plus an additional observation at 20; (dotted) additional observation at 100; (dashed) additional observations at 800, -500, and 700.

The dotted line shows how the sample myriad is affected by an additional observation of value 100. For large values of K, the myriad is very sensitive to this new observation. On the contrary, for small K, the variability of the data is assumed to be small, and the new observation is considered an outlier, not influencing significantly the value of the myriad.

More interestingly, if the additional observations are the very large data 800, -500, 700 (dashed curve), the myriad is practically unchanged for moderate values of K ($K < 10$). This behavior exhibits a very desirable outlier rejection property, not found for example in median-type estimators.

Scale-invariant Operation Unlike the sample mean or median, the operation of the sample myriad is not scale invariant, that is, for fixed values of the linearity parameter, its behavior can vary depending on the units of the data. This is formalized in the following property.

PROPERTY 8.3 (SCALE INVARIANCE) *Let* $\hat{\beta}_K(\mathbf{X})$ *denote the myriad of order* K *of the data in the vector* $\mathbf{X}$. *Then, for* $c > 0$,

$$\hat{\beta}_K(c\mathbf{X}) = c\hat{\beta}_{K/c}(\mathbf{X}). \tag{8.19}$$

Proof : Let $X_1, X_2, \ldots, X_N$ denote the data in $\mathbf{X}$. Then,

$$\hat{\beta}_K(c\mathbf{X}) = \arg\min_{\beta} \prod_{i=1}^{N} \left[K^2 + (cX_i - \beta)^2\right]$$

$$= \arg\min_{\beta} \prod_{i=1}^{N} \left[\left(\frac{K}{c}\right)^2 + \left(X_i - \frac{\beta}{c}\right)^2\right] \qquad (8.20)$$

$$= c\left(\arg\min_{\beta} \prod_{i=1}^{N} \left[\left(\frac{K}{c}\right)^2 + (X_i - \beta)^2\right]\right).$$

■

According to (8.19), a change of scale in the data is preserved in the myriad only if K experiences the same change of scale. Thus, the scale dependence of the myriad can be easily overcome if K carries the units of the data, or in other words, if K is a *scale parameter* of the data.

8.3 OPTIMALITY OF THE SAMPLE MYRIAD

Optimality In The α-Stable Model In addition to its optimality in the Cauchy distribution ($\alpha = 1$), the sample myriad presents optimality properties in the α-stable framework. First, it is well known that the sample mean is the optimal location estimator at the Gaussian model; thus, by assigning large values to the linearity parameter, the linear property guarantees the optimality of the sample myriad in the Gaussian distribution ($\alpha = 2$). The following result states the optimality of the myriad when $\alpha \to 0$, that is, when the impulsiveness of the distribution is very high. The proof of Proposition 8.1 can be found in [82].

PROPOSITION 8.1 *Let $T_{\alpha,\gamma}(X_1, X_2, \ldots, X_N)$ denote the maximum likelihood location estimator derived from a symmetric α-stable distribution with characteristic exponent α and dispersion γ. Then,*

$$\lim_{\alpha \to 0} T_{\alpha,\gamma}(X_1, X_2, \ldots, X_N) = \text{MYRIAD}\{0; X_1, X_2, \ldots, X_N\}. \qquad (8.21)$$

This proposition states that the ML estimator of location derived from an α-stable distribution with small α behaves like the sample mode-myriad. Proposition 8.1 completes what is called the *α-stable triplet* of optimality points satisfied by the myriad. On one extreme ($\alpha = 2$), when the distributions are very well-behaved, the myriad reaches optimal efficiency by making $K = \infty$. In the middle ($\alpha = 1$), the myriad reaches optimality by making $K = \gamma$, the dispersion parameter of the Cauchy distribution. On the other extreme ($\alpha \to 0$), when the distributions are extremely impulsive, the myriad reaches optimality again, this time by making $K = 0$.

The α-stable triplet demonstrates the central role played by myriad estimation in the α-stable framework. The very simple tuning of the linearity parameter empowers the myriad with good estimation capabilities under markedly different types of impulsiveness, from the very impulsive ($\alpha \to 0$) to the non impulsive ($\alpha = 2$). Since lower values of K correspond to increased resistance to impulsive noise, it is intuitively pleasant that, for *maximal* impulsiveness ($\alpha \to 0$), the optimal K takes precisely its *minimal* value, $K = 0$. The same condition occurs at the other extreme: *minimal* levels of impulsiveness ($\alpha = 2$), correspond to the *maximal* tuning value, $K = \infty$. Thus, as α is increased from 0 to 2, it is reasonable to expect, somehow, a progressive increase of the optimal K, from $K = 0$ to $K = \infty$. The following proposition provides information about the general behavior of the optimal K. Its proof is a direct consequence of Property 8.3 and the fact that γ is a scale parameter of the α-stable distribution.

PROPOSITION 8.2 *Let α and γ denote the characteristic exponent and dispersion parameter of a symmetric α-stable distribution. Let $K_o(\alpha, \gamma)$ denote the optimal tuning value of K in the sense that $\hat{\beta}_{K_o}$ minimizes a given performance criterion (usually the variance) among the class of sample myriads with non negative linearity parameter. Then,*

$$K_o(\alpha, \gamma) = K_o(\alpha, 1)\gamma. \tag{8.22}$$

Proposition 8.2 indicates a separability of K_o in terms of α and γ, reducing the optimal tuning problem to that of determining the function $K(\alpha) = K_o(\alpha, 1)$. This function is of fundamental importance for the proper operation of the myriad in the α-stable framework, and will be referred to it as the α-K *curve*. Its form is conditioned to the performance criterion chosen, and it may even depend on the sample size. In general, as discussed above, the α-K curve is expected to be monotonically increasing, with $K(0) = 0$ (very impulsive point) and $K(2) = \infty$ (Gaussian point). If the performance criterion is the asymptotic variance for example, then $K(1) = 1$, corresponding to the Cauchy point of the α-stable triplet. The exact computation of the α-K curve for α-stable distributions is still not determined. A simple empirical form that has consistently provided efficient results in a variety of conditions is

$$K(\alpha) = \sqrt{\frac{\alpha}{2 - \alpha}}, \tag{8.23}$$

which is plotted in Figure 8.15.

The α-K curve is a valuable tool for estimation and filtering problems that must *adapt* to the impulsiveness conditions of the environment. α-K curves in the α-stable framework have been used, for example, to develop myriad-based adaptive detectors for channels with uncertain impulsiveness [84].

Optimality in the Generalized t Model   The family of generalized t distributions was introduced by Hall in 1966 as an empirical model for atmospheric radio noise [90]. These distributions have been found to provide accurate fits to different types of atmospheric noise found in practice. Because of its simplicity and parsimony, it has been used by Middleton as a mathematically tractable approximation to

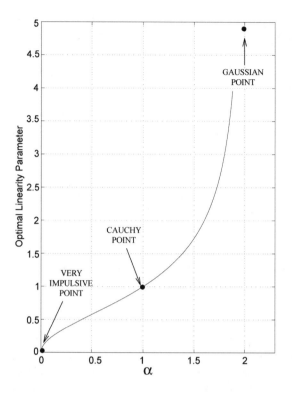

Figure 8.15 Empirical α-K curve for α-stable distributions. The curve values at $\alpha = 0, 1$, and 2 constitute the optimality points of the α-stable triplet.

his widely accepted models of electromagnetic radio noise [143]. Long before the introduction of the model by Hall, the generalized t distributions have been known in statistics as a family of heavy-tailed distributions categorized under the *type VII* of Pearson's distributional system [177].

Generalized t density functions can be conveniently parameterized as

$$f_x(x) = \frac{c}{(\alpha\sigma^2 + x^2)^{\frac{1+\alpha}{2}}}, \tag{8.24}$$

where $\sigma > 0$, $\alpha > 0$, and c is a normalizing constant given by

$$c = \frac{\Gamma(\frac{1+\alpha}{2})}{\sqrt{\pi}\Gamma(\frac{\alpha}{2})}\alpha^{\alpha/2}\sigma^{\alpha}. \tag{8.25}$$

It is easy to check that the distribution defined by (8.24) is algebraic-tailed, with tail constant α and scale parameter σ. Although α may take values larger than 2, its meaning is conceptually equivalent to the characteristic exponent of the α-stable framework. At one extreme, when $\alpha \to \infty$, the generalized t distribution is equivalent to a zero-mean Gaussian distribution with variance σ^2. As it is the case with α-stable distributions, decreased values of α correspond to increased levels of impulsiveness. For values of $\alpha \leq 2$, the impulsiveness becomes high enough to make the variance infinite, and when $\alpha = 1$, the model corresponds to the Cauchy distribution. At the other extreme, when $\alpha \to 0$, the distribution exhibits the highest levels of impulsiveness.

The maximum likelihood estimator of location derived from the t density in (8.24) is precisely the sample myriad with linearity parameter

$$K = \sqrt{\alpha}\sigma. \tag{8.26}$$

The optimality of the myriad for *all* the distributions in the generalized t family indicates its adequateness along a wide variety of noise environments, from the very impulsive ($\alpha \to 0$) to the well-behaved Gaussian ($\alpha \to \infty$). Expression (8.26) gives the optimal tuning law as a function of α and σ (note the close similarity with Equation (8.22) for α-stable distributions). Making $\sigma = 1$, the α-K curve for generalized t distributions is obtained with $K(\alpha) = \sqrt{\alpha}$. Like the α-K curve of α-stable distributions, this curve is also monotonic increasing, and contains the optimality points of the α-stable triplet, namely the Gaussian point ($K(\infty) = \infty$), the Cauchy point ($K(1) = 1$), and the very impulsive point ($K(0) = 0$).

The generalized t model provides a simple framework to assess the performance of the sample myriad as the impulsiveness of the distributions is changed. It can be proven that the normalized asymptotic variance[3] of the optimal sample myriad at the generalized t model is (for a derivation, see for example [81]):

$$V_{\text{myr}} = \frac{\alpha + 3}{\alpha + 1}\sigma^2. \tag{8.27}$$

A plot of V_{myr} vs. α is shown in Figure 8.16 for $\sigma = 1$. The asymptotic variances of the sample mean (V_{mean}) and sample median (V_{med}) are also included for comparison [81]. The superiority of the sample myriad over both mean and median in the generalized t distribution model is evident from the figure.

8.4 WEIGHTED MYRIAD SMOOTHERS

The sample myriad can be generalized to the weighted myriad smoother by assigning positive weights to the input samples (observations); the weights reflect the varying levels of reliability. To this end, the observations are assumed to be drawn from independent Cauchy random variables that are, however, *not* identically distributed. Given N observations $\{X_i\}_{i=1}^N$ and nonnegative weights $\{W_i \geq 0\}_{i=1}^N$, let the input and weight vectors be defined as $\mathbf{X} \triangleq [X_1, X_2, \ldots, X_N]^T$ and $\mathbf{W} \triangleq [W_1, W_2, \ldots, W_N]^T$, respectively. For a given *nominal* scale factor K,

---

[3]Let $V_T(N)$ be the variance of the estimator T when the sample size is N. Then, the normalized asymptotic variance V is defined as $V = \lim_{N \to \infty} N V_T(N)$.

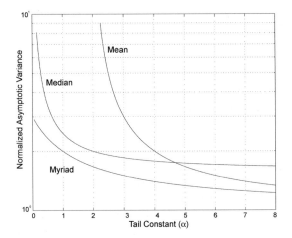

Figure 8.16 Normalized asymptotic variance of the sample mean, sample median, and optimal sample myriad in generalized t noise. The myriad outperforms the mean and median for any level of impulsiveness.

the underlying random variables are assumed to be independent and Cauchy distributed with a common location parameter β, but varying scale factors $\{S_i\}_{i=1}^{N}$: $X_i \sim \text{Cauchy}(\beta, S_i)$, where the density function of X_i has the form

$$f_{X_i}(X_i; \beta, S_i) = \frac{1}{\pi} \frac{S_i}{S_i^2 + (X_i - \beta)^2}, \quad -\infty < X_i < \infty, \tag{8.28}$$

and where

$$S_i \triangleq \frac{K}{\sqrt{W_i}} > 0, \ i = 1, 2, \dots, N. \tag{8.29}$$

A larger value for the weight W_i (smaller scale S_i) makes the distribution of X_i more concentrated around β, thus increasing the reliability of the sample X_i. Note that the special case when all the weights are equal to unity corresponds to the sample myriad at the nominal scale factor K, with all the scale factors reducing to $S_i = K$.

Again, the location estimation problem being considered here is closely related to the problem of smoothing a time-series $\{X(n)\}$ using a sliding window. The output $Y(n)$, at time n, can be interpreted as an estimate of location based on the input samples $\{X_1(n), X_2(n), \dots, X_N(n)\}$. Further, the aforementioned model of independent but not identically distributed samples synthesizes the temporal correlations usually present among the input samples. To see this, note that the output $Y(n)$, as an estimate of location, would rely more on (give more weight to) the sample $X(n)$, when compared with samples that are further away in time. By assigning varying scale factors in modeling the input samples, leading to different weights (reliabilities), their temporal correlations can be effectively accounted for.

The weighted myriad smoother output $\hat{\beta}_K(\mathbf{W}, \mathbf{X})$ is defined as the value β that maximizes the likelihood function $\prod_{i=1}^N f_{X_i}(X_i; \beta, S_i)$. Using (8.28) for $f_{X_i}(X_i; \beta, S_i)$ leads to

$$\hat{\beta}_K(\mathbf{W}, \mathbf{X}) = \arg\max_\beta \prod_{i=1}^N \frac{S_i}{S_i^2 + (X_i - \beta)^2}, \qquad (8.30)$$

which is equivalent to

$$\hat{\beta}_K(\mathbf{W}, \mathbf{X}) \triangleq \arg\min_\beta \prod_{i=1}^N \left[1 + \left(\frac{X_i - \beta}{S_i} \right)^2 \right]$$

$$= \arg\min_\beta \prod_{i=1}^N \left[K^2 + W_i (X_i - \beta)^2 \right] \qquad (8.31)$$

$$\triangleq \arg\min_\beta P(\beta); \qquad (8.32)$$

Alternatively, we can write $\hat{\beta}_K(\mathbf{W}, \mathbf{X}) \triangleq \hat{\beta}_K$ as

$$\hat{\beta}_K = \arg\min_\beta Q(\beta) \triangleq \arg\min_\beta \sum_{i=1}^N \log\left[K^2 + W_i (X_i - \beta)^2 \right]; \qquad (8.33)$$

thus $\hat{\beta}_K$ is the global minimizer of $P(\beta)$ as well as of $Q(\beta) \triangleq \log(P(\beta))$. Depending on the context, we refer to either of the functions $P(\beta)$ and $Q(\beta)$ as the *weighted myriad smoother objective function*. Note that when $W_i = 0$, the corresponding term drops out of $P(\beta)$ and $Q(\beta)$; thus a sample X_i is effectively ignored if its weight is zero.

The definition of the weighted myriad is then formally stated as

DEFINITION 8.2 (WEIGHTED MYRIAD) *Let* $\mathbf{W} = [W_1, W_2, \ldots, W_N]$ *be a vector of nonnegative weights. Given* $K > 0$, *the weighted myriad of order* K *for the data* $X_1, X_2, \ldots, X_N$ *is defined as*

$$\hat{\beta}_K = \text{MYRIAD}\{K; W_1 \circ X_1, \ldots, W_N \circ X_N\}$$

$$= \arg\min_\beta \sum_{i=1}^N \log\left[K^2 + W_i(X_i - \beta)^2 \right], \qquad (8.34)$$

where $W_i \circ X_i$ *represents the weighting operation in (8.34). In some situations, the following equivalent expression can be computationally more convenient*

$$\hat{\beta}_K = \arg\min_\beta \prod_{i=1}^N \left[K^2 + W_i(X_i - \beta)^2 \right]. \qquad (8.35)$$

It is important to note that the weighted myriad has only N independent parameters (even though there are N weights and the parameter K). Using (8.35), it can be inferred that if the value of K is changed, the same smoother output can be obtained provided the smoother weights are appropriately scaled. Thus, the following is true

$$\hat{\beta}_K(\mathbf{W}, \mathbf{X}) = \hat{\beta}_1\left(\frac{\mathbf{W}}{K^2}, \mathbf{X}\right) \tag{8.36}$$

since

$$
\begin{aligned}
\hat{\beta}_K(\mathbf{W}, \mathbf{X}) &= \arg\min_{\beta} \prod_{i=1}^{N}\left[K^2 + W_i(X_i - \beta)^2\right] \\
&= \arg\min_{\beta} \frac{1}{K^{2N}} \prod_{i=1}^{N}\left[K^2 + W_i(X_i - \beta)^2\right] \\
&= \arg\min_{\beta} \prod_{i=1}^{N}\left[1 + \frac{W_i}{K^2}(X_i - \beta)^2\right] \\
&= \hat{\beta}_1\left(\frac{\mathbf{W}}{K^2}, \mathbf{X}\right).
\end{aligned}
\tag{8.37}
$$

Equivalently:

$$\hat{\beta}_{K_1}(\mathbf{W}_1, \mathbf{X}) = \hat{\beta}_{K_2}(\mathbf{W}_2, \mathbf{X}) \quad \text{iff} \quad \frac{\mathbf{W}_1}{K_1^2} = \frac{\mathbf{W}_2}{K_2^2}. \tag{8.38}$$

Hence, the output depends only on $\frac{\mathbf{W}}{K^2}$.

The objective function $P(\beta)$ in (8.35) is a polynomial in β of degree $2N$, with well-defined derivatives of all orders. Therefore, $P(\beta)$ (and the equivalent objective function $Q(\beta)$) can have at most $(2N - 1)$ local extremes. The output is thus one of the local minima of $Q(\beta)$:

$$Q'(\hat{\beta}) = 0. \tag{8.39}$$

Figure 8.17 depicts a typical objective function $Q(\beta)$ for various values of K and different sets of weights. Note in the figure that the number of local minima in the objective function $Q(\beta)$ depends on the value of the parameter K. Note that the effect of the weight of the outlier on the cost functions (dashed lines on Fig. 8.17) is minimal for large values of K, but severe for large K. While the minimum is the same for both sets of weights using the small value of K, the minimum is shifted towards the outlier in the other case.

As K gets larger, the number of local minima of $G(\beta)$ decreases. In fact, it can be proved [111] (by examining the second derivative $G''(\beta)$) that a *sufficient* (but not *necessary*) condition for $G(\beta)$ (and $\log(G(\beta))$) to be convex and, therefore, have a unique local minimum, is that $K > \sqrt{\max\{W_j\}_{j=1}^{N}}\,(X_{(N)} - X_{(1)})$. This condition is however not necessary; the onset of convexity could be at a much lower K.

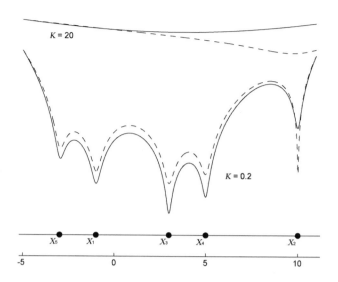

Figure 8.17 Sketch of a typical weighted myriad objective function $Q(\beta)$ for the weights $[1,\ 2,\ 3,\ 2,\ 1]$ (solid line), and $[1,\ 100,\ 3,\ 2,\ 1]$ (dashed line), and the sample set $[-1,\ 10,\ 3,\ 5,\ -3]$.

As stated in the next property, in the limit as $K \to \infty$, with the weights $\{W_i\}$ held constant, it can be shown that $Q(\beta)$ exhibits a single local extremum. The proof is a generalized form of that used to prove the linear property of the unweighted sample myriad.

PROPERTY 8.4 (LINEAR PROPERTY) *In the limit as $K \to \infty$, the weighted myriad reduces to the normalized linear estimate*

$$\lim_{K \to \infty} \hat{\beta}_K = \frac{\sum_{i=1}^{N} W_i X_i}{\sum_{i=1}^{N} W_i}. \tag{8.40}$$

Again, because of the linear structure of the weighted myriad as $K \to \infty$, the name "linearity parameter" is used for the parameter K. Equation (8.40) provides the link between the weighted myriad and a constrained linear FIR filter: the weighted myriad smoother is analogous to the weighted mean smoother having its weights *constrained* to be nonnegative and normalized (summing to unity).

Figure 8.18 also depicts that the output $\hat{\beta}$ is restricted to the dynamic range of the input of the weighted myriad smoother. In consequence, this smoother is unable to amplify the dynamic range of an input signal.

PROPERTY 8.5 (NO UNDERSHOOT/OVERSHOOT) *The output of a weighted myriad smoother is always bracketed by*

$$X_{(1)} \leq \hat{\beta}_K(\mathbf{W}; X_1, X_2, \ldots, X_N) \leq X_{(N)}, \tag{8.41}$$

where $X_{(1)}$ and $X_{(N)}$ denote the minimum and maximum samples in the input window.

Proof : For $\beta < X_{(1)}$,

$$K^2 + W_i(X_i - X_{(1)})^2 < K^2 + W_i(X_i - \beta)^2,$$

and consequently

$$\prod_{i=1}^{N}[K^2 + W_i(X_i - X_{(1)}^2)] < \prod_{i=1}^{N}[K^2 + W_i(X_i - \beta)^2].$$

This implies that any value of β smaller than $X_{(1)}$ leads to a larger value of the myriad objective function. Therefore, the weighted myriad cannot be less than $X_{(1)}$. A similar argument can be constructed for $X_{(N)}$, leading to the conclusion that the weighted myriad cannot be larger than $X_{(N)}$. ∎

At the other extreme of linearity values ($K \rightarrow 0$), the weighted myriad becomes what is referred to as the *weighted mode-myriad*. Weighted mode-myriad smoothers maintain the same mode-like behavior of the unweighted mode-myriad as stated in the following.

PROPERTY 8.6 (MODE PROPERTY) *Given a vector of positive weights,* $\mathbf{W} = [W_1, \ldots, W_N]$, *the weighted mode-myriad* $\hat{\beta}_0$ *is always equal to one of the most repeated values in the sample. Furthermore,*

$$\hat{\beta}_0 = \arg\min_{X_j \in \mathcal{M}} \left(\frac{1}{W_j}\right)^{\frac{r}{2}} \prod_{i=1, X_i \neq X_j}^{N} |X_i - X_j|, \tag{8.42}$$

where $\mathcal{M}$ is the set of most repeated values, and r is the number of times a member of $\mathcal{M}$ is repeated in the sample set.

Proof : Following the steps of the proof for the unweighted version, it is straightforward that

$$\hat{\beta}_0 = \arg\min_{X_j \in \mathcal{M}} \prod_{i=1, X_i \neq X_j}^{N} W_i(X_i - X_j)^2. \tag{8.43}$$

Dividing by $\prod_{i=1}^{N} W_i$, and applying square root to the expression to be minimized, the desired result is obtained. ∎

PROPERTY 8.7 (OUTLIER REJECTION PROPERTY) *Let $K < \infty$, and let $\mathbf{W}$ denote a vector of positive and finite weights. The outlier rejection property states that:*

$$\lim_{X_N \rightarrow \pm\infty} \hat{\beta}_K(\mathbf{W}; X_1, X_2, \ldots, X_N) = \hat{\beta}_K(\mathbf{W}; X_1, X_2, \ldots, X_{N-1}). \tag{8.44}$$

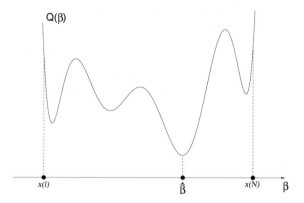

Figure 8.18 Sketch of a typical weighted myriad objective function $Q(\beta)$.

This result can be shown as follows:

$$
\begin{aligned}
\lim_{X_N \to \pm\infty} \hat{\beta}_K(\mathbf{W}; X_1, \ldots, X_N) &= \lim_{X_N \to \pm\infty} \arg\min_{\beta} \prod_{i=1}^{N} \left[K^2 + W_i(X_i - \beta)^2 \right] \\
&= \lim_{X_N \to \pm\infty} \arg\min_{\beta} \frac{1}{X_N^2} \prod_{i=1}^{N} \left[K^2 + W_i(X_i - \beta)^2 \right] \\
&= \lim_{X_N \to \pm\infty} \arg\min_{\beta} \left(\frac{K^2 + W_N(X_N - \beta)^2}{X_N^2} \right. \qquad (8.45) \\
&\qquad\qquad \left. \times \prod_{i=1}^{N-1} \left[K^2 + W_i(X_i - \beta)^2 \right] \right).
\end{aligned}
$$

Evaluating the limit on the right

$$
\lim_{X_N \to \pm\infty} \hat{\beta}_K(\mathbf{W}; X_1, \ldots, X_N) = \arg\min_{\beta} W_N \prod_{i=1}^{N-1} \left[K^2 + W_i(X_i - \beta)^2 \right] \quad (8.46)
$$

and, since W_N is positive

$$
\begin{aligned}
\lim_{X_N \to \pm\infty} \hat{\beta}_K(\mathbf{W}; X_1, \ldots, X_N) &= \arg\min_{\beta} \prod_{i=1}^{N-1} \left[K^2 + W_i(X_i - \beta)^2 \right] \\
&= \hat{\beta}_K(\mathbf{W}; X_1, \ldots, X_{N-1}). \qquad (8.47)
\end{aligned}
$$

According to Property 8.7, large gross errors are efficiently eliminated by any weighted myriad smoother with a finite linearity parameter. Note that this is not the case for the weighted median smoother, in which large positive (negative) errors can always shift the value of the smoother to the right (left).

PROPERTY 8.8 (UNBIASEDNESS) *Let* $X_1, X_2, \ldots, X_N$ *be all independent and symmetrically distributed around the point of symmetry* c. *Then,* $\hat{\beta}_K$ *is also symmetrically distributed around* c. *In particular, if* $E\hat{\beta}_K$ *exists, then* $E\hat{\beta}_K = c$.

Proof : If X_i is symmetric about c, then $2c - X_i$ has the same distribution as X_i. It follows that $\hat{\beta}_K(X_1, X_2, \ldots, X_N)$ has the same distribution as $\hat{\beta}_K(2c - X_1, 2c - X_2, \ldots, 2c - X_N)$, which from the property stated in problem 8.8, is identical to $2c - \hat{\beta}_K(X_1, X_2, \ldots, X_N)$. It follows that $\hat{\beta}_K(X_1, X_2, \ldots, X_N)$ is symmetrically distributed about c. ∎

Geometrical Interpretation Weighted myriads as defined in (8.35) can also be interpreted in a more intuitive manner. Allow a vertical bar to run horizontally through the real line as depicted in Figure 8.19a. Then, the sample myriad, $\hat{\beta}_K$, indicates the position of the bar for which the product of distances from point A to the sample points $X_1, X_2, \ldots, X_N$ is minimum. If weights are introduced, each sample point X_i is assigned a different point A_i in the bar, as illustrated in Figure 8.19b.

The geometrical interpretation of the myriad is intuitively insightful. When K approaches 0, it gives a conceptually simple pictorial demonstration of the mode-myriad formula in (8.13).

8.5 FAST WEIGHTED MYRIAD COMPUTATION

Unlike the weighted mean or weighted median, the computation of the *weighted myriad* is not available in explicit form. Its direct computation is therefore a nontrivial task, since it involves the minimization of the *weighted myriad objective function*, $Q(\beta)$ in (8.33). The myriad objective function, however, has a number of characteristics that can be exploited to construct fast iterative methods to compute its minimum.

Recall that the weighted myriad is given by

$$\hat{\beta}_K = \arg\min_{\beta} \log(P(\beta)) \overset{\triangle}{=} \arg\min_{\beta} Q(\beta)$$

$$= \arg\min_{\beta} \sum_{i=1}^{N} \log\left[1 + \left(\frac{x_i - \beta}{S_i}\right)^2\right], \tag{8.48}$$

where $Q(\beta)$ is the *weighted myriad objective function*. Having well defined derivatives, $\hat{\beta}_K$ is one of the local minima of $Q(\beta)$, that is, the values for which $Q'(\hat{\beta}) = 0$. Since $Q(\beta) = \log(P(\beta))$, the derivative of $Q(\beta)$ can be written as

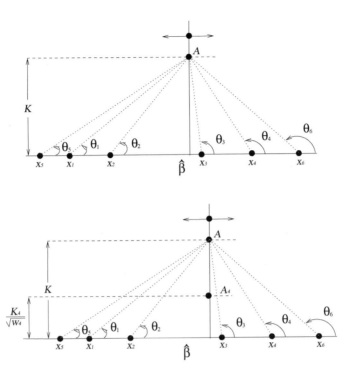

Figure 8.19 (a) The sample myriad, $\hat{\beta}_K$, indicates the position of a moving bar such that the product of distances from point A to the sample points $X_1, X_2, \ldots, X_N$ is minimum. (b) If the weight $W_4 > 1$ is introduced, the product of distances is more sensitive to the variations of the segment $\overline{X_4 A_4}$, very likely resulting in a weighted myriad $\hat{\beta}_K$ closer to X_4.

$$Q'(\beta) = \frac{P'(\beta)}{P(\beta)}. \tag{8.49}$$

where the derivative of $P(\beta) = \prod_{i=1}^{N} \left(K^2 + W_i(X_i - \beta)^2 \right)$ is given by

$$P'(\beta) = 2 \sum_{i=1}^{N} \frac{W_i \, (\beta - X_i)}{K^2 + W_i \, (X_i - \beta)^2} \, P(\beta). \tag{8.50}$$

From (8.49), it follows that

$$Q'(\beta) = 2 \sum_{i=1}^{N} \frac{W_i \, (\beta - X_i)}{K^2 + W_i \, (X_i - \beta)^2}. \tag{8.51}$$

Using the fact that $S_i = \frac{K}{\sqrt{W_i}}$, the above can be written as

$$Q'(\beta) = 2 \sum_{i=1}^{N} \frac{\left(\dfrac{\beta - X_i}{S_i^2}\right)}{1 + \left(\dfrac{X_i - \beta}{S_i}\right)^2}. \tag{8.52}$$

Defining

$$\psi(v) \triangleq \frac{2v}{1 + v^2}, \tag{8.53}$$

and referring to (8.52) the following equation is obtained for the local extremes of $Q(\beta)$:

$$Q'(\beta) = -\sum_{i=1}^{N} \frac{1}{S_i} \cdot \psi\left(\frac{X_i - \beta}{S_i}\right) = 0. \tag{8.54}$$

By introducing the *positive* functions

$$h_i(\beta) \triangleq \frac{1}{S_i^2} \cdot \varphi\left(\frac{X_i - \beta}{S_i}\right) > 0, \tag{8.55}$$

for $i = 1, 2, \ldots, N$, where

$$\varphi(v) \triangleq \frac{\psi(v)}{v} = \frac{2}{1 + v^2}, \tag{8.56}$$

the local extremes of $Q(\beta)$ in (8.54) can be formulated as

$$Q'(\beta) = -\sum_{i=1}^{N} h_i(\beta) \cdot (X_i - \beta) = 0. \tag{8.57}$$

This formulation implies that the *sum of weighted deviations* of the samples is zero, with the (positive) weights themselves being functions of β. This property, in turn, leads to a simple iterative approach to compute the weighted myriad as detailed next.

Fixed Point Formulation Equation (8.57) can be written as

$$\beta = \frac{\displaystyle\sum_{i=1}^{N} h_i(\beta) \cdot X_i}{\displaystyle\sum_{i=1}^{N} h_i(\beta)} \tag{8.58}$$

where it can be seen that each local extremum of $Q(\beta)$, including the weighted myriad $\hat{\beta}$, can be written as a *weighted mean* of the input samples X_i. Since the weights $h_i(\beta)$ are always positive, the right hand side of (8.58) is in $(X_{(1)}, X_{(N)})$, confirming that all the local extremes lie within the range of the input samples. By defining the mapping

$$T(\beta) \triangleq \frac{\displaystyle\sum_{i=1}^{N} h_i(\beta) \cdot X_i}{\displaystyle\sum_{i=1}^{N} h_i(\beta)}, \tag{8.59}$$

the local extremes of $Q(\beta)$, or the roots of $Q'(\beta)$, are seen to be the *fixed points* of $T(\cdot)$:

$$\beta^* = T(\beta^*). \tag{8.60}$$

The following *fixed point iteration* results in an efficient algorithm to compute these fixed points:

$$\beta_{m+1} \triangleq T(\beta_m) = \frac{\displaystyle\sum_{i=1}^{N} h_i(\beta_m) \cdot X_i}{\displaystyle\sum_{i=1}^{N} h_i(\beta_m)}. \tag{8.61}$$

In the classical literature, this is also called the *method of successive approximation* for the solution of the equation $\beta = T(\beta)$ [112]. It has been proven that the iterative method of (8.61) converges to a fixed point of $T(\cdot)$; thus,

$$\lim_{m \to \infty} \beta_m = \beta^* = T(\beta^*). \tag{8.62}$$

The recursion of (8.61) can be benchmarked against the update in Newton's method [112] for the solution of the equation $Q'(\beta) = 0$:

$$\beta'_{m+1} \triangleq \beta_m - \frac{Q'(\beta_m)}{Q''(\beta_m)}, \tag{8.63}$$

which is interpreted by considering the *tangent* of $Q'(\beta)$ at $\beta = \beta_m$:

$$Z(\beta) \triangleq Q'(\beta_m) + Q''(\beta_m) (\beta - \beta_m).$$

Here, $Z(\beta)$ is used as a linear approximation of $Q'(\beta)$ around the point β_m, and β'_{m+1} is the point at which the tangent $Z(\beta)$ crosses the β axis: $Q'(\beta'_{m+1}) \approx Z(\beta'_{m+1}) = 0$.

Although Newton's method can have fast (quadratic) convergence, its major disadvantage is that it may converge only if the initial value β_0 is sufficiently close to the solution β^* [112]. Thus, only local convergence is guaranteed. On the other hand, Kalluri and Arce [112] have shown that the fixed point iteration method of (8.61) decreases the objective function $Q(\beta)$ continuously at each step, leading to *global convergence* (convergence from an arbitrary starting point).

Table 8.1   Summary of the fast iterative weighted myriad search.

Parameters: Select	L = number of iterations in search
Initial Point:	$\hat{\beta}_0 = \arg\min_{X_i} P(X_i)$
Computation:	For $m = 0, 1, \ldots, L$. $\hat{\beta}_{m+1} = T(\hat{\beta}_m).$

The speed of convergence of the iterative algorithm (8.61) depends on the initial value β_0. A simple approach of selecting $\hat{\beta}_0$ is to assign its the value equal to that of the input sample X_i which that leads to the smallest cost $P(X_i)$.

Fixed Point Weighted Myriad Search

Step 1: Select the initial point $\hat{\beta}_0$ among the values of the input samples:
$$\hat{\beta}_0 = \arg\min_{X_i} P(X_i).$$

Step 2: Using $\hat{\beta}_0$ as the initial value, perform L iterations of the fixed point recursion $\beta_{m+1} = T(\beta_m)$ of (8.61). The final value of these iterations is then chosen as the weighted myriad: $\hat{\beta}_{FP} = T^{(L)}(\hat{\beta}_0).$

This algorithm can be compactly written as

$$\hat{\beta}_{FP} = T^{(L)}\left(\arg\min_{X_i} P(X_i)\right). \tag{8.64}$$

Note that for the special case $L = 0$ (meaning that no fixed point iterations are performed), the above algorithm computes the selection weighted myriad. Table 8.1 summarizes the iterative weighted myriad search algorithm.

8.6   WEIGHTED MYRIAD SMOOTHER DESIGN

8.6.1   Center-Weighted Myriads for Image Denoising

Median smoothers are effective at image denoising, especially for impulsive noise, which often results from bit errors in the transmission stage and/or in the acquisition stage. As a subset of traditional weighted median smoothers, center-weighted median (CWM) smoothers provide similar performance with much less complexity. In CW medians, only the center sample in the processing window is assigned weight, and all other samples are treated equally without emphasis, that is, are assigned a weight of 1. The larger the center weight, the less smoothing is achieved. Increasing the center weight beyond a certain threshold will turn the CW median into an identity operation. On the other hand, when the center weight is set to unity (the same as other weights), the CW median becomes a sample median operation.

The same notion of center weighting can be applied to the myriad structure as well, thus leading to the center-weighted myriad smoother (CWMy) defined as:

$$Y = \text{MYRIAD}\{K; X_1, \ldots, W_c \circ X_c, \ldots, X_N\}. \tag{8.65}$$

The cost function in (8.33) is now modified to

$$Q(\beta) = \log\left[K^2 + W_c(X_c - \beta)^2\right] + \sum_{X_i \neq X_c} \log\left[K^2 + (X_i - \beta)^2\right]. \tag{8.66}$$

While similar to a CW median, the above center weighted myriad smoother has significant differences. First, in addition to the center weight W_c, the CWMy has the free parameter (K) that controls the impulsiveness rejection. This provides a simple mechanism to attain better smoothing performance. Second, the center weight in the CWMy smoother is inevitably data dependent, according to the definition of the objective function in (8.66). For different applications, the center weight should be adjusted accordingly based on their data ranges.

For grayscale image denoising applications where pixel values are normalized between 0 and 1, the two parameters of the CWMy smoother can be chosen as follows:

(1) Choose $K = (X_{(U)} + X_{(L)})/2$, where $1 \leq L < U \leq N$, with $X_{(U)}$ being the Uth smallest sample in the window and $X_{(L)}$ the Lth smallest sample.

(2) Set $W_c = 10,000.$[4]

The linear parameter K is dynamically calculated based on the samples in the processing window. When there is "salt" noise in the window (outliers having large values), the myriad structure assures that they are deemphasized because of the outlier rejection property of K. The center weight W_c is chosen to achieve balance between outlier rejection and detail preservation. It should be large enough to emphasize the center sample and preserve signal details, but small enough so it does not let impulsive noise through.

It can also be shown that [129], the CWMy smoother with K and W_c defined as above, has the capability of rejecting "pepper" type noise (having values close to 0). This can be seen as follows. For a single "pepper" outlier sample, the cost function (8.66) evaluated at $\beta = K$ will always be smaller than that at $\beta = 0$.

---

[4]This is an empirical value. A larger value of the center weight will retain details on the image, but it will also cause some of the impulses to show in the output. A smaller value will eliminate this impulses but it might cause some loss in detail. This characteristics will be shown in Figure 8.22.

Thus, "pepper" noise will never go through the smoother if the parameters K and W_c are chosen as indicated. Denote $\mathbf{X}$ as the corrupted image, $\mathbf{Y}$ the output smoothed image, and CWMy the smoother operation. A 2-pass CWMy smoother can be defined as follows:

$$\mathbf{Y} = 1 - \text{CWMy}(1 - \text{CWMy}(\mathbf{X})). \qquad (8.67)$$

Figure 8.20 depicts the results of the algorithm defined in (8.67). Figure 8.20*a* shows the original image. Figure 8.20*b* the same image corrupted by 5% salt and pepper noise. The impulses occur randomly and were generated with MATLAB's imnoise function. Figure 8.20*c* is the output of a 5×5 CWM smoother with $W_c = 15$, *d* is the CWMy smoother output with $W_c = 10,000$ and $K = (X_{(21)} + X_{(5)})/2$. The superior performance of the CWMy smoother can be readily seen in this figure. The CWMy smoother preserves the original image features significantly better than the CWM smoother. The mean square error of the CWMy output is consistently less than half of that of the CWM output for this particular image.

The effect of the center weight can be appreciated in Figures 8.22 and 8.23. A low value of the center weight will remove all the impulsive noise at the expense of smoothing the image too much. On the other hand, a value higher than the recommended will maintain the details of the image, but some of the impulsive noise begins to leak to the output of the filter.

8.6.2 Myriadization

The linear property indicates that for very large values of K, the weighted myriad smoother reduces to a constrained linear FIR smoother. The meaning of K suggests that a linear smoother can be provided with resistance to impulsive noise by simply reducing the linearity parameter from $K = \infty$ to a finite value. This would transform the linear smoother into a myriad smoother with the same weights. In the same way as the term *linearization* is commonly used to denote the transformation of an operator into a linear one, the above transformation is referred to as *myriadization*.

Myriadization is a simple but powerful technique that brings impulse resistance to constrained linear filters. It also provides a simple methodology to design suboptimal myriad smoothers in impulsive environments. Basically, a constrained linear smoother can be designed for Gaussian or noiseless environments using FIR filter (smoother) design techniques, and then provide it with impulse resistance capabilities by means of myriadization. The value to which K is to be reduced can be designed according to the impulsiveness of the environment, for example by means of an α-K curve.

It must be taken into account that a linear smoother has to be in constrained form before myriadization can be applied. This means that the smoother coefficients W_i must be nonnegative and satisfy the normalization condition $\sum_{i=1}^{N} W_i = 1$. A smoother for which $\sum_{i=1}^{N} W_i \neq 1$, must be first decomposed into the cascade of its

Figure 8.20 (*a*) Original image, (*b*) Image with 5% salt-and-pepper noise (PSNR=17.75dB), (*c*) smoothed with 5×5 center weighted median with $W_c = 15$(PSNR=37.48dB), (*d*) smoothed with 5×5 center weighted myriad with $W_c = 10,000$ and $K = (X_{(21)} + X_{(5)})/2$ (PSNR=39.98dB)

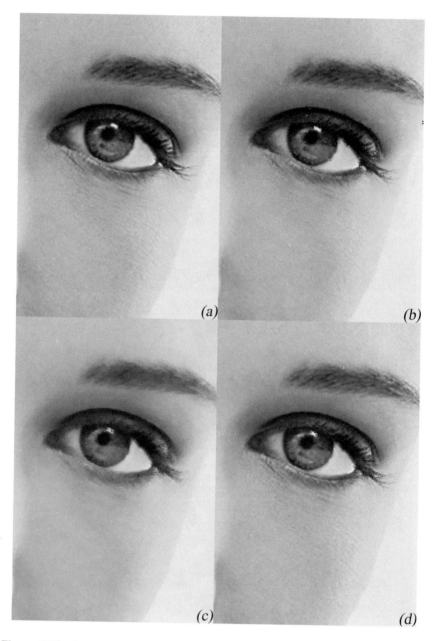

Figure 8.21 Comparison of different filtering schemes (Enlarged). (*a*) Original Image, (*b*) Image smoothed with a center weighted median (PSNR=37.48dB), (*c*) Image smoothed with a 5 × 5 permutation weighted median (PSNR=35.55dB), (*d*) Image smoothed with the center weighted myriad (PSNR=39.98dB).

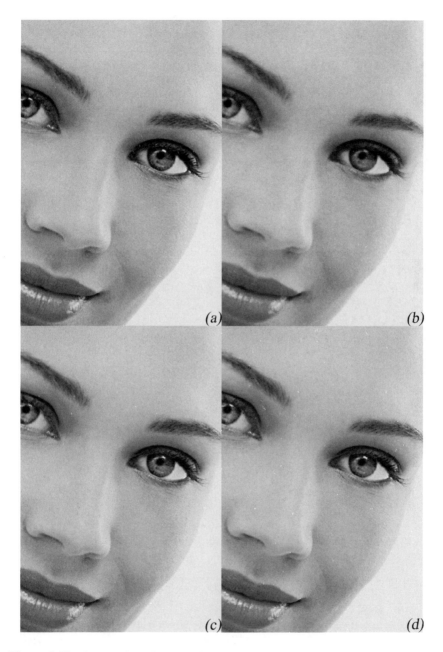

Figure 8.22 Output of the Center weighted myriad smoother for different values of the center weight W_c (*a*) Original image, (*b*) 100 (PSNR=36.74dB), (*c*) 10,000 (PSNR=39.98dB), (*d*) 1,000,000 (PSNR=38.15dB).

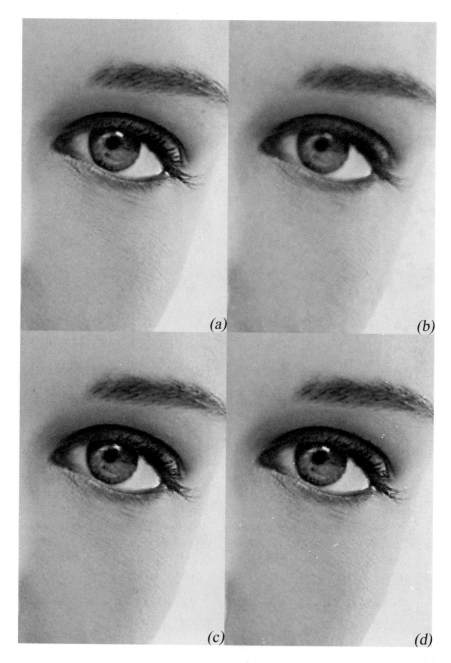

Figure 8.23 Output of the center-weighted myriad smoother for different values of the center weight W_c (enlarged) (*a*) Original image, (*b*) 100, (*c*) 10,000, (*d*) 1,000,000.

normalized version with an amplifier of gain $\sum_{i=1}^{N} W_i$. Design by myriadization is illustrated in the following example.

EXAMPLE 8.6 (ROBUST LOW PASS FILTER DESIGN)

Figure 8.24a depicts a unit-amplitude linearly swept-frequency cosine signal spanning instantaneous frequencies ranging from 0 to 400 Hz. The chirp was generated with MATLAB's chirp function having a sampling interval of 0.0005 seconds. Figure 8.24b shows the chirp immersed in additive Cauchy noise ($\gamma = 1$). The plot is truncated to the same scale as the other signals in the figure. A low-pass linear FIR smoother with 30 coefficients processes the chirp with the goal of retaining its low-frequency components. The FIR low-pass smoother weights were designed with MATLAB's fir1 function with a normalized frequency cutoff of 0.05. Under ideal, no-noise conditions, the output of the linear smoother would be that of Figure 8.24c. However, the impulsive nature of the noise introduces severe distortions to the actual output, as depicted in Figure 8.24d. Myriadizing the linear smoother by reducing K to a finite value of 0.5 , significantly improves the smoother performance (see Fig. 8.24e,f). Further reduction of K to 0.2 drives the myriad closer to a selection mode where some distortion on the smoother output under ideal conditions can be seen (see Fig. 8.24g). The output under the noisy conditions is not improved by further reducing K to 0.2, or lower, as the smoother in this case is driven to a selection operation mode.

■

EXAMPLE 8.7 (MYRIADIZATION OF PHASE LOCK LOOP FILTERS)

First-order Phase-Locked Loop (PLL) systems, depicted in Figure 8.25, are widely used for recovering carrier phase in coherent demodulators [146]. Conventional PLL utilize a linear FIR low-pass filter intended to let pass only the low frequencies generated by the multiplier.

The output of the low-pass filter represents the phase error between the incoming carrier and the recovered tone provided by the controlled oscillator. The system is working properly (i.e., achieving synchronism), whenever the output of the low-pass filter is close to zero. To test the PLL mechanisms, the FIR filter synthesized used 13 normalized coefficients where the incoming signal is a sinusoid of high frequency and unitary amplitude, immersed in additive white Gaussian noise of variance 10^{-3}, yielding a signal-to-noise ratio (SNR) of 30 dB. The parameters of the system, including (linear) filter weights and oscillator gain, were manually adjusted so that the error signal had minimum variance. Three different scenarios, corresponding to three different low-pass filter structures were simulated. The incoming and noise signals were identical for the three systems. At three arbitrary time points ($t \approx$ 400, 820, 1040), short bursts of high-power Gaussian noise were added to the noise

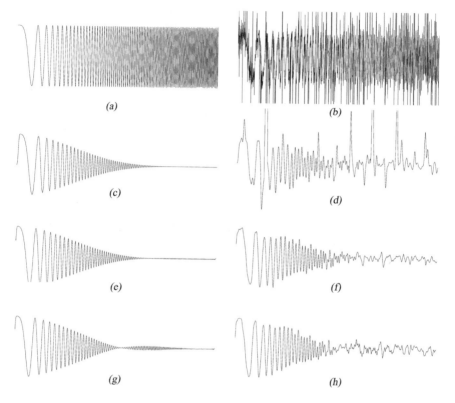

Figure 8.24 Myriadizing a linear low-pass smoother in an impulsive environment: (*a*) chirp signal, (*b*) chirp in additive impulsive noise, (*c*) ideal (no noise) myriad smoother output with $K = \infty$, (*e*) $K = 0.5$, and (*g*) $K = 0.2$; Myriad smoother output in the presence of noise with (*d*) $K = \infty$, (*f*) $K = 0.5$, and (*h*) $K = 0.2$.

signal. The length of the bursts was relatively short (between 4 and 10 sampling times) compared to the length of the filter impulse response (12 sampling times). The SNR during burst periods was very low (about -10 dB's), making the noise look heavy impulsive. Figure 8.26 shows the phase error in time when the standard linear filter was used. It is evident from the figure that this system is very likely to lose synchronism after a heavy burst. Figure 8.26*b* shows the phase error of a second scenario in which a weighted median filter has been designed to imitate the low-pass characteristics of the original linear filter [6, 201]. Although the short noise bursts do not affect the estimate of the phase, the variance of the estimate is very large. This noise amplification behavior can be explained from the inefficiency introduced by the selection property of the median, that is, the fact that the filter output is always constrained to be one of its inputs. Finally, Figure 8.26*c* shows the phase after the low-pass filter has been myriadized using a parameter K equal to half the carrier amplitude. Although the phase error is increased during the bursts, the performance of the myriadized PLL is not degraded, and the system does not lose synchronism.

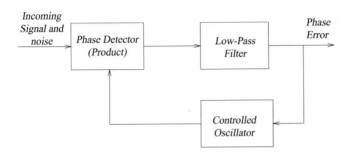

Figure 8.25 Block diagram of the Phase-Locked Loop system.

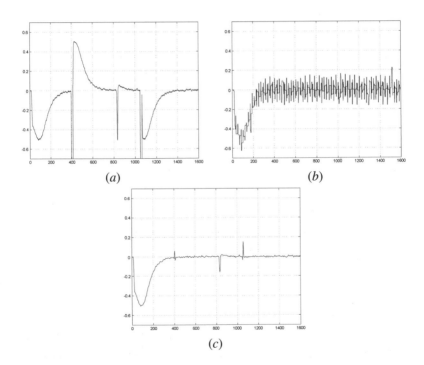

Figure 8.26 Phase error plot for the PLL with (*a*) a linear FIR filter; (*b*) an optimal weighted median filter; and (*c*) a *myriadized* version of the linear filter.

Problems

8.1   Show that:

(a) A FLOM smoother with $p < 1$ is selection type.

(b) The gamma smoother with $p = 2$ outputs the sample that is closest to the mean of the input samples.

8.2   Prove that the sample mode-myriad is shift and scale invariant. Thus, given $Z_i = aX_i + b$, for $i = 1, \ldots, N$, show that

$$\hat{\beta}_0(Z_1, \ldots, Z_N) = a\hat{\beta}_0(X_1, \ldots, X_N) + b. \tag{8.68}$$

8.3   Prove that the sample mode-myriad satisfies the "no overshoot/undershoot" property. That is $\hat{\beta}_0$ is always bounded by

$$X_{(2)} \leq \hat{\beta}_0 \leq X_{(N-1)}, \tag{8.69}$$

where $X_{(i)}$ denotes the ith-order statistic of the sample.

8.4   Show that if $N = 3$, $\hat{\beta}_0$ is equivalent to the sample median.

8.5   Show that for a Cauchy distribution with dispersion parameter K, the interquartile range is the value of K.

8.6   For the weighted median smoother defined in (8.34) show that:

$$\hat{\beta}_{K_1}(\mathbf{W}_1, \mathbf{X}) = \hat{\beta}_{K_2}(\mathbf{W}_2, \mathbf{X}) \quad \text{iff} \quad \frac{\mathbf{W}_1}{K_1^2} = \frac{\mathbf{W}_2}{K_2^2}. \tag{8.70}$$

8.7   Prove Property 8.40, the linear property of the weighted myriad.

8.8   (Shift and sign invariance properties of the weighted myriad) Let $Z_i = X_i + b$. Then, for any K and $\mathbf{W}$,

(a) $\hat{\beta}_K(Z_1, \ldots, Z_N) = \hat{\beta}_K(X_1, \ldots, X_N) + b$;

(b) $\hat{\beta}_K(-Z_1, \ldots, -Z_N) = -\hat{\beta}_K(Z_1, \ldots, Z_N)$.

8.9   (Gravitational Property of the Weighted Myriad) Let $\mathbf{W}$ denote a vector of positive finite weights, show that there always exists a sample X_i such that

$$|X_i - \beta_K| \leq \frac{K}{W_i}, \tag{8.71}$$

where W_i is the weight assigned to X_i.

9

Weighted Myriad Filters

Myriad smoothers admitting positive weights only are in essence low-pass-type filters. Weighted myriad smoothers are thus analogous to normalized linear FIR filters with nonnegative weights summing to unity. There is a clear need to extend these smoothers into a general filter structure, comparable to linear FIR filters, that admit real-valued weights. In the same way that weighted median smoothers are extended to the weighted median filter, a generalized weighted myriad filter structure that admits real-valued weights is feasible. This chapter describes the structure and properties of such class of filters admitting positive as well as negative weights. Adaptive optimization algorithms are presented. As would be expected, weighted myriad filters reduce to weighted myriad smoothers whenever the filter coefficients are constrained to be positive.

9.1 WEIGHTED MYRIAD FILTERS WITH REAL-VALUED WEIGHTS

The approach used to generalize median smoothers to a general class of median filters can be used to develop a generalized class of weighted myriad filters. To this end, the set of real-valued weights are first decoupled in their sign and magnitude. The sign of each weight is then attached to the corresponding input sample and the weight magnitude is used as a positive weight in the weighted myriad smoother structure. Starting from the definition of the weighted myriad smoothers (Def. 8.2), the class of weighted myriad filters admitting real-valued weights emerges as follows:

DEFINITION 9.1 (WEIGHTED MYRIAD FILTERS) *Given a set of N real valued weights $(W_1, W_2, \ldots, W_N)$ and the observation vector $\mathbf{X} = [X_1, X_2, \ldots, X_N]^T$, the weighted myriad filter output is defined as*

$$
\begin{aligned}
\hat{\beta}_K(\mathbf{W}, \mathbf{X}) &\triangleq MYRIAD\left(|W_i| \circ \mathrm{sgn}(W_i)X_i\right)|_{i=1}^N \\
&= \arg\min_{\beta} Q(\beta),
\end{aligned}
\tag{9.1}
$$

where

$$
Q(\beta) \triangleq \sum_{i=1}^N \log\left[K^2 + |W_i| \cdot (\mathrm{sgn}(W_i)X_i - \beta)^2\right]
\tag{9.2}
$$

is the objective function of the weighted myriad filter.

Since the log function is monotonic, the weighted myriad filter is also defined by the equivalent expression

$$
\begin{aligned}
\hat{\beta}_K(\mathbf{W}, \mathbf{X}) &\triangleq \arg\min_{\beta} \prod_{i=1}^N \left[K^2 + |W_i|\,(\mathrm{sgn}(W_i)X_i - \beta)^2\right] \tag{9.3} \\
&\triangleq \arg\min_{\beta} P(\beta); \tag{9.4}
\end{aligned}
$$

thus $\hat{\beta}_K$ is the global minimizer of $P(\beta)$ as well as of $Q(\beta) \triangleq \log(P(\beta))$.

Like the weighted myriad smoother, the weighted myriad filter also has only N independent parameters. Using (9.2), it can be inferred that if the value of K is changed, the same filter output can be obtained provided the filter weights are appropriately scaled. The following is true

$$
\hat{\beta}_K(\mathbf{W}, \mathbf{X}) = \hat{\beta}_1\left(\frac{\mathbf{W}}{K^2}, \mathbf{X}\right)
\tag{9.5}
$$

or

$$
\hat{\beta}_{K_1}(\mathbf{W}_1, \mathbf{X}) = \hat{\beta}_{K_2}(\mathbf{W}_2, \mathbf{X}) \quad \text{iff} \quad \frac{\mathbf{W}_1}{K_1^2} = \frac{\mathbf{W}_2}{K_2^2}.
\tag{9.6}
$$

Hence, the output depends only on $\frac{\mathbf{W}}{K^2}$.

The objective function $P(\beta)$ in (9.4) is a polynomial in β of degree $2N$, with well-defined derivatives of all orders. Therefore, $P(\beta)$ (and the equivalent objective function $Q(\beta)$) can have at most $(2N - 1)$ local extremes. The output is thus one of the local minima of $Q(\beta)$:

$$
Q'(\hat{\beta}) = 0.
\tag{9.7}
$$

Figure 9.1 depicts a typical objective function $Q(\beta)$, for various values of K. The effect of assigning a negative weight on the cost function is illustrated in this figure. As in the smoother case, the number of local minima in the objective function $Q(\beta)$ depends on the value of the parameter K. When K is very large only one extremum exists.

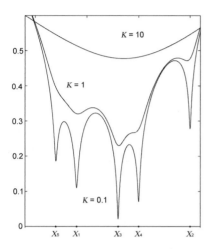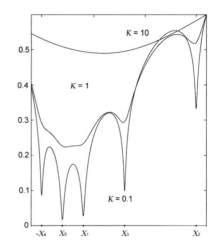

Figure 9.1 Weighted myriad cost function for the sample set $X = [-1, \ 10, \ 3, \ 5, \ -3]$ with weights $W = [1, \ 2, \ 3, \ \pm 2, \ 1]$ for $k = 0.1, \ 1, \ 10$

In the limit as $K \to \infty$, with the weights $\{W_i\}$ held constant, it can be shown that $Q(\beta)$ exhibits a single local extremum. The proof is a generalized form of that used to prove the linear property of weighted myriad smoothers.

PROPERTY 9.1 (LINEAR PROPERTY) *In the limit as $K \to \infty$, the weighted myriad filter reduces to the normalized linear FIR filter*

$$\lim_{K \to \infty} \hat{\beta}_K = \frac{\sum_{i=1}^{N} W_i X_i}{\sum_{i=1}^{N} |W_i|}. \tag{9.8}$$

Once again, the name 'linearity parameter' is used for the parameter K.

At the other extreme of linearity values ($K \to 0$), the weighted myriad filter maintains a mode-like behavior as stated in the following.

PROPERTY 9.2 (MODE PROPERTY) *Given a vector of real-valued weights, $\mathbf{W} = [W_1, \ldots, W_N]$, the weighted mode-myriad $\hat{\beta}_0$ is always equal to one of the most repeated values in the signed sample set $\{\mathrm{sgn}(W_1)X_1, \ \mathrm{sgn}(W_2)X_2, \ldots, \ \mathrm{sgn}(W_N) X_N\}$. Furthermore,*

$$\hat{\beta}_0 = \arg \min_{\mathrm{sgn}(W_j)X_j \in \mathcal{M}} \left(\frac{1}{|W_j|} \right)^{\frac{r}{2}} \prod_{i=1, X_i \neq X_j}^{N} |\mathrm{sgn}(W_i)X_i - \mathrm{sgn}(W_j)X_j|, \tag{9.9}$$

where $\mathcal{M}$ is the set of most repeated signed values, and r is the number of times a member of $\mathcal{M}$ is repeated in the signed sample set.

9.2   FAST REAL-VALUED WEIGHTED MYRIAD COMPUTATION

Using the same technique developed in Section 8.5, the fast computation for the real-valued weighted myriad can be derived. Recall that the myriad objective function now is

$$Q(\beta) = \sum_{i=1}^{N} \log \left[K^2 + |W_i| \cdot (\text{sgn}(W_i)X_i - \beta)^2 \right]. \tag{9.10}$$

Its derivative with respect to β can be easily found as

$$Q'(\beta) = 2 \sum_{i=1}^{N} \frac{|W_i| \, (\beta - \text{sgn}(W_i)X_i)}{K^2 + |W_i| \, (\text{sgn}(W_i)X_i - \beta)^2}. \tag{9.11}$$

By introducing the *positive* functions

$$h_i(\beta) \triangleq \frac{2|W_i|}{K^2 + |W_i| \, (\text{sgn}(W_i)X_i - \beta)^2} > 0, \tag{9.12}$$

for $i = 1, 2, \ldots, N$, the local extremes of $Q(\beta)$ satisfy the following condition

$$Q'(\beta) = - \sum_{i=1}^{N} h_i(\beta) \cdot (\text{sgn}(W_i)X_i - \beta) = 0. \tag{9.13}$$

Since $\hat{\beta}_K$ is one of the local minima of $Q(\beta)$, we have $Q'(\hat{\beta}_K) = 0$. Equation (9.13) can be further written as *weighted mean* of the signed samples

$$\beta = \frac{\displaystyle\sum_{i=1}^{N} h_i(\beta) \cdot \text{sgn}(W_i)X_i}{\displaystyle\sum_{i=1}^{N} h_i(\beta)}. \tag{9.14}$$

By defining the mapping

$$T(\beta) \triangleq \frac{\displaystyle\sum_{i=1}^{N} h_i(\beta) \cdot \text{sgn}(W_i)X_i}{\displaystyle\sum_{i=1}^{N} h_i(\beta)}, \tag{9.15}$$

the local extremes of $Q(\beta)$, or the roots of $Q'(\beta)$, are seen to be the *fixed points* of $T(\cdot)$:

$$\beta^* = T(\beta^*). \tag{9.16}$$

The following efficient *fixed point iteration* algorithm is used to compute the fixed points:

$$\beta_{m+1} \triangleq T(\beta_m) = \frac{\sum_{i=1}^{N} h_i(\beta_m) \cdot \text{sgn}(W_i)X_i}{\sum_{i=1}^{N} h_i(\beta_m)}. \tag{9.17}$$

In the limit, the above iteration will converge to one of the fixed points of $T(\cdot)$

$$\lim_{m \to \infty} \beta_m = \beta^* = T(\beta^*). \tag{9.18}$$

It is clear that the global convergence feature of this fixed point iteration can be assured from the analysis in Section 8.5, since the only difference is the sample set.

Fixed Point Weighted Myriad Search Algorithm

Step 1: Couple signs of the weights and samples to form the signed sample vector $[\text{sgn}(W_1)X_1, \text{sgn}(W_2)X_2, \ldots, \text{sgn}(W_N)X_N]$.

Step 2: Compute the selection weighted myriad: $\hat{\beta}_0 = \underset{\beta \in \{\text{sgn}(W_i)X_i\}}{\arg\min} P(\text{sgn}(W_i)X_i)$.

Step 3: Using $\hat{\beta}_0$ as the initial value, perform L iterations of the fixed point recursion $\beta_{m+1} = T(\beta_m)$ of (9.17). The final value of these iterations is then chosen as the weighted myriad: $\hat{\beta}_{\text{FP}} = T^{(L)}(\hat{\beta}_0)$.

The compact expression of the above algorithm is

$$\hat{\beta}_{\text{FP}} = T^{(L)} \left(\underset{\beta \in \{\text{sgn}(W_i)X_i\}}{\arg\min} P(\text{sgn}(W_i)X_i) \right). \tag{9.19}$$

9.3 WEIGHTED MYRIAD FILTER DESIGN

9.3.1 Myriadization

The linear property indicates that for very large values of K, the weighted myriad filter reduces to a constrained linear FIR filter. This characteristic of K suggests that a linear FIR filter can be provided with resistance to impulsive noise by simply reducing the linearity parameter from $K = \infty$ to a finite value. This would transform the linear FIR filter into a myriad filter with the same weights. This transformation is referred to as *myriadization* of linear FIR filters. Myriadization is a simple but powerful technique that brings impulse resistance to linear FIR filters. It also provides a simple methodology to design suboptimal myriad filters in impulsive environments. A linear FIR filter can be first designed for Gaussian or noiseless environments using FIR filter design tools, and then provided with impulse resistance capabilities by means of myriadization. The value to which K is to be reduced can be designed according to the impulsiveness of the environment. Design by myriadization is illustrated in the following example.

Example: Robust Band-Pass Filter Design   Figure 9.2*a* depicts a unit-amplitude linearly swept-frequency cosine signal spanning instantaneous frequencies ranging from 0 to 400 Hz. The chirp was generated with MATLAB's chirp function having a sampling interval of 0.0005 seconds. Figure 9.3*a* shows the chirp immersed in additive Cauchy noise ($\gamma = 0.05$). The plot is truncated to the same scale as the other signals in the figure. A band-pass linear FIR filter with 31 coefficients processes the chirp with the goal of retaining its mid-frequency components. The FIR band-pass filter weights were designed with MATLAB's fir1 function with normalized frequency cutoffs of 0.15 and 0.25. Under ideal, no-noise conditions, the output of the FIR filter would be that of Figure 9.2*b*. However, the impulsive nature of the noise introduces severe distortions to the actual output, as depicted in Figure 9.3*b*. Myriadizing the linear filter by reducing K to a finite value of 0.5, significantly improves the filter performance (see Figs. 9.2*c* and 9.3*c*). Further reduction of K to 0.2 drives the myriad closer to a selection mode where some distortion on the filter output under ideal conditions can be seen (see Fig. 9.2*d*). The output under the noisy conditions is not improved by further reducing K to 0.2, or lower, as the filter in this case is driven to a selection operation mode (Fig. 9.3*d*).

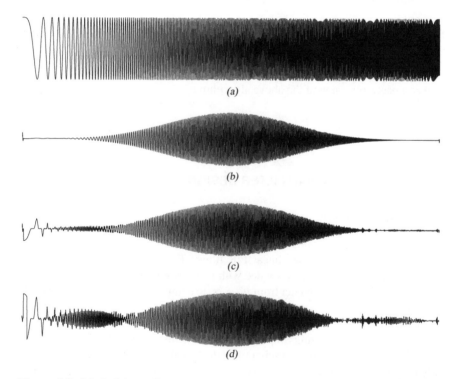

(a)

(b)

(c)

(d)

Figure 9.2   Myriadizing a linear band-pass filter in an impulsive environment: (*a*) chirp signal, (*b*) ideal (no noise) myriad smoother output with $K = \infty$, (*c*) $K = 0.5$, and (*d*) $K = 0.2$.

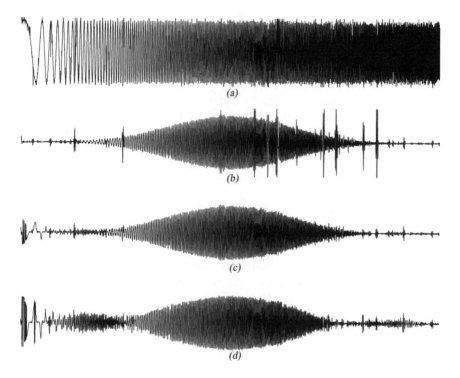

Figure 9.3 Myriadizing a linear band-pass filter in an impulsive environment (continued): (*a*) chirp in additive impulsive noise. Myriad filter output in the presence of noise with (*b*) $K = \infty$, (*c*) $K = 0.5$, and (*d*) $K = 0.2$

9.3.2 Optimization

The *optimization* of the weighted myriad filter parameters for the case when the linearity parameter K satisfies $K > 0$ was first described in [111]. The goal is to design the set of weighted myriad filter weights that optimally estimate a desired signal according to a *statistical error criterion*. Although the mean absolute error (MAE) criterion is used here, the solutions are applicable to the mean square error (MSE) criterion with simple modifications.

Given an input (observation) vector $\mathbf{X} \triangleq [X_1, X_2, \ldots, X_N]^T$, a weight vector $\mathbf{W} \triangleq [W_1, W_2, \ldots, W_N]^T$ and linearity parameter K, denote the weighted myriad filter output as $Y \equiv Y_K(\mathbf{W}, \mathbf{X})$, sometimes abbreviated as $Y(\mathbf{W}, \mathbf{X})$. The filtering error, in estimating a desired signal $D(n)$, is then defined as $e(n) = Y(n) - D(n)$. Under the mean absolute error (MAE) criterion, the cost function is defined as

$$J_1(\mathbf{W}, K) \triangleq E\{|e|\} = E\{|Y_K(\mathbf{W}, \mathbf{X}) - D|\}, \qquad (9.20)$$

where $E\{\cdot\}$ represents statistical expectation. The mean square error (MSE) is defined as

$$J_2(\mathbf{W}, K) \triangleq E\{e^2\} = E\left\{(Y_K(\mathbf{W}, \mathbf{X}) - D)^2\right\}. \qquad (9.21)$$

When the error criterion adopted is clear from the context, the cost function is written as $J(\mathbf{W}, K)$. Further, the optimal filtering action is independent of K (the filter weights can be scaled to keep the output invariant to changes in K). The cost function is therefore sometimes written simply as $J(\mathbf{W})$, with an assumed arbitrary choice of K. Obtaining conditions for a global minimum that are both necessary and sufficient is quite a formidable task. Necessary conditions, on the other hand, can be attained by setting the gradient of the cost function equal to zero. The *necessary* conditions to be satisfied by the optimal filter parameters are obtained as:

$$\frac{\partial J(\mathbf{W})}{\partial W_i} = 2 E\left\{e \frac{\partial Y}{\partial W_i}\right\} = 0, \ i = 1, 2, \dots, N. \qquad (9.22)$$

The nonlinear nature of the equations in (9.22) prevents a closed-form solution for the optimal parameters. The *method of steepest descent* is thus applied, which continually updates the filter parameters in an attempt to converge to the global minimum of the cost function $J(\mathbf{W})$:

$$W_i(n+1) = W_i(n) - \frac{1}{2}\mu \frac{\partial J}{\partial W_i}(n), \ i = 1, 2, \dots, N, \qquad . (9.23)$$

where $W_i(n)$ is the ith parameter at iteration n, $\mu > 0$ is the *step-size* of the update, and the gradient at the nth iteration is given by

$$\frac{\partial J}{\partial W_i}(n) = 2 E\left\{e(n) \frac{\partial Y}{\partial W_i}(n)\right\}, \ i = 1, 2, \dots, N. \qquad (9.24)$$

When the underlying signal statistics are unavailable, *instantaneous estimates* for the gradient are used, since the expectation in (9.24) cannot be evaluated. Thus, removing the expectation operator in (9.24) and using the result in (9.23), the following weight update is found

$$W_i(n+1) = W_i(n) - \mu\, e(n) \frac{\partial Y}{\partial W_i}(n), \ i = 1, 2, \dots, N. \qquad (9.25)$$

All that remains is to find an expression for

$$\frac{\partial Y(n)}{\partial W_i} = \frac{\partial}{\partial W_i} \hat{\beta}_K(\mathbf{W}, \mathbf{X}), \ i = 1, 2, \dots, N. \qquad (9.26)$$

Recall that the output of the weighted myriad filter is $\hat{\beta}_K(\mathbf{W}, \mathbf{X}) = \arg\min\limits_{\beta} Q(\beta)$, where $Q(\beta)$ is given by

$$Q(\beta) = \sum_{i=1}^{N} \log\left[K^2 + |W_i| \cdot (\mathrm{sgn}(W_i)X_i - \beta)^2\right]. \qquad (9.27)$$

The derivative of $\hat{\beta}_K(\mathbf{W}, \mathbf{X})$ with respect to the weight W_i, holding all other quantities constant must be evaluated. Since $\hat{\beta}_K(\mathbf{W}, \mathbf{X}) \equiv \hat{\beta}$ is one of the local minima of $Q(\beta)$, it follows that

$$Q'(\hat{\beta}) = 0. \tag{9.28}$$

Differentiating (9.27) and substituting into (9.28), results in

$$G(\hat{\beta}, W_i) \triangleq Q'(\hat{\beta}) = 2 \sum_{j=1}^{N} \frac{|W_j| \cdot (\hat{\beta} - \text{sgn}(W_j)X_j)}{K^2 + |W_j| \cdot \left(\hat{\beta} - \text{sgn}(W_j)X_j\right)^2} = 0, \tag{9.29}$$

where the function $G(\cdot, \cdot)$ is introduced to emphasize the *implicit* dependency of the output $\hat{\beta}$ on the weight W_i, since we are interested in evaluating $\frac{\partial \hat{\beta}}{\partial W_i}$ while holding all other quantities constant. Implicit differentiation of (9.29) with respect to W_i leads to

$$\left(\frac{\partial G}{\partial \hat{\beta}}\right) \cdot \left(\frac{\partial \hat{\beta}}{\partial W_i}\right) + \left(\frac{\partial G}{\partial W_i}\right) = 0, \tag{9.30}$$

from which $\frac{\partial \hat{\beta}}{\partial W_i}$ can be found once $\frac{\partial G}{\partial \hat{\beta}}$ and $\frac{\partial G}{\partial W_i}$ are evaluated. Using (9.29), it is straightforward to show that

$$\frac{\partial G}{\partial \hat{\beta}} = 2 \sum_{j=1}^{N} |W_j| \frac{K^2 - |W_j| \cdot \left(\hat{\beta} - \text{sgn}(W_j)X_j\right)^2}{\left(K^2 + |W_j| \cdot \left(\hat{\beta} - \text{sgn}(W_j)X_j\right)^2\right)^2}. \tag{9.31}$$

Evaluation of $\frac{\partial G}{\partial W_i}$ from (9.29) is, however, a more difficult task. The difficulty arises from the term $\text{sgn}(W_i)$ that occurs in the expression for $G(\hat{\beta}, W_i)$. It would seem impossible to differentiate $G(\hat{\beta}, W_i)$ with respect to W_i, since this would involve the quantity $\frac{\partial}{\partial W_i} \text{sgn}(W_i)$, which clearly cannot be found. Fortunately, this problem can be circumvented by rewriting the expression for $G(\hat{\beta}, W_i)$ in (9.29), expanding it as follows:

$$G(\hat{\beta}, W_i) = 2 \sum_{j=1}^{N} \frac{|W_j| \hat{\beta} - W_j X_j}{K^2 + |W_j| \left(\hat{\beta}^2 + X_j^2\right) - 2W_j \hat{\beta} X_j}, \tag{9.32}$$

where the fact that $|W_j| \cdot \text{sgn}(W_j) = W_j$ was used. Equation (9.32) presents no mathematical difficulties in differentiating it with respect to W_i, since the term $\text{sgn}(W_i)$ is no longer present. Differentiating with respect to W_i, and performing some straightforward manipulations, leads to

$$\frac{\partial G}{\partial W_i} = 2\left[\left(\text{sgn}(W_j)\hat{\beta} - X_j\right)\left(K^2 + |W_j|\left(\hat{\beta}^2 + X_j^2\right) - 2W_j\hat{\beta}X_j\right) \\ - \left(|W_j|\hat{\beta} - W_jX_j\right)\left(\text{sgn}(W_j)\left(\hat{\beta}^2 + X_j^2\right) - 2\hat{\beta}X_j\right)\right] \\ \times \left(K^2 + |W_i| \cdot \left(\hat{\beta} - \text{sgn}(W_i)X_i\right)^2\right)^{-2}$$

(9.33)

multiplying and cancelling common terms reduces to:

$$\frac{\partial G}{\partial W_i} = \frac{2\,K^2\,\text{sgn}(W_i)\left(\hat{\beta} - \text{sgn}(W_i)X_i\right)}{\left(K^2 + |W_i| \cdot \left(\hat{\beta} - \text{sgn}(W_i)X_i\right)^2\right)^2}.$$

(9.34)

Substituting (9.31) and (9.34) into (9.30), the following expression for $\frac{\partial\hat{\beta}}{\partial W_i}$ is obtained:

$$\frac{\partial}{\partial W_i}\hat{\beta}_K(\mathbf{W},\mathbf{X}) = \frac{-\left[\dfrac{K^2\,\text{sgn}(W_i)\left(\hat{\beta} - \text{sgn}(W_i)X_i\right)}{\left(K^2 + |W_i| \cdot \left(\hat{\beta} - \text{sgn}(W_i)X_i\right)^2\right)^2}\right]}{\left[\displaystyle\sum_{j=1}^{N} |W_j|\,\dfrac{K^2 - |W_j| \cdot \left(\hat{\beta} - \text{sgn}(W_j)X_j\right)^2}{\left(K^2 + |W_j| \cdot \left(\hat{\beta} - \text{sgn}(W_j)X_j\right)^2\right)^2}\right]}.$$

(9.35)

Using (9.25), the following adaptive algorithm is obtained to update the weights $\{W_i\}_{i=1}^{N}$:

$$W_i(n+1) = W_i(n) - \mu\,e(n)\frac{\partial\hat{\beta}}{\partial W_i}(n),$$

(9.36)

with $\frac{\partial\hat{\beta}}{\partial W_i}(n)$ given by (9.35).

Considerable simplification of the algorithm can be achieved by just removing the denominator from the update term above; this does not change the direction of the gradient estimate or the values of the final weights. This leads to the following computationally attractive algorithm:

$$W_i(n+1) = W_i(n) - \mu e(n)\left[\frac{K^2\,\text{sgn}(W_i)\left(\hat{\beta} - \text{sgn}(W_i)X_i\right)}{\left(K^2 + |W_i| \cdot \left(\hat{\beta} - \text{sgn}(W_i)X_i\right)^2\right)^2}\right].$$

(9.37)

It is important to note that the optimal filtering action is independent of the choice of K; the filter only depends on the value of $\frac{\mathbf{w}}{K^2}$. In this context, one might ask how the algorithm scales as the value of K is changed and how the step size μ and the initial weight vector $\mathbf{w}(0)$ should be changed as K is varied. To answer this, let $\mathbf{g}_o \overset{\triangle}{=} \mathbf{w}_{o,1}$ denote the optimal weight vector for $K = 1$. Then, from (9.6), $\frac{\mathbf{w}_{o,K}}{K^2} = \frac{\mathbf{g}_o}{(1)^2}$ or $\mathbf{g}_o = \frac{\mathbf{w}_{o,K}}{K^2}$. Now consider two situations. In the first, the algorithm in (9.37) is used with $K = 1$, step size $\mu = \mu_1$, weights denoted as $g_i(n)$ and initial weight vector $\mathbf{g}(0)$. This is expected to converge to the weights $\mathbf{g}_o$. In the second, the algorithm uses a general value of K, step size $\mu = \mu_K$ and initial weight vector $\mathbf{w}_K(0)$. Rewrite (9.37) by dividing throughout by K^2 and writing the algorithm in terms of an update of $\frac{w_i}{K^2}$. This is expected to converge to $\frac{\mathbf{w}_{o,K}}{K^2}$ since (9.37) should converge to $\mathbf{w}_{o,K}$. Since $\mathbf{g}_o = \frac{\mathbf{w}_{o,K}}{K^2}$, the above two situations can be compared and the initial weight vector $\mathbf{w}_K(0)$ and the step size μ_K can be chosen such that the algorithms have the *same behavior* in both cases and converge, as a result, to the *same filter*. This means that $g_i(n) = \frac{w_i(n)}{K^2}$ at each iteration n. It can be shown that this results in

$$\mu_K = K^4 \mu_1 \text{ and } \mathbf{w}_K(0) = K^2 \mathbf{w}_1(0). \tag{9.38}$$

This also implies that if K is changed from K_1 to K_2, the new parameters should satisfy

$$\mu_{K_2} = \left(\frac{K_2}{K_1}\right)^4 \mu_{K_1} \text{ and } \mathbf{w}_{K_2}(0) = \left(\frac{K_2}{K_1}\right)^2 \mathbf{w}_{K_1}(0). \tag{9.39}$$

EXAMPLE 9.1 (ROBUST HIGH-PASS FILTER DESIGN)

Figure 9.4 illustrates some highpass filtering operations with various filter structures over a two-tone signal corrupted by impulsive noise. The signal has two sinusoidal components with normalized frequency 0.02 and 0.4 respectively. The sampling frequency is 1000Hz. Figure 9.4a shows the two-tone signal in stable noise with exponent parameter $\alpha = 1.4$, and dispersion $\gamma = 0.1$. The result of filtering through a high-pass linear FIR filter with 30 taps is depicted in Figure 9.4b. The FIR high-pass filter coefficients are designed with MATLAB's fir1 function with a normalized cutoff frequency of 0.2. It is clear to see that the impulsive noise has strong effect on the linear filter output, and the high frequency component of the original two-tone signal is severely distorted. The myriadization of the above linear filter gives a little better performance as shown in Figure 9.4c in the sense that the impulsiveness is greatly reduced, but the frequency characteristics are still not satisfactorily recovered. Figure 9.4e is the result of the optimal weighted myriad filtering with step size $\mu = 3$. As a comparison, the optimal weighted median filtering result is shown in Figure 9.4d, step size $\mu = 0.15$. Though these two nonlinear optimal filters perform significantly better than the linear filter, the optimal weighted myriad filter performs best since its output has no impulsiveness presence and no perceptual distortion (except magnitude fluctuations) as in the myriadization and the optimal weighted median cases. Moreover, signal details are better preserved in the optimal weighted myriad realization as well.

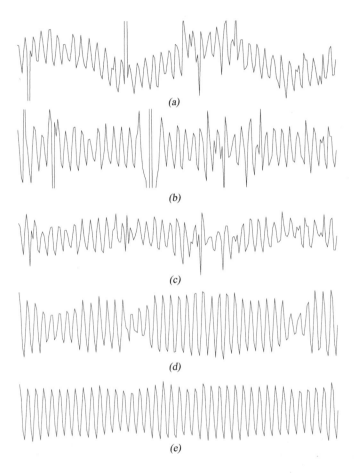

Figure 9.4 Robust high-pass weighted myriad filter: (*a*) two-tone signal corrupted by α-stable noise, $\alpha = 1.4$, $\gamma = 0.1$, (*b*) output of the 30-tap linear FIR filter, (*c*) myriadization, (*d*) optimal weighted median filter, (*e*) optimal weighted myriad filter.

Another interesting observation is depicted in Figure 9.5, where ensemble performances are compared. Though both median and myriad filters will converge faster when the step size is large and slower when the step size is small, they reach the same convergence rate with different step sizes. This is expected from their different filter structures. As shown in the plot, when the step size μ is chosen to be 0.15 for the median, the comparable performance can be found when the step size μ is in the vicinity of 3 for the myriad. The slight performance improvement can be seen from the plot in the stable region where the myriad has lower excess error floor than the median.

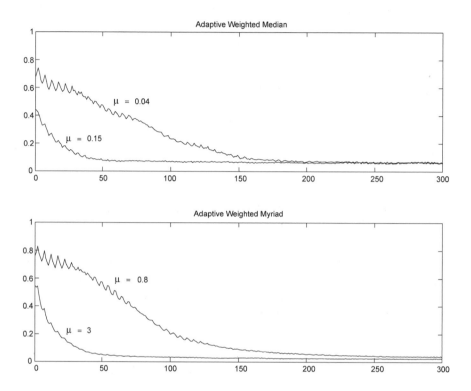

Figure 9.5 Comparison of convergence rate of the optimal weighted median and the optimal weighted myriad: (*a*) optimal weighted median at $\mu = 0.04, 0.15$, (*b*) optimal weighted myriad at $\mu = 0.8, 3$.

■

EXAMPLE 9.2 (ROBUST BLIND EQUALIZATION)

The *constant modulus algorithm (CMA)* may be the most analyzed and deployed blind equalization algorithm. The CMA is often regarded as the workhorse for blind channel equalization, just as the least mean square (LMS) is the benchmark used for supervised adaptive filtering [101]. Communication technologies such as digital cable TV, DSL, and the like are ideally suited for blind equalization implementations, mainly because in their structures, training is extremely costly if not impossible. In CMA applications, the linear FIR filter structure is assumed by default. In applications such as DSL, however, it has been shown that impulsive noise is prevalent [204], where inevitably, CMA blind equalization using FIR structure collapses. Here we describe a real-valued blind equalization algorithm that is the combination of the constant modulus criterion and the weighted myriad filter structure. Using myriad filters, one should expect a performance close to that of linear filters when the linear parameter K is set to values far bigger than that of the data samples. When the noise contains impulses, by reducing K to a suitable level, one can manage to remove their influence without losing the capability of keeping the communication eye open.

Consider a pulse amplitude modulation (PAM) communication system, where the signal and channel are both real. The constant modulus cost function is defined as follows:

$$J(\mathbf{W}, K) \triangleq \frac{1}{4} E\left\{\left(|Y(n)|^2 - R_2\right)^2\right\}, \tag{9.40}$$

where

$$R_2 = \frac{E|S(n)|^4}{E|S(n)|^2}$$

$S(n)$ is the signal constellation, $Y(n)$ the filter output. The gradient of the above cost function can be calculated out as

$$\nabla_{\mathbf{W}} J(\mathbf{W}, K) = E\left\{Y(n)\left(|Y(n)|^2 - R_2\right)\frac{\partial Y(n)}{\partial \mathbf{W}}\right\}. \tag{9.41}$$

A scaled version of a real valued weighted myriad filter is used where the sum of the magnitudes of the filter weights are used as the scaling factor.

$$Y(n) = \left(\sum_{i=1}^{N} |W_i|\right) \cdot \text{MYRIAD}\left(|W_i| \circ \text{sgn}(W_i) X_i \mid_{i=1}^{N}; K\right) \tag{9.42}$$

$$= \left(\sum_{i=1}^{N} |W_i|\right) \cdot \hat{\beta}. \tag{9.43}$$

This is particularly important for equalization applications since the signal energy needs to be considered in these cases. The derivative of the filter output with respect to a single weight is expressed as

$$\frac{\partial Y}{\partial W_i} = \text{sgn}(W_i)\hat{\beta} + \left(\sum_{i=1}^{N} |W_i|\right)\frac{\partial \hat{\beta}}{\partial W_i}, \tag{9.44}$$

and the derivative of $\hat{\beta}$ has already been shown in (9.35). Finally, the weight update can be carried out using the following equation

$$W_i(n+1) = W_i(n) + \mu Y(n)\left(|Y(n)|^2 - R_2\right)\left(\text{sgn}(W_i)\hat{\beta} + \left(\sum_{i=1}^{N} |W_i|\right)\frac{\partial \hat{\beta}}{\partial W_i}\right). \tag{9.45}$$

Unlike the regular WMy filters having only N independent parameters, as described in [113], all $N + 1$ parameters of the scaled WMy filters, that is N weights and one linear parameter, are independent. Thus, to best exploit the proposed structure, K needs to be updated adaptively as well. This time, we need to reconsider the objective function of the weighted myriad, since K is now a free parameter:

$$\hat{\beta} \triangleq \text{MYRIAD} \left(|W_i| \circ \text{sgn}(W_i) X_i \mid_{i=1}^{N} ; K \right)$$

$$= \arg\max_{\beta} \prod \frac{K}{K^2 + |W_i| \cdot (\text{sgn}(W_i) X_i - \beta)^2}$$

$$= \arg\min_{\beta} \sum \left[\log \left(K^2 + |W_i| \cdot (\text{sgn}(W_i) X_i - \beta)^2 \right) - \log K \right] \quad (9.46)$$

$$= \arg\min_{\beta} Q(\beta, K).$$

Denote

$$G(\hat{\beta}, K) \triangleq Q'(\hat{\beta}, K)$$

$$= 2 \sum_{i=1}^{N} \frac{|W_i| (\hat{\beta} - \text{sgn}(W_i) X_i)}{K^2 + |W_i| (\text{sgn}(W_i) X_i - \hat{\beta})^2} = 0. \quad (9.47)$$

Following a similar analysis as in the weight update, one can develop an update algorithm for K. However, two reasons make it more attractive to update the *squared linearity parameter* $\mathcal{K} \triangleq K^2$ instead of K itself. First, in myriad filters, K always occurs in its squared form. Second, the adaptive algorithm for K might have an ambiguity problem in determining the sign of K. Rewrite (9.47) as

$$G(\hat{\beta}, \mathcal{K}) = 2 \sum_{i=1}^{N} \frac{|W_i| (\hat{\beta} - \text{sgn}(W_i) X_i)}{\mathcal{K} + |W_i| (\text{sgn}(W_i) X_i - \hat{\beta})^2} = 0. \quad (9.48)$$

Implicitly differentiating both sides with respect to $\mathcal{K}$, leads to

$$\left(\frac{\partial G}{\partial \hat{\beta}} \right) \cdot \left(\frac{\partial \hat{\beta}}{\partial \mathcal{K}} \right) + \left(\frac{\partial G}{\partial \mathcal{K}} \right) = 0. \quad (9.49)$$

Thus,

$$\frac{\partial \hat{\beta}}{\partial \mathcal{K}} = - \frac{\frac{\partial G}{\partial \mathcal{K}}}{\frac{\partial G}{\partial \hat{\beta}}}. \quad (9.50)$$

Finally, the update for $\mathcal{K}$ can be expressed as

$$\mathcal{K}_i(n+1) = \mathcal{K}_i(n) - \frac{1}{2} \mu \frac{\partial J(\mathbf{W}, \mathcal{K})}{\partial \mathcal{K}} (n)$$

$$= \mathcal{K}(n) - \mu E \left\{ e(n) \frac{\partial \hat{\beta}_\mathcal{K}}{\partial \mathcal{K}} (n) \right\}. \quad (9.51)$$

Figure 9.6 depicts a blind equalization experiment where the constellation of the signal is BPSK, and the channel impulse response is simply [1 0.5]. Additive stable noise with $\alpha = 1.5$, $\gamma = 0.002$ corrupts the transmitted data. Figure 9.6a is the traditional linear CMA equalization while Figure 9.6b is the myriad CMA equalization. It is can be seen that, under the influence of impulsive noise, the linear equalizer diverge, but the myriad equalizer is more robust and still gives very good performance. Figure 9.7 shows the adaptation of parameter K in the corresponding realization.

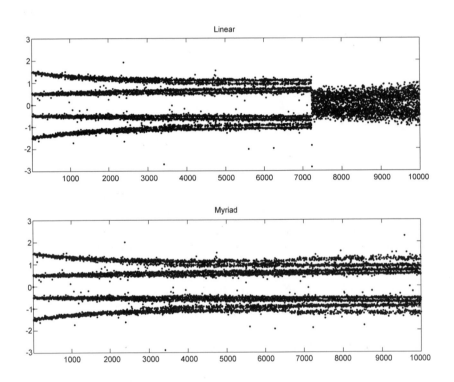

Figure 9.6 Blind equalization of a BPSK signal using (a) linear blind equalization, (b) myriad blind equalization.

■

Problems

9.1 Prove the scale relationships established in Equations (9.5) and (9.6).

9.2 Prove the linear property of the weighted myriad filters (Property 9.1).

9.3 Prove the mode property of the weighted myriad filters (Property 9.2).

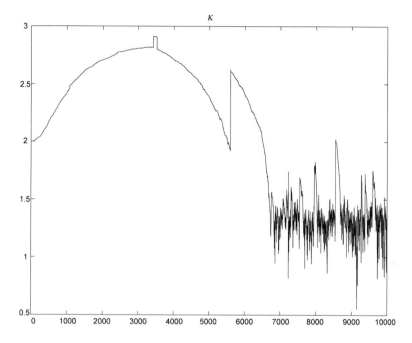

Figure 9.7 Adaptation of K

9.4 Prove if weighted myriad filters satisfy or not the shift and sign invariance properties described in problem 8.8.

References

1. G. Aloupis, M. Soss, and G. Toussaint. On the computation of the bivariate median and a fermat-torricelli problem. Technical Reports SOCS-01.2, School of Computer Science, McGill University, Montreal, Canada, February 2001.

2. S. Ambike and D. Hatzinakos. A new filter for highly impulsive α-stable noise. In *Proc. of the 1995 Int. Workshop on Nonlinear Signal and Image Proc.*, Halkidiki, Greece, June 1995.

3. D. Andrews, D. Bickel, P. Hampel, F. Huber, P. Rogers, and J. Tukey. *Robust Estimates of Location: Survey and Advances*. Princeton University Press, Princeton, NJ, 1972.

4. A. Antoniou. *Digital Filter. Analysis, Design and Applications*. McGraw-Hill, Inc, New Jersey, U.S.A., 1993.

5. G. R. Arce. Statistical threshold decomposition for recursive and nonrecursive median filters. *IEEE Trans. on Information Theory*, IT-32(2):243–253, March 1986.

6. G. R. Arce. A general weighted median filter structure admitting negative weights. *IEEE Trans. on Signal Proc.*, 46(12), December 1998.

7. G. R. Arce and R. E. Foster. Detail-preserving ranked-order based filters for image processing. *IEEE Trans. on Acoustics, Speech, and Signal Proc.*, 37(1):83–98, January 1989. Also reprinted in Digital Image Processing, by R. Chellappa, IEEE Press 1992.

8. G. R. Arce and N. C. Gallagher. Stochastic analysis of the recursive median filter process. *IEEE Trans. on Information Theory*, IT-34(4), July 1988.

9. G. R. Arce and N. C. Gallagher, Jr. State description of the root set of median filters. *IEEE Trans. on Acoustics, Speech, and Signal Proc.*, 30(6):894–902, December 1982.

10. G. R. Arce, T. A. Hall, and K. E. Barner. Permutation weighted order statistic filters. *IEEE Trans. on Image Proc.*, 4, August 1995.

11. G. R. Arce and R. Hasan. Elimination of interference terms in the wigner ville distribution using nonlinear filtering. *IEEE Trans. on Signal Proc.*, 48(8):2321–2331, August 2000.

12. G. R. Arce and Y. Li. Median power and median correlation theory. *IEEE Trans. on Signal Proc.*, 50(11):2768–2776, November 2002.

13. G. R. Arce and J. Paredes. Image enhancement with weighted medians. In S. Mitra and G. Sicuranza, editors, *Nonlinear Image Processing*, pages 27–67. Academic Press, 2000.

14. G. R. Arce and J. L. Paredes. Recursive weighted median filters admitting negative weights and their optimization. *IEEE Trans. on Signal Proc.*, 48(3):768–779, March 2000.

15. R. D. Armstrong, E. L. Frome, and D. S. Kung. A revised simplex algorithm for the absolute deviation curve fitting problem. *Communications in Statistics*, B8:175 – 190, 1979.

16. B. C. Arnold, N. Balakrishnan, and H. N. Nagaraja. *A First Course in Order Statistics*. John Wiley & Sons, New York, NY, 1992.

17. A. Asano, K. Itoh, and Y. Ichioka. Rondo: Rank-order based nonlinear differential operator. *Pattern Recognition*, 25(9):1043–1059, 1992.

18. A. Asano, K. Itoh, and Y. Ichioka. A generalization of the weighted median filter. In *Proc. 20th Joint Conference on Image Technology*, Tokio, Japan, 2003.

19. J. Astola, P. Haavisto, and Y. Neuvo. Vector median filters. *Proc. of the IEEE*, 78(4):678–689, April 1990.

20. J. Astola, P. Heinonen, and Y. Neuvo. Linear median hybrid filters. *IEEE Trans. on Circuits and Systems*, CAS-36:1430–1438, November 1989.

21. J. Astola and P. Kuosmanen. *Nonlinear Digital Filtering*. CRC Press, New York, NY, 1997.

22. J. Astola and P. Kuosmanen. Representation and optimization of stack filters. In E. Dougherty and J. Astola, editors, *Nonlinear Filters for Image Processing*, chapter 7. SPIE, 1999.

23. J. Astola and Y. Neuvo. Optimal median type filters for exponential noise distributions. *Signal Proc.*, 17:95–104, June 1989.

24. N. Balakrishnan and C. Rao. Order statistics: Applications. In *Handbook of Statistics*, volume 17. Elsevier Science, Amsterdam, 1998.

25. A. Bangham. Properties of a series of nested median filters, namely the data sieve. *IEEE Trans. on Signal Proc.*, 41:31–42, 1993.

26. K. E. Barner and G. R. Arce. Coloring schemes for the design of permutation filters: A group theoretic method for filter class reduction. *IEEE Trans. on Circuits and Systems*, July 1996.

27. V. Barnett. The ordering of multivariate data. *J. R. Stat. Soc. A*, 139:331–354, 1976.

28. V. Barnett and T. Lewis. *Outliers in Statistical Data*. John Wiley & Sons, New York, second edition, 1984.

29. I. Barrodale and F. D. K. Roberts. An improved algorithm for discrete l_1 linear approximation. *SIAM J. Numer. Anal.*, 10(5):839 – 848, October 1973.

30. J. B. Bednar and T. L. Watt. Alpha-trimmed means and their relationship to median filters. *IEEE Trans. on Acoustics, Speech, and Signal Proc.*, ASSP-32(2), February 1984.

31. J. Beran, R. Sherman, M. S. Taqqu, and W. Willinger. Long-range dependance in variable bit-rate video traffic. *IEEE Trans. on Communications*, 43:1566–1579, 1995.

32. P. Bickel and K. Doksum. *Mathematical Statistics*. Holden-Day, San Francisco, CA, 1977.

33. G. Blom. Nearly best linear estimates of location and scale parameters. In A. E. Sarhan and B. G. Greenberg, editors, *Contributions to Order Statistics*, pages 34–46. John Wiley & Sons, New York, NY, 1962.

34. P. Bloomfield and W. Steiger. Least absolute deviations curve-fitting. *SIAM J. Sci. Statist. Comput.*, 1:290 – 301, 1980.

35. P. Bloomfield and W. L. Steiger. *Least Absolute Deviations: Theory, Applications, and Algorithms*. Birkhauser, Boston, MA, 1983.

36. C. G. Boncelet, Jr. Algorithms to compute order statistic distributions. *SIAM J. Sci. Stat. Comput.*, 8(5), September 1987.

37. A. C. Bovik. Streaking in median filtered images. *IEEE Trans. on Acoustics, Speech, and Signal Proc.*, ASSP-35, April 1987.

38. A. C. Bovik, T. S. Huang, and Jr. D.C. Munson. A generalization of median filtering using linear combinations of order statistics. *IEEE Trans. on Acoustics, Speech, and Signal Proc.*, 31(12), December 1983.

39. K. L. Boyer, M. J. Mirza, and G. Ganguly. The robust sequential estimator: A general approach and its application to surface organization in range data. *IEEE Trans. on Pattern Analysis and Machine Intelligence*, 16(10):987–1001, October 1994.

40. B. Wade Brorsen and S. R. Yang. Maximum likelihood estimates of symmetric stable distribution parameters. *Commun. Statist.-Simula.*, 19(4):1459–1464, 1990.

41. D. R. K. Brownrigg. The weighted median filter. *Communications of the ACM*, 27(8):807–818, August 1984.

42. O. Cappé, E Moulines, J. C. Pesquet, A. Petropulu, and X. Yang. Long-range dependence and heavy-tail modeling for teletraffic data. *IEEE Signal Proc. Magazine*, May 2002.

43. J. M. Chambers, C. Mallows, and B. W. Stuck. A method for simulating stable random variables. *J. Amer. Stat. Association*, 71(354):340–344, 1976.

44. A. Charnes, W. W. Cooper, and R. O. Ferguson. Optimal estimation of executive compensation by linear programming. *Management Science*, 1(2):138 – 151, January 1955.

45. H.-I. Choi and W. J. Williams. Improved time-frequency representation of multicomponent signals using exponential kernels. *IEEE Trans. on Acoustics, Speech, and Signal Proc.*, 37:862–871, 1989.

46. K.S. Choi, A.W. Morales, and S.J. Ko. Design of linear combination of weighted medians. *IEEE Trans. on Signal Proc.*, 49(9):1940–1952, September 2001.

47. T. C. Chuah, B. S Sharif, and O. R. Hinton. Nonlinear decorrelator for multiuser detection in nongaussian impulsive environments. *IEE Electronic Letters*, 36(10):920–922, 2000.

48. T. C. Chuah, B. S Sharif, and O. R. Hinton. Nonlinear space-time decorrelator for multiuser detection in nongaussian channels. *IEE Electronic Letters*, 36(24):2041–2043, 2000.

49. T. C. Chuah, B. S Sharif, and O. R. Hinton. Robust adaptive spread-spectrum receiver with neural-net preprocessing in nongaussian noise. *IEEE TRans. on Neural Networks*, 12:546–558, 2001.

50. T. C. Chuah, B. S Sharif, and O. R. Hinton. Robust decorrelating decision-feedback multiuser detection in nongaussian channels. *Signal Processing*, 81(2001):1997–2004, 2001.

51. G. A. Churchill. Fundamentals of experimental design for cdna microarrays. *Nature Genetics*, 32:490 – 495, 2002.

52. P. M. Clarkson and G. A. Williamson. Minimum variance signal estimation with adaptive order statistic filters. In *Proc. of the 1992 IEEE International Conference on Acoustics, Speech and Signal Processing*, volume 4, pages 253–256, 1992.

53. L. Cohen. *Time-Frequency Analysis*. Prentice-Hall, Upper Saddle River, NJ, 1st edition, 1995.

54. E. J. Coyle. Rank order operators and the mean absolute error criterion. *IEEE Trans. on Acoustics, Speech, and Signal Proc.*, 36(1):63–76, January 1988.

55. E. J. Coyle and J.-H. Lin. Stack filters and the mean absolute error criterion. *IEEE Trans. on Acoustics, Speech, and Signal Proc.*, 36(8), August 1988.

56. H. Cramer. *Mathematical Methods of Statistics*. Princeton University Press, New Jersey, 1946.

57. E. L. Crow and M. M. Siddiqui. Robust estimation of location. *J. Amer. Statist. Association*, 62:353–389, 1967.

58. H. A. David. *Order Statistics*. Wiley Interscience, New York, 1981.

59. L. Davis and A. Rosenfeld. Noise cleaning by iterated local averaging. *IEEE Trans. on Systems, Man, and Cybernetics*, 8:705–710, September 1978.

60. Y. Dodge, editor. *Statistical Data Analysis: Based on the L_1-Norm and Related Methods*. Elsevier Science, The Netherlands, 1987.

61. Y. Dodge, editor. *L_1-Statistical Analysis and Related Methods*. North-Holland, The Netherlands, 1992.

62. Y. Dodge, editor. *L_1-Statistical Procedures and Related Topics*. Institute of Mathematical Statistics, 1997.

63. Y. Dodge and W. Falconer, editors. *Statistical Data Analysis Based on the L_1-Norm and Related Methods*. Barika Photography & Productions, 2002.

64. D. L. Donoho, I. M. Johnstone, G. Kerkyacharian, and D. Picard. Wavelet shrinkage: Asymptopia? *J. R. Statist. Soc. B.*, 57(2):301–337, 1995.

65. R. Durrett. *Probability: Theory and Examples*. Duxbury Press, New York, second edition, 1996.

66. F. Y. Edgeworth. A new method of reducing observations relating to several quantities. *Phil. Mag. (Fifth Series)*, 24:222 – 223, 1887.

67. R. J. Adler et al., editor. *A Practical Guide to Heavy Tails: Statistical Techniques and Applications*. Birkhauser, Boston, MA, 2002.

68. E. F. Fama and R. Roll. J. amer. stat. association. *J. Amer. Stat. Association*, 66(6), June 1971.

69. W. Feller. *An Introduction to Probability Theory and its Applications*, volume I of *Wiley Series in Probability and Mathematical Statistics*. John Wiley & Sons, New York, 1970.

70. R. A. Fisher. On the mathematical foundation of theoretical statistics. *Phylosophical Trans. of the Royal Society of London*, 222(Series A), 1922.

71. J. Fitch. Software and VLSI algorithms for generalized ranked order filtering. *IEEE Trans. on Circuits and Systems*, 34(5), May 1987.

72. J. P. Fitch, E. J. Coyle, and N. C. Gallagher. Median filtering by threshold decomposition. *IEEE Trans. on Acoustics, Speech, and Signal Proc.*, 32(12), December 1984.

73. A. Flaig, G. R. Arce, and K. E. Barner. Affine order statistic filters: Medianization of fir filters. *IEEE Trans. on Signal Proc.*, 46(8):2101–2112, August 1998.

74. M. Fréchet. Sur la loi des erreurs d'observation. *Matematicheskii Sbornik*, 32:1–8, 1924.

75. M. S. Taqqu G. Samorodnitsky. *Stable Non-Gaussian Random Processes: Stochastic Models with Infinite Variance*. Chapman & Hall, New York, 1994.

76. M. Gabbouj and E. J. Coyle. Minimum mean absolute error stack filtering with structural constraints and goals. *IEEE Trans. on Acoustics, Speech, and Signal Proc.*, 38(6):955–968, June 1990.

77. J. Galambos. *The Asymptotic Theory of Extreme Order Statistics*. John Wiley & Sons, New York, 1978.

78. N. C. Gallagher, Jr. and G. L. Wise. A theoretical analysis of the properties of median filters. *IEEE Trans. on Acoustics, Speech, and Signal Proc.*, 29(12), December 1981.

79. P. Ghandi and S. A. Kassam. Design and performance of combination filters. *IEEE Trans. on Signal Proc.*, 39(7), July 1991.

80. C. Gini and L. Galvani. Di talune estensioni dei concetti di media ai caratteri qualitative. *Metron*, 8:3–209, 1929.

81. J. G. Gonzalez. *Robust Techniques for Wireless Communications in Non-Gaussian Environments*. Ph. D. dissertation, Dept. of Electrical and Computer Engineering, University of Delaware, Newark, DE, 1997.

82. J. G. Gonzalez and G. R. Arce. Weighted myriad filters: A robust filtering framework derived from alpha-stable distributions. In *Proceedings of the 1996*

IEEE International Conference on Acoustics, Speech, and Signal Processing, Atlanta, Georgia, May 1996.

83. J. G. Gonzalez and G. R. Arce. Optimality of the myriad filter in practical impulsive-noise environments. *IEEE Trans. on Signal Proc.*, 49:438–441, February 2001.

84. J. G. Gonzalez, D. W. Griffith, Jr., A. B. Cooper, III, and G. R. Arce. Adaptive reception in impulsive noise. In *Proc. IEEE Int. Symp. on Information Theory*, Ulm, Germany, June 1997.

85. J. G. Gonzalez, D. Griffith, and G. R. Arce. Zero order statistics. In *Proc. of the 1997 Workshop on Higher-Order Statistics*, Banff, Canada, July 1997.

86. J. G. Gonzalez and D. L. Lau. The closest-to-mean filter: an edge preserving smoother for gaussian environments. In *Proc. IEEE ICASSP 1997*, Munich, Germany, April 1997.

87. C. Goodall. M-estimators of Location: An Outline of the Theory. In D. C. Hoaglin et al., editor, *Understanding Robust and Exploratory Data Analysis*, chapter 11, pages 339–403. John Wiley & Sons, New York, 1983.

88. C. Großand T. Strempel. On generalizations of conics and on a generalization of the fermat-torricelli problem. *American Mathematical Monthly*, 105(8):732–743, 1998.

89. J. B. S. Haldane. Note on the median on a multivariate distribution. *Biometrika*, 35(3/4):414–415, December 1948.

90. H. M. Hall. A new model for "impulsive" phenomena: Application to atmosheric noise communication channels. Technical Reports 3412-8 and 7050-7, Stanford Electronics Lab., Stanford University, August 1966.

91. F. Hampel, E. Ronchetti, P. Rousseeuw, and W. Stahel. *Robust Statistics: The Approach Based on Influence Functions*. John Wiley & Sons, New York, NY, 1986.

92. R. Hardie and G. R. Arce. Ranking in r^p and its use in multivariate image estimation. *IEEE Trans. on Video Technology*, 1(2), June 1991.

93. R. C. Hardie and K. E. Barner. Rank conditioned rank selection filters for signal restoration. *IEEE Trans. on Image Proc.*, 3(2), March 1994.

94. R. C. Hardie and C. G. Boncelet, Jr. LUM filters: A class rank order based filters for smoothing and sharpening. *IEEE Trans. on Signal Proc.*, 41(5), May 1993.

95. T. E. Harris. Regression using minimum absolute deviations. *American Statistician*, 4(1):14 – 15, Feb. 1950.

96. H. L. Harter. *Order Statistics and Their Use in Testing and Estimation*, volume 1 & 2. U.S. Government Printing Office, Washington D.C., 1970.

97. R. W. Hawley and N. C. Gallagher. On edgeworth's method for minimum error linear regression. *IEEE Trans. on Signal Proc.*, 43(8), August 1994.

98. J. F. Hayford. What is the center of an area or the center of a population? *J. Amer. Stat. Association*, 8(58):47–58, 1902.

99. S. Haykin. *Adaptive Filter Theory*. Prentice Hall, New Jersey, 1995.

100. S. Haykin. *Neural Networks: A Comprehensive Foundation*. Prentice Hall, New Jersey, 1998.

101. S. Haykin. *Unsupervised Adaptive Filtering, Volume I and Volume II*. Wiley-Interscience, 2000.

102. J. R. M. Hosking. L-estimation. In C. R. Rao and N. Balakrishnan, editors, *Order Statistics and Their Applications*, Handbook of Statistics. Elsevier, 1998.

103. S. Hoyos, J. Bacca, and G. R. Arce. Spectral design of weighted median filters: A general iterative approach. *IEEE Trans. on Signal Proc.*, 2004. Accepted for Publication.

104. S. Hoyos, Y. Li, J. Bacca, and G. R. Arce. Weighted median filters admitting complex-valued weights and their optimization. *IEEE Trans. on Signal Proc.*, 2004. Accepted for Publication.

105. P. J. Huber. *Robust Statistics*. John Wiley & Sons, New York, 1981.

106. J. Ilow and D. Hatzinakos. Analytic alpha-stable noise modeling in a Poisson field of interferers or scatterers. *IEEE Trans. on Signal Proc.*, 46(6):1601–1611, June 1998.

107. A. K. Jain. *Fundamentals of Digital Image Processing*. Prentice Hall, New Jersey, 1989.

108. D. L. Jones and R. G. Baraniuk. A signal dependent time-frequency representation: Optimal kernel design. *IEEE Trans. on Signal Proc.*, 41:1589–1601, April 1993.

109. D. L. Jones and R. G. Baraniuk. An adaptive optimal-kernel time-frequency representation. *IEEE Trans. on Signal Proc.*, 43:2361–2371, October 1995.

110. T. Kailath. *Linear Systems*. Prentice-Hall, New Jersey, 1980.

111. S. Kalluri and G. R. Arce. Adaptive weighted myriad filter optimization for robust signal processing. In *Proc. of the 1996 Conference on Information Science and Systems*, Princeton, NJ, 1996.

112. S. Kalluri and G. R. Arce. Fast algorithms for weighted myriad computation by fixed point search. *IEEE Trans. on Signal Proc.*, 48:159–171, January 2000.

113. S. Kalluri and G. R. Arce. Robust frequency-selective filtering using weighted myriad filters admitting real–valued weights. *IEEE Trans. on Signal Proc.*, 49(11):2721–2733, November 2001.

114. Y.-T. Kim and G. R. Arce. Permutation filter lattices: a general order statistic filtering framework. *IEEE Trans. on Signal Proc.*, 42(9), September 1994.

115. S.-J. Ko and Y. H. Lee. Center weighted median filters and their applications to image enhancement. *IEEE Trans. on Circuits and Systems*, 38(9), September 1991.

116. V. Koivunen. Nonlinear filtering of multivariate images under robust error criterion. *IEEE Trans. on Image Proc.*, 5, June 1996.

117. C. Kotropoulos and I. Pitas. Constrained adaptive LMS L-filters. *Signal Proc.*, 26(3):335–358, 1992.

118. C. Kotropoulos and I. Pitas. Adaptive LMS L-filters for noise suppression in images. *IEEE Trans. on Image Proc.*, 5(12):1596–1609, December 1996.

119. P. Kuosmanen and J. Astola. Optimal stack filters under rank selection and structural constraints. *Signal Proc.*, 41, February 1995.

120. P. Kuosmanen, P. Koivisto, P. Huttunen, and J. Astola. Shape preservation criteria and optimal soft morphological filtering. *J. Math. Imag. Vis.*, 5, December 1995.

121. E. E. Kuruoğlu. Density parameter estimation of skewed α-stable distributions. *IEEE Trans. on Signal Proc.*, 49(10), 2001.

122. P. S. Laplace. Mémoire sur la probabilité des causes par les évènemens. *Mémoires de Mathématique et de Physique*, 6, 1774.

123. D. L. Lau, G. R. Arce, and N. C. Gallagher. Robust image wavelet shrinkage for denoising. In *Proceedings International Conference on Image Processing 1996*, volume 1, pages 371–374, 1996.

124. J. S. Lee. Digital image smoothing and the sigma filter. *Computer Vision, Graphics, and Image Processing*, 24:255–269, November 1983.

125. Y. H. Lee and S. A. Kassam. Generalized median filtering and related nonlinear filtering techniques. *IEEE Trans. on Acoustics, Speech, and Signal Proc.*, 33(6), June 1985.

126. E. L. Lehmann. *Theory of Point Estimation*. J Wiley & Sons, New York, NY, 1983.

127. W. E. Leland, M. S. Taqqu, W. Willinger, and D. V. Wilson. On the self-similar nature of ethernet traffic (extended version). *IEEE/ACM Trans. on Networking*, 2(1):1–15, February 1994.

128. P. Lévy. *Calcul des probabilités*. Gauthier-Villars, Paris, 1925.

129. Y. Li and G. R. Arce. Center weighted myriad smoother in image denoising. Technical report, Dept. of Elect. Comp. Eng., University of Delaware, February 2003.

130. Y. Li, G. R. Arce, and J. Bacca. Generalized vector medians for correlated channels. In *Proceedings of the 2004 EUSIPCO European Signal Processing Conference*, Vienna, Austria, September 2004.

131. J. Lin and Y.-T. Kim. Fast algorithms for training stack filters. *IEEE Trans. on Signal Proc.*, 42(4), April 1994.

132. J.-H. Lin, T. M. Sellke, and E. J. Coyle. Adaptive stack filtering under the mean absolute error criterion. *IEEE Trans. on Acoustics, Speech, and Signal Proc.*, 38(6), June 1990.

133. R. Y. Liu. On a notion of data depth based on random simplices. *Annals of Statistics*, 18(1):405–414, March 1990.

134. E. H. Lloyd. Least–squares estimation of location and scale parameter using order statistics. *Biometrika*, 39, 1952.

135. X. Ma and C. L. Nikias. Parameter estimation and blind channel identification in impulsive signal environments. *IEEE Trans. on Signal Proc.*, 43(12):2884–2897, December 1995.

136. X. Ma and C. L. Nikias. Joint estimation of time delay and frequency delay in impulsive noise using Fractional Lower-Order Statistics. *IEEE Trans. on Signal Proc.*, 44(11):2669–2687, November 1996.

137. C. L. Mallows. Some theory of nonlinear smoothers. *Annals of Statistics*, 8(4), 1980.

138. B. Mandelbrot. Long-run linearity, locally Gaussian processes, H-spectra, and infinite variances. *Interant. Econ. Rev.*, 10:82–111, 1969.

139. I. Mann, S. McLaughlin, W. Henkel, R. Kirkby, and T. Kessler. Impulse generation with appropriate amplitude, length, interarrival, and spectral characteristics. *IEEE Journal on Selected Areas in Communications*, 20(5):901–912, June 2002.

140. P. A. Maragos and R. W. Schafer. Morphological filters - Part I: Their set theoretic analysis and relations to linear shift invariant filters, and Morphological filters - Part II: Their relations to median, order–statistic, and stack filters. *IEEE Trans. on Acoustics, Speech, and Signal Proc.*, 35(8):1153–1169, August 1987.

141. V. J. Mathews and G. Sicuranza. *Polynomial Signal Processing*. John Wiley & Sons, New York, NY, 1994.

142. J. H. McCulloch. Simple consistent estimators of stable distribution parameters. *Communications in Statistics–Simulation and Computation*, 1986.

143. D. Middleton. Statistical-physical models of electromagnetic interference. *IEEE Trans. on Electromagnetic Compatibility*, EMC-19:106–127, 1977.

144. S. K. Mitra. *Digital Signal Processing: A Computer-Based Approach, 2e with DSP Laboratory using MATLAB*. McGraw-Hill Higher Education, 2nd edition, 2001.

145. S. Muroga. *Threshold Logic and Its Applications*. John Wiley & Sons, New York, NY, 1971.

146. H. Myer and G. Ascheid. *Synchronization in Digital Communications*. John Wiley & Sons, New York, NY, 1990.

147. A. Nieminen, P. Heinonen, and Y. Neuvo. A new class of detail-preserving filters for image processing. *IEEE Trans. on Pattern Analysis and Machine Intelligence*, 9(1), January 1987.

148. C. L. Nikias and A. T. Petropulu. *Higher-Order Spectra Analyisis: A Nonlinear Signal Processing Framework*. Prentice-Hall, New Jersey, 1993.

149. C. L. Nikias and M. Shao. *Signal Processing with Alpha-Stable Distributions and Applications*. Wiley-Interscience, New York, 1995.

150. T. A. Nodes and N. C. Gallagher, Jr. Median filters: some modifications and their properties. *IEEE Trans. on Acoustics, Speech, and Signal Proc.*, 30(2):739–746, October 1982.

151. John P. Nolan. *Stable Distributions*. Birkhauser, first edition, June 2002.

152. H. Oja. Descriptive statistics for multivariate distributions. *Statistics & Probability Letters*, 1(6):327–332, October 1983.

153. R. Oten and R.J.P. de Figueiredo. An efficient method for l-filter design. *IEEE Trans. on Signal Proc.*, 51:193–203, January 2003.

154. J. M. Paez-Borrallo and S. Zazo-Bello. On the joint statistical characterization of the input and output of an order statistic filter. *IEEE Trans. on Signal Proc.*, 42(2):456–459, February 1994.

155. F. Palmieri and C. G. Boncelet, Jr. Ll-filters–a new class of order statistic filters. *IEEE Trans. on Acoustics, Speech, and Signal Proc.*, 37(5), May 1989.

156. F. Palmieri and C. G. Boncelet, Jr. Frequency analysis and synthesis of a class of nonlinear filters. *IEEE Trans. on Acoustics, Speech, and Signal Proc.*, 38(8), August 1990.

157. A. Papoulis. *Probability, Random Variables, and Stochastic Processes.* McGraw-Hill, New York, NY, 1991.

158. J. L. Paredes and G. R. Arce. Stack filters, stack smoothers, and mirrored threshold decomposition. *IEEE Trans. on Signal Proc.*, 47:2757–2767, October 1999.

159. V. Paxson and S. Floyd. Wide-area traffic: The failure of poisson modeling. *IEEE-ACM Trans. on Networking*, 3:226–244, 1995.

160. I. Pitas and A. N. Venetsanopoulos. *Non-linear Filters.* Kluwer, 1989.

161. I. Pitas and A. N. Venetsanopoulos. Adaptive filters based on order statistics. *IEEE Trans. on Signal Proc.*, 39(2), February 1991.

162. I. Pitas and A. N. Venetsanopoulos. Order statistics in digital image processing. *Proceedings of the IEEE*, 80(12):1893–1921, December 1991.

163. I. Pitas and S. Vougioukas. LMS order statistic filters adaptation by back propagation. *Signal Processing*, 25(3), December 1991.

164. C. A. Pomalaza-Raez and C. D. McGillem. An adaptive, nonlinear edge-preserving filter. *IEEE Trans. on Acoustics, Speech, and Signal Proc.*, 32(3):571–576, June 1984.

165. M. K. Prasad and Y. H. Lee. Stack filters and selection probabilities. *IEEE TRansactions on Signal Processing*, 42:2628–2642, October 1994.

166. J. G. Proakis and D.G. Manolakis. *Digital Signal Processing. Principles, Algorithms and Applications.* Prentice Hall, third edition, 1996.

167. S. Rappaport and L. Kurz. An optimal nonlinear detector for digital data transmission through non-gaussian channels. *IEEE Trans. on Communications*, 14(3), March 1966.

168. E. A. Robinson. *Multichannel Time Series Analysis.* Goose Pond Press, Houston, Texas, 1983.

169. S. Ross. *A First Course in Probability.* Prentice Hall, Upper Saddle River, New Jersey, 1998.

170. P. J. Rousseeuw and M. Hubert. Depth in an arrangement of hyperplanes. *Discrete and Computational Geometry*, 22(2):167 – 176, September 1999.

171. E. Sarhan and B. G. Greenberg. *Contributions to Order Statistics.* John Wiley & Sons, New York, NY, 1962.

172. M. I. Shamos. Geometry and statistics: Problems at the interface. In J. F. Traub, editor, *Algorithms and Complexity: New Directions and Recent Results.* Academic Press, Inc., New York, 1976.

173. Y. Shen and K. Barner. Fast optimization of weighted vector median filters. *IEEE Trans. on Signal Proc.*, 2004. Submitted.

174. C. L. Sheng. *Threshold Logic*. The Ryerson Press, Ontario, Canada, 1969.

175. I. Shmulevich and G. R. Arce. Spectral design of weighted median filters admitting negative weights. *Signal Processing Letters, IEEE*, 8(12):313–316, December 2001.

176. J. Shynk. Adaptive IIR filtering. *IEEE ASSP Magazine*, 6(2):4–21, April 1989.

177. H. S. Sichel. The method of frequency moments and its application to Type VII populations. *Biometrika*, pages 404–425, 1949.

178. C. Small. A survey of multidimensional medians. *Int. Stat. Rev.*, 58(3), 1990.

179. E. Souza. Performance of a spread spectrum packet radio network in a Poisson field of interferences. *IEEE Trans. on Information Theory*, IT-38(6):1743–1754, November 1992.

180. K. S. Srikantan. Recurrence relations between the pdfs of order statistics and some applications. *Annals of Mathematical Statistics*, 33:169–177, 1962.

181. F. Steiner. Most frequent value and cohesion of probability distributions. *Acta Geod., geophys. et Mont. Acad. Sci. Hung.*, 8(3-4):381–395, 1973.

182. B. W. Stuck. Minimum error dispersion linear filtering of scalar symmetric stable processes. *IEEE Trans. on Automatic Control*, 23(3):507–509, 1978.

183. M. T. Subbotin. On the law of frequency of errors. *Matematicheskii Sbornik*, 31:296–301, 1923.

184. T. Sun, M. Gabbouj, and Y. Neuvo. Center weighted median filters: Some properties and their applications in image processing. *Signal Proc.*, 35:213–229, 1994.

185. P. E. Trahanias and A. N. Venetsanopoulos. Directional processing of color images: Theory and experimental results. *IEEE Trans. on Image Proc.*, 5, June 1996.

186. I. Tăbuces, D. Petrescu, and M Gabbouj. A training framework for stack and boolean filtering fast optimal design procedurtes and robustness case study. *IEEE Trans. on Image Proc., Special Issue on Nonlinear Image Processing*, 5:809–826, June 1996.

187. J. Tukey. A problem of berkson and minimum variance orderly estimators. *Annals of Mathematical Statistics*, 29:588–592, 1958.

188. J. W. Tukey. A survey of sampling from contaminated distributions. In I. Olkin, S. G. Ghurye, W. Hoeffding, W. G. Madow, and H. B. Mann, editors, *Contributions to Probability and Statistics, Essays in Honor of Harold Hotelling*, pages 448–485. Stanford University Press, Stanford, CA, 1960.

189. J. W. Tukey. Nonlinear (nonsuperimposable) methods for smoothing data. In *Conf. Rec.*, Eascon, 1974.

190. J. W. Tukey. Mathematics and the picturing of data. In *Int Congress Math.*, pages 523–531, Vancouver, 1975.

191. V. V. Uchaikin and V. M. Zolotarev. *Chance and Stability, Stable Distributions and Their Applications (Modern Probability and Statistics)*. VSP, Utrecht, Netherlands, 1999.

192. P. D. Welch. The use of fast fourier transforms for the estimation of power spectrum: A method based on time averaging over short modified periodograms. *IEEE Trans. on Audio and Electroacustics*, 15:70–73, June 1967.

193. P. D. Wendt. Nonrecursive and recursive stack filters and their filtering behavior. *IEEE Trans. on Acoustics, Speech, and Signal Proc.*, 33, August 1986.

194. P. D. Wendt, E. J. Coyle, and N. C. Gallagher, Jr. Some convergence properties of median filters. *IEEE Trans. on Circuits and Systems*, 33(3), March 1986.

195. P. D. Wendt, E. J. Coyle, and N. C. Gallagher, Jr. Stack filters. *IEEE Trans. on Acoustics, Speech, and Signal Proc.*, 34(8), August 1986.

196. G. O. Wesolowsky. A new descent algorithm for the least absolute value regression. *Commun. Statist.*, B10(5):479 – 491, 1981.

197. B. Widrow and S. D. Stearns. *Adaptive Signal Processing*. Prentice Hall, Englewood Cliffs, NJ, 1985.

198. R. Yang, L. Yin, M. Gabbouj, J. Astola, and Y. Neuvo. Optimal weighted median filters under structural constraints. *IEEE Trans. on Signal Proc.*, 43(6), June 1995.

199. L. Yin, J. Astola, and Y. Neuvo. Adaptive stack filtering with applications to image processing. *IEEE Trans. on Acoustics, Speech, and Signal Proc.*, 41(1), January 1993.

200. L. Yin and Y. Neuvo. Fast adaptation and performance characteristics of fir–wos hybrid filters. *IEEE Trans. on Signal Proc.*, 42(7), July 1994.

201. L. Yin, R. Yang, M. Gabbouj, and Y. Neuvo. Weighted median filters: a tutorial. *Trans. on Circuits and Systems II*, 41, May 1996.

202. O. Yli-Harja, J. Astola, and Y. Neuvo. Analysis of the properties of median and weighted median filters using threshold logic and stack filter representation. *IEEE Trans. on Acoustics, Speech, and Signal Proc.*, 39(2), February 1991.

203. P. T. Yu and E. Coyle. On the existence and design of the best stack filter based on associative memory. *IEEE Trans. on Circuits and Systems*, 39(3), March 1992.

204. W. Yu, D. Toumpakaris, J. Coffi, D. Gardan, and F. Gauthier. Performance of asymmetric digitial subscriber lines (adsl) in an impulse noise environment. *IEEE Trans. on Communications*, 51(10):1653–1657, October 2003.

205. B. Zeng. Optimal median-type filtering under structural constraints. *IEEE Trans. on Image Proc.*, pages 921–931, July 1995.

206. Y. Zhang. Primal-dual interior point approach for computing l_1-solutions and l_∞-solutions of overdetermined linear systems. *J. Optim. Theory Appl.*, 77(2):323 – 341, May 1993.

207. V. M. Zolotarev. *One-dimensional Stable Distributions*. American Mathematical Society, Providence, Rhode Island, 1986. Translation from the Russian, 1983.

Appendix A
Software Guide

Chapter 2

astable	Generate an α-stable distributed data set.
astable_flom	Estimate the density parameters of α-stable distributions.
astable_logmom	Estimate the density parameters of α-stable distributions.
contgaussrnd	Generates a random data set with a contaminated Gaussian distribution function.
Laplacian	Generates a data set with a Laplacian distribution.
parestssd	Parameter estimates for Symmetric Stable Distributions

Chapter 4

t_mean	Calculates the trimmed mean of a data vector.
w_mean	Calculates the windsorized mean of a data vector.

Chapter 5

g3hvd5filt	Smooths an image keeping horizontal, vertical and diagonal lines.
pwmedfilt2	Two dimensional permutation weighted median filtering of a sequence.
wmedfilt	One-dimensional weighted median filtering.
wmedfilt2	Two-dimensional weighted median filtering.
wmedian	Compute the weighted median of an observation vector.

Chapter 6

LMS_MPCCWM	Designs an optimal marginal phase coupled complex weighted median filter.
marginalWMMI	Multivariate weighted median filtering for stationary correlated channels.
mcwmedfilt	One-dimensional marginal phase coupled complex weighted median filtering.
mcwmedian	Calculates the marginal phase coupled complex weighted median of an observation vector.
optmarginalWMMI	Optimization algorithm for the marginal WMM I.
optVmedfilt	Calculates the optimum weights for the weighted vector median filter.
optWMMII	Optimum weights for the weighted multivariate median filter II.
rwmedfilt	One-dimensional recursive weighted median filter.
rwmedfilt2	Two-dimensional recursive weighted median filter.
rwmedopt	Design one-dimensional recursive weighted median filters using the fast "recursive decoupling" adaptive optimization algorithm.

Chapter 6

rwmedopt2	Design two-dimensional recursive weighted median filters using the fast "recursive decoupling" adaptive optimization algorithm.
SSP2MPCCWM	Finds the closest marginal phase coupled complex weighted median filter to a given complex valued linear filter.
SSP2WM	Finds the closest weighted median filter to a given linear FIR filter.
Vwmedfilt	Weighted vector median filtering.
Vwmedian	Weighted vector median of an observation window.
WM2SSP_real	Finds the linear filter closest in the MSE sense to a given real valued weighted median filter.
WM2SSP_realfd	Same as the previous including also the first derivative of the cost function in (6.36)
wmedopt	Design one-dimensional weighted median filters using a fast adaptive optimization algorithm.
wmedopt2	Design two-dimensional weighted median filters using a fast adaptive optimization algorithm.
WMMII	Weighted multivariate median II filtering.
wmsharpener	Sharpens a gray-scale image using permutation high pass median filters.

Chapter 7

LCWM	One-dimensional LCWM filtering.
LCWM_design	Designs a LCWM filter based on a given linear filter.
LCWMsymmetric	Designs a LCWM filter based on a symmetric linear filter.
Lfilter	Performs L-filtering of the sequence X
Llfilter	Performs Ll-filtering of the sequence X.
median_affine	Performs median affine filtering of the vector X.
opt_median_affine	Designs an optimal median affine filter (linear weights and dispersion parameter) with an adaptive algorithm.
opt_weights_L	Finds the optimum weights for a location estimator using L-filters.
opt_weights_Ll	Performs Ll-filtering of the sequence X.

Chapter 8	
cwmyrfilt2	Smoothing of images with the use of weighted myriad filters.
fwmyriad	Compute the fast weighted myriad of an observation vector.
wmyrfilt	One-dimensional weighted myriad filter.
wmyriad	Compute the fast weighted myriad of an observation vector.

Additional	
impoint	Pointillize image
medcor	Sample median correlation.
medcov	Sample median covariation.
medpow	Sample median power.

astable

Purpose   Generate an α-stable distributed data set.

Syntax   x = astable(m,n,alpha,delta,gamma,beta).

Description *astable* returns a $m \times n$ dimensional data set satisfying an α-stable distribution described by the parameters: α, δ, γ and β. The index of stability $\alpha \in (0, 2]$ measures the thickness of the tails and, therefore, the impulsiveness of the distribution. The symmetry parameter $\delta \in [-1, 1]$ sets the skewness of the distribution. The scale parameter $\gamma > 0$, also called the dispersion, is similar to the variance. The location parameter $\beta \in (-\infty, \infty)$ sets the shift of the probability distribution function.

Example   Generate a α-stable distributed data sequence

$$x = astable(10, 2, 1.5, 0.5, 1, 20);$$

The result is

$x =$	22.8381	21.5816
	20.0192	18.7585
	18.4857	15.4310
	20.7009	19.8038
	20.0218	16.9714
	19.4775	20.4440
	20.3570	20.2192
	20.3778	21.1832
	20.6206	19.8470
	20.8048	19.4305

Algorithm   Refer to [43] for details of the algorithm.

See Also   astable_logmom, astable_flom, Section 2.2.4.

astable_flom

Purpose Estimate the density parameters of α-stable distributions.

Syntax [alpha,delta,gamma,beta] = astable_flom(x)

Description astable_flom returns the density parameters α, δ, γ and β of the skewed α-stable distribution of a sample set x calculated by fractional lower order moment methods. The index of stability $\alpha \in (0, 2]$ measures the thickness of the tails and, therefore, the impulsiveness of the distribution. The symmetry parameter $\delta \in [-1, 1]$ sets the skewness of the distribution. The scale parameter $\gamma > 0$, also called the dispersion, is similar to the variance. The location parameter $\beta \in (-\infty, \infty)$ sets the shift of the probability distribution function.

Example Generate an α-stable distributed data sequence

$$x = astable(10000, 1, 1.5, 0.5, 1, 20);$$

Then estimate the density parameters α, γ, δ and β of the distribution

$$[alpha, delta, gamma, beta] = astable\_flom(x);$$

The result is
$$
\begin{aligned}
alpha &= \quad 1.5270 \\
delta &= \quad 0.4493 \\
gamma &= \quad 1.0609 \\
beta &= \quad 20.4428
\end{aligned}
$$

Algorithm The algorithm is based on the properties of the fractional lower-order moments of α-stable distribution. Refer to [121] for details.

See Also astable, astable_logmom, Section 2.3.3.

astable_logmom

Purpose   Estimate the density parameters of α-stable distributions.

Syntax   [alpha,delta,gamma,beta] = astable_logmom(x)

Description   astable_logmom returns the density parameters α, δ, γ and β of
                   the skewed α-stable distribution of a data set x. The parameters
                   are calculated by logarithmic moment methods. The index of
                   stability $\alpha \in (0, 2]$ measures the thickness of the tails and,
                   therefore, the impulsiveness of the distribution. The symmetry
                   parameter $\delta \in [-1, 1]$ sets the skewness of the distribution. The
                   scale parameter $\gamma > 0$, also called the dispersion, is similar to
                   the variance. The location parameter $\beta \in (-\infty, \infty)$ sets the
                   shift of the probability distribution function.

Example   Generate an α-stable distributed data sequence

$$x = astable(10000, 1, 1.5, 0.5, 1, 20);$$

Then estimate the density parameters α, γ, δ and β of the distribution

$$[alpha, delta, gamma, beta] = astable\_logmom(x);$$

The result is
$$
\begin{aligned}
alpha &= & 1.4772 \\
delta &= & 0.4564 \\
gamma &= & 0.9790 \\
beta &= & 20.4522
\end{aligned}
$$

Algorithm   The algorithm is based on the properties of the logarithmic
                  moments of α-stable distribution. Refer to [121] for details.

See Also   astable, astable_flom, Section 2.3.3.

contgaussrnd

Purpose

Generates a random data set with a contaminated Gaussian distribution function.

Syntax

Y = contgaussrnd(M, N, mean, sigma1, sigma2, p)

Description

contgaussrnd returns a $M \times N$ data set that satisfies a contaminated Gaussian distribution with parameters mean, sigma1, sigma2 and p. Mean is the mean of the distribution, sigma1 is the standard deviation of the original Gaussian distribution, sigma2 is the standard deviation of the contaminating Gaussian distribution and p is the proportion of contaminated samples.

Example

Generate a contaminated Gaussian data sequence

$$x = contgaussrnd(10, 2, 20, 1, 10, 0.1);$$

The result is

$x =$	20.8013	20.6137
	19.9186	17.8818
	20.4586	17.1818
	19.9364	20.4960
	21.5763	18.7519
	19.9550	20.4427
	19.0683	18.2451
	17.8455	11.2404
	20.7491	20.4547
	19.7286	18.6845

Algorithm

Initially a $M \times N$ vector of uniformly distributes random variables is generated. This vector is read component by component. Every time a component of the vector is greater than p, a Gaussian random variable with standard deviation sigma1 is generated. Otherwise, a Gaussian random variable with standard deviation sigma2 is generated.

See Also

Laplacian and Section 2.1.

cwmyrfilt2

Purpose   Smoothing of images with the use of weighted myriad filters.

Syntax   y = cwmyrfilt2(X, Wc, N, L, U)

Description   y = cwmyrfilt2 performs a double center weighted myriad operation on the image X to remove impulsive noise. The input parameters are:

X Input data vector.

Wc Center weight (A value of 10,000 is recommended for gray scale images).

N Window size.

L Lower order statistic used in the calculation of K.

U Upper order statistic used in the calculation of K.

Example   cwmyrfilt2 is used to clean an image corrupted with 5% salt-and-pepper noise. The clean, noisy and filtered images are shown below.

Algorithm The output of the algorithm is obtained as $\mathbf{Y} = 1 - \text{CWMy}(1 - \text{CWMy}(\mathbf{X}))$, where $CWMy$ represents the center weighted myriad with $K = \frac{X_U + X_L}{2}$.

See Also wmyrfilt, Section 8.6.1 and [129].

fwmyriad

Purpose   Compute the fast weighted myriad of an observation vector.

Syntax   Owmy = fwmyriad(x, w, k)

Description fwmyriad(x,w,k) finds the approximate value of the weighted myriad of a vector x using a fast algorithm. w is an N-component vector that contains the real-valued weights and k is the linearity parameter that takes on values greater than zero. The default values for the weights and the linearity parameter are the all-one vector and one respectively.

Example   x = [3 2 4 5 8];
w = [0.15 -0.2 0.3 -0.25 0.1];
Owmf = wmyriad(x,w,1);
Owmf = 3.1953

Algorithm   fwmyriad(x,w,k) implements a fast algorithm introduced by S. Kalluri et al. to compute an approximate value for the weighted myriad of an observation vector.

See Also   wmyriad, wmyrfilt, wmyropt, Section 8.5 and [112].

g3hvd5filt

Purpose   Smooths an image keeping horizontal, vertical, and diagonal lines.

Syntax   y = g3hvd5filt(X)

Description   y = g3hvd5filt(X) slides a 5×5 running window over the black and white image X and performs a smoothing operation with a mask that keeps the horizontal, vertical, and diagonal lines of length 3.

Example   The following shows a clean image, the same image contaminated with salt & pepper noise and the output of the stack smoother applied to the noisy signal.

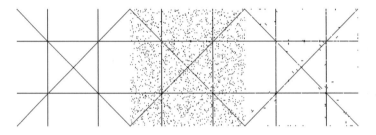

References   See Section 5.3.1 and [205].

impoint

Purpose   Pointillize image.

Syntax   impoint(n,'infilename','outfilename')

Description   impoint first scales the image 'infilename' by a factor of n. It then applies brushstrokes to the image to give it a "painted" feeling. Smaller 7×7 pixel strokes are used near high frequency edges of the image. Larger 14×10 pixel strokes are used on the rest of the image. The red, green, and blue components of each brushstroke are independently leveled to 0, 64, 128, 192, or 255, so there are at most $5^3 = 125$ colors used in the finished image, which is written out to 'outfilename'.

Example   The algorithm was applied to the following picture:

Figure A.1 Luminance of the original and pointillized images.

Laplacian

Purpose   Generates a data set with a Laplacian distribution.

Syntax   Y = Laplacian(M, N, lambda)

Description   Laplacian returns a $M \times N$ of random variables with a Laplacian distribution with parameter lambda.

Example $x = laplacernd(10, 2, 1)$; The result is
$x =$
0.5424	-0.7765
0.0290	-0.6927
0.3497	-1.7152
0.1406	1.1511
-0.4915	2.3975
0.2874	-0.4506
1.9228	0.3298
-2.7274	0.0151
0.2739	-1.4150
-0.4697	-0.2175

Algorithm   The program generates a $M \times N$ vector (U) of uniformly distributed random variables between 0 and 1 and applies:

$$Y = F^{-1}(U)$$

where F^{-1} represents the inverse of the distribution function of a Laplacian random variable with parameter lambda.

See Also   contgaussrnd, Section 2.1 and [169].

LCWM

Purpose	One-dimensional LCWM filtering.
Syntax	y = LCWM(alpha,Bp,x)
Description	LCWM(alpha,Bp,x) filters the sequence x using the LCWM filter described by the vector of linear coefficients $alpha$ and the matrix of median weights Bp.
Algorithm	The program performs weighted median operations over the input sequence x using the weighted medians described by the rows of the matrix Bp. The outputs of the medians are scaled using the coefficients in the vector $alpha$ and added together to obtain the final output of the filter.
See Also	combmat, rowsearch, LCWMsymmetric, LCWM_design, Section 7.6.1 and [46].

LCWM_design

Purpose

Designs a LCWM filter based on a given linear filter.

Syntax

[alpha,Bp] = LCWM_design(h,M)

Description

LCWM_design(h,M) designs a LCWM filter based on medians of size M and the linear filter h. The matrix Bp contains the weights for the medians and the vector $alpha$ the coefficients for the linear combination.

Example

Design a LCWM filter from a high-pass, 7-tap linear filter with a cutoff frequency of 0.5 using medians of length 3.

$$h = [0.0087, 0, -0.2518, 05138, -0.2518, 0, 0.0087]^T$$
$$\vec{\alpha} = [-0.7844, 1.5124, -0.7555, 0, 0.0262, 0.0275, 0.0014]$$

$$\mathbf{B}_p = \begin{bmatrix} 1 & 1 & 1 & 0 & 0 & 0 & 0 \\ 1 & 1 & 0 & 1 & 0 & 0 & 0 \\ 1 & 1 & 0 & 0 & 1 & 0 & 0 \\ 1 & 1 & 0 & 0 & 0 & 1 & 0 \\ 1 & 1 & 0 & 0 & 0 & 0 & 1 \\ 1 & 0 & 1 & 1 & 0 & 0 & 0 \\ 0 & 1 & 1 & 1 & 0 & 0 & 0 \end{bmatrix}$$

The spectra of the original linear filter and the LCWM filter are shown in Figure A.2.

Algorithm

The program uses the routines combmat and rowsearch to generate a set of linearly independent combinations of m elements from a set of n elements (n is the length of h). This combinations are grouped in a matrix that is used to calculate the coefficients of the linear combination.

See Also

combmat, rowsearch, LCWM, LCWMsymmetric, Section 7.6.2 and [46].

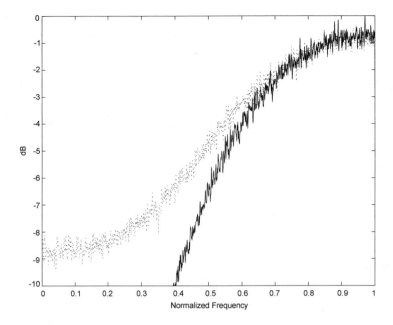

Figure A.2 Approximated spectra of a LCWM filter (dotted) and the linear filter used as a reference to design it (solid).

LCWMsymmetric

Purpose

Designs a LCWM filter based on a symmetric linear filter.

Syntax

[alpha,Bp] = LCWMsymmetric(h,M)

Description

LCWMsymmetric(h,M) designs a LCWM filter based on medians of size $2M + 1$ and $2(M + 1)$ and the symmetric linear filter h. The matrix Bp contains the weights for the medians and the vector $alpha$ the coefficients for the linear combination.

Example

Design a LCWM filter from a high-pass, 7-tap linear filter with a cutoff frequency of 0.5 using medians of length 3 and 4 ($M = 1$).

$$h = [0.0087,\ 0,\ -0.2518,\ 05138,\ -0.2518,\ 0,\ 0.0087]^T$$
$$\vec{\alpha} = [0.3798,\ 1.1353,\ 0.0262,\ -1.5138]$$
$$\mathbf{B}_p = \begin{bmatrix} 0 & 0 & 1 & 1 & 1 & 0 & 0 \\ 0 & 1 & 0 & 1 & 0 & 1 & 0 \\ 1 & 0 & 0 & 1 & 0 & 0 & 1 \\ 0 & 1 & 1 & 0 & 1 & 1 & 0 \end{bmatrix}$$

The spectra of the original linear filter and the LCWM filter are shown in Figure A.3.

Algorithm

The program uses the routines **combmat** and **rowsearch** to generate a set of linearly independent combinations of m elements from a set of n elements ($n = \frac{l}{2}$ or $n = \frac{l+1}{2}$, depending if l is even or odd, l is the length of h). This combinations are grouped in a matrix that is used to calculate the coefficients of the linear combination.

See Also

combmat, rowsearch, LCWM, LCWM_design, Section 7.6.3 and [46].

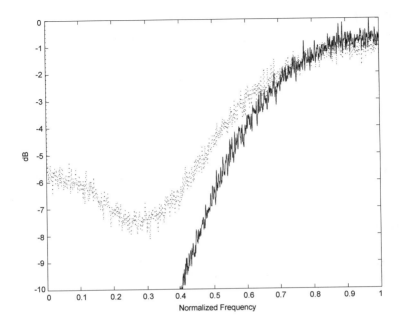

Figure A.3 Approximated spectra of a symmetric LCWM filter (dotted) and the linear filter used as a reference to design it (solid).

Lfilter

Purpose   Performs L-filtering of the sequence X

Syntax   y=Lfilter(X,W)

Description   y=Lfilter(X,W) filters the data in vector X with the Lfilter described by the weight vector W.

Algorithm

$$Y(n) = \sum_{i=1}^{N} W_i X_{(i)}.$$

See Also   opt_weights_L, Llfilter, Section 7.2 and [16, 58, 102].

Llfilter

Purpose   Performs Ll-filtering of the sequence X.

Syntax   y=Llfilter(X,W)

Description   y=Llfilter(X,W) filters the data in vector X using the Ll-filter described by the weight vector W.

Algorithm   Llfilter passes a window of length N (the length of W is N^2) over the data X. With the data in the window it generates the Lℓ vector $X_{L\ell}$ and multiplies it by W to obtain the output.

See Also   Lfilter, opt_weights_Ll, Section 7.3 and [79, 155].

LMS_MPCCWM

Purpose
Designs an optimal marginal phase coupled complex weighted median filter.

Syntax
[Wmpc,e,y]=LMS_MPCCWM(mu,taps,reference,received)

Description
LMS_MPCCWM(mu,taps,reference,received) Uses an adaptive LMS algorithm to design an optimum marginal phase coupled complex weighted median. The filter is optimum in the sense that the MSE between its output and a reference signal is minimized. The input parameters are:

- mu: Step size of the adaptive algorithm.
- taps: Number of weights of the filter to be designed.
- reference: Reference signal (desired output) for the LMS algorithm.
- received: Received signal. Input to the complex weighted median being designed.

Example
Design a high-pass complex weighted median filter with cut-off frequencies equal to ± 0.3 and test its robustness against impulsive noise.

```
Fs=2000;                                          % Sampling freq.
t=0:1/Fs:1;
d=exp(2*pi*i*400*t);                              % Desired signal
x=exp(2*pi*i*20*t)+d;                             % Training signal
n=41;                                             % Number of
                                                  % filter coefficients
mu=0.001;                                         % Step size
h=cremez(n-1,[-1 -.3 -.2 .2 .3 1],'highpass');    % Linear coeff.
[Wmpc,e,y]=LMS_MPCCWM(mu,n,d,x);                  % Training stage
yo=mcwmedfilt(x,Wmpc);                            % WM filter output
yl=filter(h,1,x);                                 % FIR output
Tn=x+astable(1, length(t), 1.5, 0, 0.5,0)
        +j*astable(1,length(t),1.5, 0, 0.5, 0);   % Noisy signal
yon=mcwmedfilt(Tn,Wmpc);                          % WM filtering of Tn
yln=filter(h,1,Tn);                               % FIR filtering of Tn
```

The next figure shows the results:

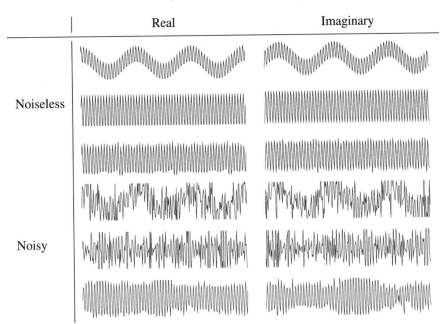

Figure A.4 Filtering of a sum of exponentials in a noiseless and noisy environment. The first plot represents the original signal, the second the signal filtered with a FIR filter and the third the signal filtered with an optimal marginal phase coupled complex weighted median filter.

Algorithm The weights are updated according to the equation:

$$
\begin{aligned}
W_i(n+1) &= W_i(n) + \mu\{-\nabla J(n)\} \\
&\approx W_i(n) + \mu e^{j\theta_i} \Big\{ e_R(n)\mathrm{sgn}(P_{R_i}(n) - \hat{\beta}_R(n)) \\
&\quad + e_I(n)\mathrm{sgn}(P_{I_i}(n) - \hat{\beta}_I(n)) \\
&\quad + 2j e_R(n)(P_{I_i}(n)\delta(P_{R_i}(n) - \hat{\beta}_R(n)) \\
&\quad + 2j e_I(n)(P_{R_i}(n)\delta(P_{I_i}(n) - \hat{\beta}_I(n)) \Big\}
\end{aligned}
$$

$$(A.1)$$

where $P_{R_i}(n)$ and $P_{I_i}(n)$ represent the ith phase-coupled sample at time n[1], $\hat{\beta}_R(n)$ and $\hat{\beta}_I(n)$ are the output of the filter at

---

[1] The subindexes R and I represent real and imaginary parts respectively

time n, $e_R(n)$ and $e_I(n)$ represent the difference between the output of the filter and the desired output at time n.

See Also   mcwmedfilt, mcwmedian, Section 6.6.4 and [104].

marginalWMMI

Purpose   Performs marginal multivariate weighted median filtering of stationary cross-channel correlated signals.

Syntax   [y] = marginalWMMI(X,W,V,wsize)

Description   [y] = marginalWMMI(X,W,V,wsize) filters the data in vector X with the marginal multivariate weighted median filter I described by the N-dimensional vector V and the $M \times M$ matrix W. N is the number of samples in the observation window and M is the dimension of the components of X. The window size is specified by the parameter wsize. This parameter has the form $[m\ n]$, where $m \times n = N$.

Algorithm   marginalWMMI pases a window over the data in X and computes, for each component of each sample in the window, the weighted median described by the columns of W. After that it calculates the output of the marginal weighted median described by V applied to the outputs of the previous operation.

See Also   optmarginalWMMI, WMMII and Section 6.7.3.

mcwmedfilt

Purpose

One-dimensional marginal phase-coupled complex weighted median filtering.

Syntax

y = mcwmedfilt(X,W)

Description

y=mcwmedfilt(X,W) filters the data in vector X with the complex weighted median filter described by the vector W using the marginal phase coupling algorithm. The window size is specified by the dimensions of the vector W.

Algorithm

mcwmedfilt pases a window over the data in X and computes, at each instant n, the marginal phase-coupled complex weighted median of the samples $X\left(n - \frac{N-1}{2}\right), \ldots, X(n), \ldots, X\left(n + \frac{N-1}{2}\right)$, where $N = \text{length}(W)$. The resultant value is the filter output at instant n.

See Also

mcwmedian, Section 6.6.2 and [104].

mcwmedian

Purpose   Calculates the marginal phase-coupled complex weighted median of an observation vector.

Syntax   y = mcwmedian(X, W)

Description mcwmedian computes the phase-coupled complex-weighted median of an observation vector X. W is a complex-valued vector of the same length of X containing the filter weights.

Example

$$X = [-0.7128 + 0.5547i, \ 0.8858 - 0.3101i, \ 0.5131$$
$$-0.2349i, \ 0.9311 + 0.4257i, \ -0.8017 - 0.6254i]$$
$$W = [-0.0430 + 0.0592i, \ 0.3776 - 0.1924i, \ 0.6461,$$
$$0.3776 + 0.1924i, \ -0.0430 - 0.0592i]$$
$$P = [-0.0297 - 0.9027i, \ 0.6484 - 0.6785i, \ 0.5131$$
$$-0.2349i, \ 0.6363 + 0.8020i, \ -0.0347 + 1.0162i]$$
$$y = mcwmedian(X, W) = 0.6363 - 0.2349i \qquad \text{(A.2)}$$

Algorithm   Given a set of N complex-valued weights $\langle W_1, W_2, \ldots, W_N \rangle$ and an observation vector $\mathbf{X} = [X_1, X_2, \ldots, X_N]^T$, the output of the marginal phase coupled complex-weighted median filter is defined as:

$$\hat{\beta}_{marginal} = \text{MEDIAN}\left(|W_{R_i}| \lozenge \text{sgn}(W_{R_i}) X_{R_i} |_{i=1}^N\right)$$
$$+ j\text{MEDIAN}\left(|W_{I_1}| \lozenge \text{sgn}(W_{I_1}) X_{I_1} |_{i=1}^N\right) \text{(A.3)}$$

See Also   mcwmedfilt, wmedian, Section 6.6.2 and [104].

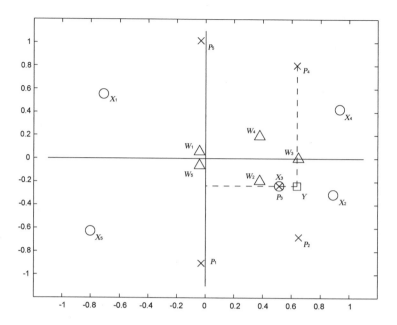

Figure A.5 Marginal phase-coupled complex-weighted median. The samples are represented by ○, the weights by △, the phase-coupled samples by ×, and the output by □. The dashed lines show how the output is composed by the real part of P_4 and the imaginary part of P_3.

medcor

Purpose   Sample median correlation.

Syntax   mcor = medcor(x,y)

Description   medcor(x,y) returns the sample median correlation between the sequences x and y. Both sequences should have the same length.

Algorithm   Given the observation sets $\{x_i(\cdot)\}$ and $\{y_i(\cdot)\}$ taken from two joint random variables x and y, the sample median correlations is defined as

$$\tilde{R}_{xy} = \left(\frac{1}{n} \sum_{i=1}^{n} |y_i| \right) \cdot \mathrm{MEDIAN}\left(|y_i| \diamond \mathrm{sgn}(y_i)x_i \,|_{i=1}^{n} \right)$$

where n is the length of the observation sequences and $|y_i| \diamond \mathrm{sgn}(y_i)x_i \,|_{i=1}^{n} = |y_1| \diamond \mathrm{sgn}(y_1)x_1, |y_2| \diamond \mathrm{sgn}(y_2)x_2, \ldots, |y_n| \diamond \mathrm{sgn}(y_n)x_n$.

See Also   medpow, medcov and [12].

medcov

Purpose

Sample median covariation.

Syntax

mcov = medcov(x,y)

Description

medcov(x,y) computes the sample median covariation between the sequences x and y. Both sequences should have the same length.

Algorithm

Given the observation sets $\{x_i(\cdot)\}$ and $\{y_i(\cdot)\}$ taken from two joint random variables x and y, the sample median correlations is given by

$$\hat{R}_{xy} = \text{MEDIAN}(|y_i|\ |_{i=1}^{n}) \cdot \text{MEDIAN}\left(|y_i| \diamond \text{sgn}(y_i)x_i\ |_{i=1}^{n}\right)$$

where n is the length of the observation sequences and $|y_i| \diamond \text{sgn}(y_i)x_i\ |_{i=1}^{n} = |y_1| \diamond \text{sgn}(y_1)x_1, |y_2| \diamond \text{sgn}(y_2)x_2, \ldots, |y_n| \diamond \text{sgn}(y_n)x_n$.

See Also

medpow, medcor and [12].

median_affine

Purpose   Performs median affine filtering of the vector X.

Syntax   y=median_affine(x,W,C,gamma)

Description   median_affine filters the data in vector X using the median weights C to calculate the reference, the parameter γ to calculate the affinity and the weights W to perform the linear combination.

Algorithm   median_affine passes a window over the data vector X that selects $N = length(W)$ samples. Then it takes a window of size $M = length(C)$ around the center sample and calculates the reference as the weighted median of this samples with the weights in C. Once the reference point is obtained, the exponential affinity function is calculated for each sample in the original window. With the weights and the affinity function, the normalization constant K is calculated. with all this elements the final value of the median affine can be calculated as:

$$Y_\gamma(n) = K(n) \sum_{i=1}^{N} \exp\left(-\frac{(X_i - \mu(n))^2}{\gamma}\right) W_i X_i$$

References   See Section 7.5.1 and [73].

medpow

Purpose

Sample median power.

Syntax

mpow = medpow(x)
mcor = medpow(x,type)

Description

medpow returns the sample median power of the observation sequence x. medpow(x,type) uses a second input parameter to specify the type of median power to be returned.

- If type = cor, medpow(x,type) returns the sample median correlation power of x.
- If type = cov, medpow(x,type) returns the sample median covariation power of x.

By default, medpow(x) returns the sample median correlation power.

Algorithm

For an observation sequence $\{x(\cdot)\}$ of length n, the sample median correlation power and the sample median covariation power are respectively defined as

$$\tilde{R}_{xx} = \left(\frac{1}{n} \sum_{i=1}^{n} |x_i| \right) \cdot \text{MEDIAN} \left(|x_i| \diamond |x_i| \,|_{i=1}^{n} \right)$$

$$\hat{R}_{xx} = \text{MEDIAN}(|x_i| \,|_{i=1}^{n}) \cdot \text{MEDIAN} \left(|y_i| \diamond |x_i| \,|_{i=1}^{n} \right)$$

where $|x_i| \diamond |x_i| \,|_{i=1}^{n} = |x_1| \diamond |x_1|, \ |x_2| \diamond |x_2|, \ \ldots, \ |x_n| \diamond |x_n|$.

See Also

medcov, medcor and [12].

Opt_marginalWMMI

Purpose   Finds the optimal weights for the marginal multivariate weighted median filter.

Syntax   [v,w,y,e] = Opt_marginalWMMI(x,wsize,d,muv, muw)

Description   Opt_marginalWMMI finds the optimum weights for the marginal weighted multivariate median filter I using an adaptive LMA algorithm. The parameters of the function are:

x is the input signal to be filtered.

wsize is the window size of the filter in a vector form ($[m\ n]$).

d reference signal for the adaptive algorithm.

muv step size used in the calculation of the outer weights v.

muw step size used in the calculation of the inner weights w.

v N-dimensional vector of optimum outer weights ($N = m \times n$).

w $M \times M$ matrix of optimum inner weights (M) is the dimension of the space.

y Output signal.

e Absolute error of the output signal.

Example   Filter a color image contaminated with 10% salt and pepper noise using the marginal WMM filter I and a 5×5 window.

See Also   marginalWMMI, Opt_WMMII and 6.7.4.

Figure A.6 Luminance of the noiseless image, noisy image, and output of the marginal WMM filter I.

opt_median_affine

Purpose   Designs an optimal median affine filter (linear weights and dispersion parameter) with an adaptive algorithm.

Syntax   [Wopt, gopt, y] = opt_median_affine(x, Wini, C, gini, stepW, stepg, yd)

Description opt_median_affine calculates the optimal linear weights and dispersion parameter for a median affine filter using a given median operator and the exponential distance as affinity function. The input parameters are:

x is the input signal to be filtered.

Wini is the initial value for the linear weights of the filter.

C contains the median weights of the filter.

gini is the initial value for the dispersion parameter.

stepW is the step size for the adaptive algorithm that calculates the weights. It should be much smaller than the step size used in the adaptive algorithm of the dispersion parameter.

stepg is the step size for the adaptive algorithm of the dispersion parameter.

yd is the reference signal.

Example   Design a median affine filter to recover a sinusoidal contaminated with α-stable noise.

```
t=1:1:1000;
yd=sin(2*pi*t/50);
x=yd+astable(1, 1000, 1, 0, 0.2, 0);
( Wopt, gopt, y) = opt_median_affine(x, ones(1,9),
ones(1,5), 1, 0.001, 0.1, yd);
y=median_affine(x,Wopt,ones(1,5),gopt);
```

The original signal, the noisy signal and the output of the filter are shown below

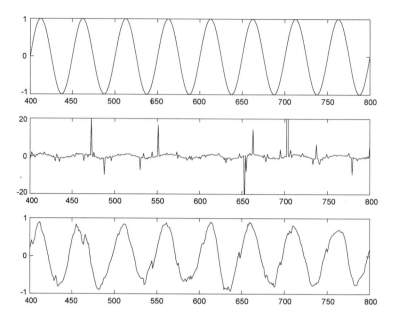

Figure A.7   Reference signal, input and output of a median affine filter.

Algorithm

The weights and the dispersion parameter are updated according to the following equations:

$$W_i(n+1) = W_i(n) + \nu_W e(n) \left(g_i \sum_{k=1}^{N} W_k g_k (\text{sgn}(W_k) X_i - \tanh(W_i) X_k) \right)$$

$$\gamma(n+1) = \gamma(n) + \\ \nu_\gamma e(n) \left(\sum_{i=1}^{N} (X_i - Y_\gamma(n)) W_i \frac{(X_i - \mu(n))^2}{\gamma^2} e^{-\frac{(X_i - \mu(n))^2}{\gamma}} \right)$$

where g_i stands for the abbreviated affinity function of the ith sample.

See Also

median_affine, Section 7.5.1, and [73].

opt_Vwmedfilt

Purpose

Finds the optimum weights for the vector-weighted median filter.

Syntax

[w, wCurve] = Opt_Vwmedfilt(I_n, I, w0, mu)

Description

Opt_Vwmedfilt calculates the optimum weights for a vector-weighted median filter using a fast adaptive greedy algorithm. The parameters of the algorithm are:

I_n Input signal.

I Reference signal.

w0 Initialization of the weights (usually an all ones $m \times n$ matrix).

mu Step size for the adaptive algorithm.

w Matrix of optimum weights with the same dimensions as w0.

wCurve Values of the weights during all the adaptive algorithm

Example

Filter a color image contaminated with 10% salt-and-pepper noise using the vector-weighted median filter and a 5×5 window.

Algorithm

The weights are optimized according to:

$$W_i(n+1) = W_i(n) + \mu \Delta W_i, \quad i = 1, 2, \cdots, N, \quad \text{(A.4)}$$

where

$$\Delta W_i = \frac{d(\vec{X}_{e_{min}}) - d(\vec{X}_{j_0})}{||\vec{X}_{j_0} - \vec{X}_i|| - ||\vec{X}_{e_{min}} - \vec{X}_i||}, \quad i = 1, 2, \cdots, N. \quad \text{(A.5)}$$

where $d(\vec{X}_j) = \sum_{i=1}^{N} W_i ||\vec{X}_j - \vec{X}_i||$ and $\vec{X}_{j_0}$ is the current filter output.

See Also

Vwmedfilt, Vwmedian, Section 6.7.2 and [173].

Figure A.8 Luminance of the noiseless image, noisy image, and output of the VWM filter.

opt_weights_L

Purpose   Finds the optimum weights for a location estimator using *L*-filters.

Syntax   wopt=opt_weights_L(x,w,d)

Description   wopt=opt_weights_L(x,w,d) calculates a $w \times w$ correlation matrix and the expected values of the order statistics of the vector x and uses them to calculate the optimum weights for the location estimator. w is the filter window size and d is the reference signal.

Example   Suppose a 3V DC signal is embedded in alpha-stable noise with parameters $\alpha = 1$, $\delta = 0$, $\gamma = 0.2$, $\beta = 3$. find the optimum 9-tap *L*-filter to estimate the DC signal.

x = astable(1, 1000, 1, 0, 0.2, 3)
wopt = opt_weights_L(x,9,3)
y = Lfilter (x,wopt).

The resulting weights are: $wopt = [-0.012, -0.0209, -0.0464, 0.3290, 0.5528, 0.1493, 0.1010, -0.0630, -0.0012]$. The noisy signal has an average value of 4.2894 and a MSE of 1.036×10^3. The filtered signal has an average value of 2.9806 and a MSE of 0.0562.

Algorithm

$$wopt = R^{-1}d\mu^T$$

where R is the correlation matrix of the order statistics of X and μ is the vector of expected values of the order statistics.

See Also   Lfilter, opt_weights_Ll, Section 7.3.1 and [38].

opt_weights_Ll

Purpose

Finds the optimum weights for a location estimator using $L\ell$ filters.

Syntax

wopt=opt_weights_Ll(x,w,d)

Description

opt_weights_Ll(x,w,d) calculates a $w^2 \times w^2$ correlation matrix and the expected values of the vector $X_{L\ell}$ and uses them to calculate the optimum weights for the location estimator. w is the window size and d is the desired signal.

Example

Suppose a 3V DC signal is embedded in alpha-stable noise with parameters $\alpha = 1$, $\delta = 0$, $\gamma = 0.2$, $\beta = 3$. find the optimum 9-tap $L\ell$-filter to estimate the DC signal.

x = 3 + astable(1, 1000, 1, 0, 0.2, 0)
wopt = opt_weights_Ll(x,9,3)
y = Llfilter (x,wopt).

The noisy signal has an average value of 4.2894 and a MSE of 1.036×10^3. The filtered signal has an average value of 2.9810 and a MSE of 0.0554.

Algorithm

$$wopt = R^{-1}d\mu^T$$

See Also

Llfilter, opt_weights_L, Section 7.3.1 and [79, 155].

opt_WMMII

Purpose   Finds the optimal weights for the multivariate weighted median filter II.

Syntax   [v,w,y,e] = Opt_WMMII(x,d,wsize,mu)

Description   Opt_WMMII(x,d,wsize,mu) Uses an adaptive algorithm to calculate the optimal weights **V** and **W** for the WMM filter II. The parameters of the function are:

x Input data vector.

wsize Window size in vector form ($[m\ n]$).

d desired signal.

mu Vector containing the step sizes for the outer and inner weights respectively ($[muv\ muw]$)

v M-dimensional vector containing the optimal time/space N-dimensional vector weights ($N = m \times n$).

w $M \times M$ matrix containing the optimal cross-channel weights. M is the dimension of the input samples.

y Marginal multivariate weighted median filtered sequence.

e Absolute error of the output.

See Also   WMMII and Section 6.7.4.

parestssd

Purpose   Parameter estimates for Symmetric Stable Distributions

Syntax   [alpha, disp, loc] = parestssd(x)

Description   parestssd(x) uses a method based on sample fractiles to esti-
                      mate the characteristic exponent (alpha), the dispersion (disp)
                      and the location parameter (loc) of the symmetric alpha-stable
                      distribution that outputs x.

Example   x = astable(1, 10000, 1.75, 0, 2, 5);
                      [alpha, disp, loc] = parestssd(x);
                      alpha = 1.7436
                      dips = 2.0188
                      loc   = 5.0289

Algorithm   parestssd uses a simplified method based on McCulloch's frac-
                      tile method for estimating the characteristic exponent and the
                      dispersion of a symmetric α-stable distribution. This method is
                      based on the computation of four sample quantiles and simple
                      linear interpolations of tabulated index number. The location
                      parameter (loc) is estimated as a p-percent truncated mean. A
                      75% truncated mean is used for $\alpha > 0.8$ whereas for $\alpha \leq 0.8$,
                      a 25% truncated mean is used. McCulloch's method provides
                      consistent estimators for all the parameters if $\alpha \geq 0.6$.

See Also   astable, Section 2.3.3 and [142, 149, 75].

pwmedfilt2

Purpose   Two-dimensional permutation weighted median filtering.

Syntax   y=pwmedfilt2(X,N,Tl,Tu,a)

Description   pwmedfilt2 filters X with a weighted median filter of size N^2
                   whose weights depend on the ranking of the center sample.

Example   load portrait.mat
                   X = imnoise(I,'salt & pepper',.03)
                   Y = pwmedfilt2(X, 5, 6, 20, 0)
                   imshow([I X Y])

Algorithm   pwmedfilt2 pases a window over the data in X that selects N^2
                   samples. It sorts them and finds the rank of the center sample.
                   The program performs an identity operation (i.e., the center
                   weight is set equal to the window size N^2 while all the others
                   are set to one) when the rank of the center sample is in the
                   interval [Tl Tu], and a standard median operation (i.e., all the
                   weights are set to one) when the rank of the center sample is
                   outside the interval.

See Also   wmedfilt2, Section 6.1.1 and [10, 93].

rwmedfilt

Purpose One-dimensional recursive weighted median filter.

Syntax y = rwmedfilt(x, a, b)
y = rwmedfilt(x, a, b, oper)

Description rwmedfilt(x, w) filters the one-dimensional sequence x using a recursive weighted median filter with weights a and b, where a is an N-component vector containing the feedback filter coefficients and b is an M-component vector containing the feedforward filter coefficients.

oper indicates the kind of filtering operation to be implemented. oper = 0, for low-pass filter applications whereas for high-pass or band-pass applications oper = 1. The default value for oper is 0.

See Also wmedfilt, rwmedopt, Section 6.4 and [14].

rwmedfilt2

Purpose

Two-dimensional recursive weighted median filter.

Syntax

Y = rwmedfilt2(X, W)
Y = rwmedfilt2(X, W, a)

Description

Y = rwmedfilt2(X, W) filters the data in X with the two-dimensional recursive weighted median with real-valued weights W. The weight matrix, W, contains the feedback and feedfoward filter coefficients according to the following format

$$
W = \begin{bmatrix}
A_{-m,-n,} & \cdots & A_{-m,0} & \cdots & A_{-m,n} \\
\vdots & & \vdots & & \vdots \\
A_{0,-n} & \cdots & B_{0,0} & \cdots & B_{0,n} \\
\vdots & & \vdots & & \vdots \\
A_{m,-n} & \cdots & A_{m,0} & \cdots & A_{m,n}
\end{bmatrix}
$$

where $A_{i,j}$'s are the feedback coefficients, $B_{i,j}$'s are the feed-foward coefficients, and $2m + 1 \times 2n + 1$ is the observation window size.

rwmedfilt2(X, W, a) uses a third input parameter to indicate the filtering operation to be implemented.

- a = 0 for low-pass operations
- a = 1 for band-pass or high-pass operations

Example

load portrait.mat
X = imnoise(I,'salt & pepper',.05)
W = [1 1 1;1 4 1;1 1 1]
Y = rwmedfilt2(X, W, 0)
imshow(X,[ ])
figure
imshow(Y,[ ])

Algorithm

rwmedfilt2 passes a window over the image, X that selects, at each window position, a set of samples to compromise the observation array. The observation array for a window of size

$2m+1 \times 2n+1$ positioned at the ith row and jth column is given by [Y(i-m: i-1, j-n:j+n); Y(i, j-n: j-1), X(i, j: j+n); X(i+1: i+m, j-n:j+n)] where Y(k,l) is the previous filter output.

rwmedfilt2 calls **wmedian** and passes the observation array and the weight W as defined above.

See Also   wmedian, rwmedopt2, Section 6.4 and [14].

rwmedopt

Purpose
Design one-dimensional recursive weighted median filters using the fast "recursive decoupling" adaptive optimization algorithm.

Syntax
[fb, ff] = rwmedopt(x, d, fb0, ff0, mu, a)
[fb, ff, e, y] = rwmedopt(x, d, fb0, ff0, mu, a)

Description
rwmedopt implements the fast "recursive decoupling" adaptive optimization algorithm for the design of recursive WM filters. The optimal recursive WM filter minimizes the mean absolute error between the observed process, x, and the desired signal, d. The input parameters are as follows.

- x is the training input signal.

- d is the desired signal. The algorithm assumes that both x and d are available.

- fb0 is an N-component vector containing the initial values for the feedback coefficients. It is recommended to initialize the feedback coefficients to small positive random numbers (on the order of 10^{-3}).

- ff0 is an M-component vector containing the initial values for the feedforward coefficients. It is recommended to initialize the feedforward coefficients to the values outputted by Matlab's fir1 with M taps and the same passband of interest.

- mu is the step-size of the adaptive optimization algorithm. A reliable guideline to select the algorithm step-size is to select a step-size on the order of that required for the standard LMS algorithm.

- a is the input parameter that defines the type of filtering operation to be implemented. a = 0 for low-pass applications. a = 1 for high-pass and band-pass applications.

[fb, ff] = rwmedopt(x, d, fb0, ff0, mu, a) outputs the optimal feedback and feedfoward filter coefficients. [fb, ff, e, y] = rwmedopt(x, d, fb0, ff0, mu, a) also outputs the error between the desired signal and the recursive WM filter output, (e), and the recursive WM filter output (y) as the training progresses.

Example

Design an one-dimensional band-pass recursive WM filter with pass band between 0.075 and 0.125 (normalized frequency, where 1.0 corresponds to half the sampling rate). Compare the performance of the designed recursive WM filter to that yielded by a linear IIR filter with the same number of taps and passband of interest.

```
% TRAINING STAGE
x = randn(1, 700);                          %training data
lfir = fir1(121, [0.075 0.125]);            %model
d = filter(lfir, 1, x);                     %Desired signal
fb0 = 0.001*rand(1, 31);                    %Initial weights
ff0 = fir1(31, [0.075 0.125]);
mu = 0.01;   a = 1;
[fb, ff] = rwmedopt(x, d, fb0, ff0, mu, a)  %Training
% TESTING STAGE
Fs = 2000;
t = 0: 1/Fs: 2;
z = chirp(t,0,1,400) ;                      %Test signal
```

```
% Linear IIR filter with the same passband of interest
[fbiir,ffiir]=yulewalk(30,  [0 0.075 0.075 0.125 0.125 1],
                            [0 0 1 1 0 0]);
Orwm = rwmedfilt(z, fb, ff, 1);   % RWM filter output
Oiir = filter(fbiir, ffiir, z);       % IIR filter output
```

```
figure
subplot(3,1,1);    plot(z);       axis([1 1200 -1 1]);
subplot(3,1,2);    plot(Oiir);    axis([1 1200 -1 1]);
subplot(3,1,3);    plot(Orwm);    axis([1 1200 -1 1]);
```

```
% Test stage with α-stable noise
zn = z + 0.2 astable(1, length(z),1.4);
Orwmn = rwmedfilt(zn, fb, ff, 1);  % RWM filter output
Oiirn = filter(fbiir, ffiir, zn);      % IIR filter output
```

```
figure
subplot(3,1,1);    plot(zn);      axis([1 1200 -4 4]);
subplot(3,1,2);    plot(Oiirn);   axis([1 1200 -1.5 1.5]);
subplot(3,1,3);    plot(Orwmn);   axis([1 1200 -1 1]);
```

See Also

rwmedian, wmedopt, Section 6.4.2 and [14, 176].

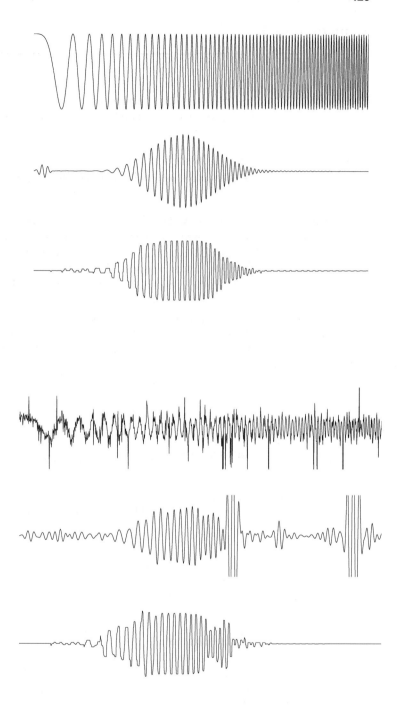

rwmedopt2

Purpose

Design two-dimensional recursive weighted median filters using the fast "recursive decoupling" adaptive optimization algorithm.

Syntax

Wopt = rwmedopt2(X, D, W0, mu, a)
[Wopt, e, Y] = rwmedopt2(X, D, W0, mu, a)

Description

rwmedopt2(X, D, W0, mu, a) implements the fast "recursive decoupling" optimization algorithm for the design of two-dimensional recursive weighted median filters. X is the input image used as training data, D is the desired image, W0 is an $m \times n$ matrix containing the initial values for the filter coefficients, mu is the step-size used in the adaptive algorithm and a is an input parameter that specifies the type of filtering operation on training. a = 0 for low-pass filtering operations, whereas for high-pass or band-pass filtering operations a = 1.

Wopt is an $m \times n$ matrix that contains the optimal feedback and feedforward filter coefficients in accordance with the following format

$$\text{feedback} = \text{Wopt}(i,j) \text{ for } \begin{cases} i = 1,2 \ldots \frac{m-1}{2} & \text{and} \quad j = 1,\ldots,n \\ i = \frac{m+1}{2} & \text{and} \quad j = 1,2 \ldots \frac{n-1}{2} \end{cases}$$

$$\text{feedforward} = \text{Wopt}(i,j) \text{ for } \begin{cases} i = \frac{m+1}{2} \text{ and } j = \frac{n+1}{2}, \ldots, n \\ i = \frac{m+3}{2} \ldots m \text{ and } j = 1, \ldots n \end{cases}$$

where n and m are assumed to be odd numbers.

[Wopt, e, Y] = rwmedopt2(X, D, W0, mu, a) outputs the optimal filter weights, the error signal, $e = d - Y$, and the recursive WM filter output, Y, as the training progresses.

Example

```
load portrait.mat
D=I;   %Desired image
Xn = 255*imnoise(D/255,'salt & pepper',0.1);   % Training
data
W0 = ones(3,3); W0(2,2) = 5;   % Initialization of filter co-
efficients
mu = 0.001;
```

```
Wopt = rwmedopt2(Xn(1:60,1:60), D(1:60,1:60), W0, mu,
0);
Y0=rwmedfilt2(Xn, W0, 0);
Yopt=rwmedfilt2(Xn, Wopt, 0);
imshow([D, Xn; Y0 Yopt], []);
```

See Also   rwmedfilt2, rwmedopt1, Section 6.4.2 and [14, 176].

SSP2MPCCWM

Purpose

Finds the closest marginal phase-coupled complex weighted median filter to a given complex-valued linear filter.

Syntax

W = SSP2MPCCWM(h,u)

Description

SSP2MPCCWM(h,u) calculates the marginal phase-coupled complex weighted median filter W that is closest in the mean square error sense to the complex valued linear filter h according to the theory of Mallows.

Example

$$h = [-0.0147 + 0.0930i, \ -0.1044 + 0.1437i,$$
$$0.3067 - 0.1563i, \ 0.5725, \ 0.3067 + 0.1563i,$$
$$-0.1044 - 0.1437i, \ -0.0147 - 0.0930i]$$
$$W = [-0.0055 + 0.0346i, \ -0.0662 + 0.0911i,$$
$$0.1857 - 0.0946i, \ 0.2878, \ 0.1857 + 0.0946i,$$
$$-0.0662 - 0.0911i, \ 0.0055 - 0.0346i]$$

Algorithm

The algorithm divides the complex weights in magnitude and phase, normalizes the magnitudes and runs **SSP2WM** with the normalized magnitudes of the weights as the input and the parameter u as the step size. The output of this algorithm is then coupled with the phases of the original linear weights to get the final median weights.

See Also

SSP2WM, Section 6.6.5 and [103].

SSP2WM

Purpose   Finds the closest weighted median filter to a given linear FIR filter.

Syntax   W = SSP2WM(h,u)

Description SSP2WM(h,u) calculates the weighted median filter W that is closest in the mean square error sense to the linear filter h according to the theory of Mallows. The code implements and adaptive algorithm that tries to minimize the MSE between the output of the reference linear filter h and the tentative weighted median filter W. The weights of the median filter are adjusted according to the first derivative of the MSE and the step size u.

Example

$$h = [0.0548, -0.1214, 0.1881, -0.2714,$$
$$0.1881, -0.1214, 0.0584]$$
$$\text{SSP2WM}(P, 0.01) = [0.0544, -0.1223, 0.1891, -0.2685,$$
$$0.1891, -0.1223, 0.0544]$$

Algorithm   The weights of the median filter are updated according to the equation

$$W_l(n+1) = W_l(n) + \mu \left(-\frac{\partial}{\partial W_l} \mathbf{J}(\mathbf{W}) \right).$$

The derivative $\frac{\partial}{\partial W_l} \mathbf{J}(\mathbf{W})$ can be found in [1].

See Also   WM2SSP_real, WM2SSP_realfd, Section 6.2 and [103].

t_mean

Purpose   Calculates the trimmed mean of a data vector.

Syntax   T = t_mean(X,alpha)

Description   t_mean sorts the samples in the input vector X, then discards
                     the highest and the lowest alpha-order statistics and calculates
                     the average of the remaining samples.

Example

$$x = [-2, 2, -1, 3, 6, 8];$$
$$\alpha = 1$$
$$\text{t\_mean}(x, \alpha) = \frac{1}{4}[-1 + 2 + 3 + 6]$$
$$= 2.5$$

See Also   w_mean, wmedian and Section 4.3.

Vwmedfilt

Purpose   Performs vector-weighted median filtering of the data X.

Syntax   [y]=Vwmedfilt(X,W)

Description   Vwmedfilt(X,W) Filters the vector valued data in X using the real-valued weights in the $m \times n$ matrix W.

Algorithm   Vwmedfilt(X,W) passes a window over the data in X and passes the data in the window to Vwmedian to calculate the output of the filter at a given instant.

See Also   Opt_Vwmedfilt, Vwmedian and Section 6.7.2.

Vwmedian

Purpose

Performs vector-weighted median filtering of an observation vector with real-valued weights

Syntax

[y,dist]=Vwmedian(X,W,dist)

Description

Vwmedian(X,W,dist) filters the vector valued data in X using the real-valued weights in the $m \times n$ matrix W. X and W should have the same size. The parameter dist is a $D \times D$ matrix ($D = m \times n$) that initializes the values of the distances between samples, it is used to avoid recalculation of distances in Vwmedfilt and, if unknown, it should be initialized to a zero matrix.

Algorithm

Vwmedian(X,W,dist) calculates, for each sample, the distances to all the other samples and obtains a weighted sum of them using the weights in W. Then it compares the results and chooses the minimum weighted sum. The output of the filter is the sample corresponding to this weighted sum.

See Also

Opt_Vwmedfilt, Vwmedfilt and Section 6.7.2.

w_mean

Purpose   Calculates the windsorized mean of a data vector.

Syntax   w_mean(X,r)

Description w_mean sorts the samples in the vector X, then removes the lowest and highest r-order statistics and replaces them with the $r + 1$st and the $N - r$th-order statistics of the vector to calculate the mean.

Example

$$x = [-2, 2, -1, 3, 6, 8];$$
$$r = 1$$
$$\text{w\_mean}(x, r) = \frac{1}{6}[2 + 3 + 2 * (-1 + 6)]$$
$$= 2.5$$

Algorithm

$$W_N(r) = \frac{1}{N}\left[\sum_{i=r+2}^{N-r-1} Z_{(i):N} + (r + 1)\left[Z_{(r+1):N} + Z_{(N-r):N}\right]\right]$$

See Also   t_mean, wmedian and Section 4.3.

WM2SSP_real

Purpose

Finds the linear filter closest in the MSE sense to a given real-valued weighted median filter.

Syntax

h = WM2SSP_real(W)

Description

WM2SSP_real(W) calculates the linear filter h that is closest in the mean square error sense to the real-valued weighted median filter W according to the theory of Mallows.

Example

$$W = \langle 1, \ -2, \ 3, \ -4, \ 3, \ -2, \ 1 \rangle$$
$$\text{WM2SSP\_real}(W) = \left[\frac{23}{420}, \ -\frac{17}{140}, \ \frac{79}{420}, \ -\frac{19}{70}, \ \frac{79}{420}, \ -\frac{17}{140}, \ \frac{23}{420} \right]$$

Algorithm

The algorithm is based on the closed form function for the samples selection probabilities developed in [1].

See Also

SSP2WM, WM2SSP_realfd, Section 6.2.2 and [103, 137].

WM2SSP_realfd

Purpose

Finds the closest linear filter to a given real-valued weighted median filter in the MSE sense and the first derivative (gradient) of the cost function

$$J(\mathbf{W}) = \|\mathbf{P}(\mathbf{W}) - \mathbf{h}\|^2 = \sum_{j=1}^{N} \left(p_j(\mathbf{W}) - h_j \right)^2$$

where $\mathbf{W}$ is a vector of median weights and $\mathbf{h}$ is a normalized linear filter.

Syntax

[h,fd] = WM2SSP_realfd(W)

Description

WM2SSP_realfd(W) calculates the linear filter h that is closest in the mean square error sense to the real-valued weighted median filter W according to the theory of Mallows. It also calculates the gradient of the cost function indicated above. This algorithm is used in the iterative calculation of the weighted median closest to a given linear filter in SSP2WM.

Algorithm

The algorithm is based on the closed form function for the samples selection probabilities developed in [1].

See Also

SSP2WM, WM2SSP_real, Sections 6.2.2, 6.2.4 and [103].

wmedfilt

Purpose

One-dimensional weighted median filtering.

Syntax

```
y = wmedfilt(x, w)
y = wmedfilt(x, w, a)
```

Description

y = wmedfilt(x, w) filters the data in vector x with the weighted median filter described by weight vector w. The window size is specified by the dimensions of the vector w.

y = wmedfilt(x, w, a) uses the third input argument to specify the filtering operation at hand. For low-pass filtering operation a is set to zero whereas for band-pass or high-pass filtering application a is set to one.

Algorithm

wmedfilt passes a window over the data x that computes, at each instant n, the weighted median value of the samples x(n-(m-1)/2), ..., x(n), ..., x(n+(m-1)/2) where m = length(w). The resultant value is the filter output at instant n.

Due to the symmetric nature of the observation window, m/2 samples are appended at the beginning and at the end of the sequence x. Those samples appended at the beginning of the data have the same value as the first signal sample and those appended at the end have the same value of the last signal sample.

See Also

wmedian, wmedopt, Lfilter, Section 6.1 and [6, 189].

wmedfilt2

Purpose

Two-dimensional weighted median filtering.

Syntax

Y = wmedfilt2(X, W, a)

Description

Y = wmedfilt2(X, W) filters the image in X with the two-dimensional weighted median filter with real-valued weights W. Each output pixel contains the weighted median value in the m-by-n neighborhood around the corresponding pixel in the input image, where [m, n] = size(W). The third input argument is set to zero for low-pass filtering and to one for band-pass or high-pass filtering. The program appends m/2 (n/2) rows(columns) at the top(left) and bottom(right) of the input image to calculate the values in the borders.

Example

```
load portrait.mat                    imshow(X,[ ])
X = imnoise(I,'salt & pepper',.03)   figure
W = [1 1 1;1 4 1;1 1 1]              imshow(Y,[ ])
Y = wmedfilt2(X, W, 0)
```

Algorithm

wmedfilt2 uses wmedian to perform the filtering operation using an m-by-n moving window.

See Also

wmedian, wmedopt2, Section 6.1 and [13, 115].

Figure A.9 Image contaminated with salt-and-pepper noise and output of the weighted median filter.

wmedian

Purpose   Compute the weighted median of an observation vector.

Syntax   y = wmedian(x, w, a)

Description   wmedian(x, w, a) computes the weighted median value of the observation vector x. w is a real-valued vector of the same length of x that contains the filter weights. For a = 0 the output is one of the signed samples, whereas for a = 1 the output is the average of two signed samples. The default value for a is zero.

Example   x = [-2, 2, -1, 3, 6, 8];
w = [0.2, 0.4, 0.6, -0.4, 0.2, 0.2];
wmedian(x, w, 0) = -1
wmedian(x, w, 1) = -1.5

See Also   wmedfilt, wmedfilt2, Section 6.1 and [6].

wmedopt

Purpose

Design one-dimensional weighted median filters using a fast adaptive optimization algorithm.

Syntax

wopt = wmedopt(x, d, w0, mu, a)
[wopt, e, y] = wmedopt(x, d, w0, mu, a)

Description

wmedopt implements the fast adaptive optimization algorithm for the design of weighted median filters. The filters are optimal in the sense that the mean absolute error between the observed process, x, and the desired signal, d, is minimized.

w0 are the initial values for the filter coefficients. As good initial values, use the filter coefficients of a linear FIR filter designed for the same application. That is, w0 = fir1(n-1,Wn) where n is the number of filter coefficients and Wn represents the frequency of interest. See MATLAB's fir1 function for further information.

mu is the step size of the adaptive optimization algorithm. A reliable guideline to select the algorithm step-size is to select a step size on the order of that required for the standard LMS.

a = 0 for low-pass filter applications, 1 otherwise.

wopt = wmedopt(x, d, w0, mu, a) returns the row vector, wopt, containing the n optimal filter coefficients. [wopt, e, y] = wmedopt(x, d, w0, mu, a) also returns the error between the desired signal and the WM filter output (e), and the WM filter output (y) as the training progresses.

Example

Design a high-pass WM filter with cutoff frequency equal to 0.3 (normalized frequency, where 1 corresponds to half the sampling frequency), and then test its robustness again impulsive noise.

```
Fs = 2000;                           % Sampling frequency
t = [0: 1/Fs: 1];
x = sin(2*pi*20*t) + sin(2*pi*400*t);   % Training signal
```

```
d = sin(2*pi*400*t);                    % Desired signal
n = 40;                                 % Number of filter coefficients
mu = 0.001;                             % Step size parameter
w0 = fir1(n, 0.3, 'high');              % Initialization of coefficients
wopt = wmedopt(x, d, w0, mu, 1);        % Training stage
Owmf = wmedfilt(Ts, wopt, 1);           % WM filter output
Ofir = filter(w0, 1, Ts);               % Linear FIR filter output
% Testing stage with α-stable noise
Tn = Ts + 0.5*astable(1, length(Ts), 1.50);
Owmfn = wmedfilt(Tn, wopt, 1);
Ofirn = filter(w0, 1, Tn);
```

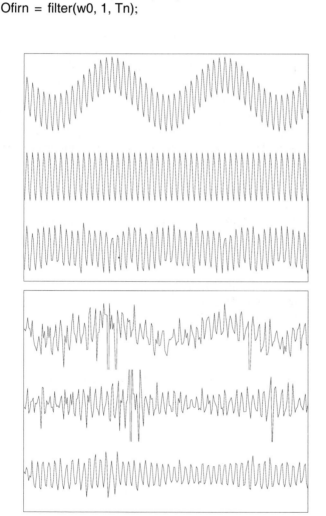

See Also   wmedian, wmedopt , Section 6.3.2 and [6, 200].

wmedopt2

Purpose   Design two-dimensional weighted median filters using a fast
                    adaptive optimization algorithm.

Syntax   Wopt = wmedopt2(X, D, W0, mu, a)
                    [Wopt, e, Y] = wmedopt2(X, D, W0, mu, a)

Description   wmedopt2(X, D, w0, mu, a) outputs the optimal two-dimensional
                    filter coefficients where X is the training input image, D is the
                    desired image, W0 is a matrix containing the initial values of
                    the weights, mu is the step size and a describes the kind of
                    WM filtering operation. Use a = 0 for low-pass operations and
                    a = 1 for band-pass or high-pass operations. wmedopt2 also
                    outputs the difference between the training input data and the
                    desired signal, and the filter output as the training progresses.

Example   Xn = 255*imnoise(D/255,'salt & pepper',0.1);   % Training
                    data
                    W0 = ones(3,3); W0(2,2) = 5;   % Initialization of filter co-
                    efficients
                    [Wopt, e, Y] = wmedopt2(Xn(1:60,1:60), D(1:60,1:60),
                    W0, 0.001, 0);
                    Y0=wmedfilt2(Xn, W0, 0);
                    Yopt=wmedfilt2(Xn, Wopt, 0);
                    imshow([D, Xn, Y0 Yopt], []);

See Also   wmedopt, wmedfilt2, Section 6.3.2 and [6, 200].

Figure A.10 Noiseless image, image contaminated with 10% salt-and-pepper noise, output of a weighted median filter, and output of an optimized weighted median filter.

WMMII

Purpose
Performs multivariate weighted median filtering of a vector valued signal.

Syntax
[y] = WMMII(X,W,V,wsize)

Description
WMMII(X,W,V,wsize) filters the data in vector X with the multivariate weighted median filter II described by the M-dimensional vector V and the $M \times M$ matrix W. Each component of V is a N-dimensional vector. N is the window size and M is the dimension of the input samples. The window size is specified by the parameter wsize in the form $[m\ n]$, where $m \times n = N$.

Algorithm
Running window over the data in X that computes, for each component of each sample in the window, the weighted median described by the columns of W. After that it calculates the output of the marginal weighted median described by each component of V applied to the outputs of the previous operation.

See Also
marginalWMMI, Opt_WMMII, section 6.7.4 and [10].

wmsharpener

Purpose   Sharpens a grayscale image using permutation high pass median filters.

Syntax   s=wmsharpener(X, N, lambda1, lambda2, L)

Description   wmsharpener performs a linear combination of the original image and two high-pass filtered versions of it (positive edges enhanced and negative edges enhanced) to obtain a sharper version of it.

Example   Y = wmsharpener(I, 3, 2, 2, 1)

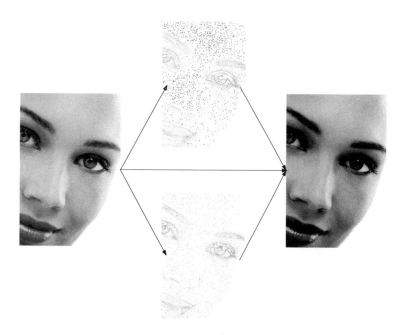

Algorithm   To obtain the two high-pass filtered images, the same permutation high-pass weighted median is applied to the original image, to enhance positive edges, and to its negative, to enhance negative edges. Once the two filtered images are obtained, they are scaled with the coefficients lambda1 and lambda2 and added to the original image to obtain the final output.

See Also   pwmedfilt2, Example 6.1 and [10].

wmyrfilt

Purpose   One-dimensional weighted myriad filter.

Syntax
y = wmyrfilt(x, w, k)
y = wmyrfilt(x, w, k, method)

Description   y = wmyrfilt(x, w, k) performs the weighted myriad filtering of the data x. w is an N-component vector that contains the myriad filter coefficients and k is the linearity parameter. The observation window size is defined by the length of the weight vector w.

The nth element of the output vector y is the weighted myriad value of observation vector [x(n-(N-1)/2), ..., x(n), ... x(n+(N-1)/2)]. Due to the symmetric nature of the observation window, wmyrfilt(x, w, k) pads the input sequence x with (N-1)/2 zeros at the beginning and at the end. If the fourth input parameter is used, it indicates the method used to compute the weighted myriad. method is a string that can have one of these values:

- 'exact' (default) uses the exact method to compute the weighted myriad value of the observation vector. At each window position, wmyrfilt(x, w, k,'exact') calls the wmyriad function.

- 'approximate' uses the approximate method to compute the weighted myriad value of the observation vector. At each window position, wmyrfilt(x, w, k,'approximate') calls the fwmyriad function.

If you omit the method argument, wmyrfilt uses the default value of 'exact'.

Example   Test the robustness properties of weighted myriad filter.

```
t = 0:.001:0.5;
x = sin(2*pi*10*t);
xn = x +.05*astable(1, length(x), 1.5);
w = fir1(7, 0.3);
linearFIR = filter(w, 1, xn);                 % FIR filter output
wmyrE = wmyrfilt(xn, w, 0.1, 'exact');        % WMy exact output
wmyrA = wmyrfilt(xn, w, 0.1, 'approximate');  % WMy approx output
figure
subplot(4,1,1);  plot(xn);        axis([0 500 -2 4]);   axis off
subplot(4,1,2);  plot(linearFIR); axis([0 500 -2 4]);   axis off
subplot(4,1,3);  plot(wmyrE);     axis([0 500 -2 4]);   axis off
subplot(4,1,4);  plot(wmyrA);     axis([0 500 -2 4]);   axis off
```

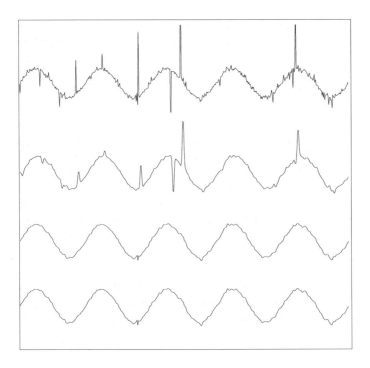

See Also   wmyropt, wmyriad, fwmyriad , Section 9.1 and [112].

wmyriad

Purpose   Compute the weighted myriad of an observation vector.

Syntax   Omyr = wmyriad(x,w,k)

Description   wmyriad(x,w,k) computes the weighted myriad of an observation vector x. The length of the observation vector defines the
observation window size, N.

- w is a N-component real-valued vector that contains the
  coefficients of the filter.

- k is the linearity parameter that takes on positive values.
  As k goes to $+\infty$, wmyriad(x,w,k) reduces to a weighted
  sum of the observation samples. If k tends to zero, the
  weighted myriad reduces to a selection filter. If k is not
  specified, it is set to one.

Example   x = [3 2 4 5 8];
w = [0.15 -0.2 0.3 -0.25 0.1];
Owmf = wmyriad(x,w,1);
Owmf = 3.3374

Algorithm   Given a set of observation samples $x_1, x_2, \ldots, x_N$, a real-valued
weights $w_1,$
$w_2, \ldots,\ w_N$ and a real parameter $k > 0$, the sample weighted
myriad of order k is defined as

$$\hat{\beta}_k = \arg\min_{\beta} \prod_{i=1}^{N} \left[k^2 + |w_i|\, (s_i - \beta)^2 \right]$$

According to the window size, wmyriad uses different algorithms to find the value of β that minimizes the above equation.
For small window size, $N \leq 11$, wmyriad treats the above equation as a polynomial function. The global minimum is found by
examining all the local extrema, which are found as the roots
of the derivative of the polynomial function. For large window

size, $N > 11$, wmyriad uses MATLAB's fmin function to find the global minimum.

See Also fwmyriad, fmin, Section 9.1 and [83].

Index

CUSTOMER NOTE: IF THIS BOOK IS ACCOMPANIED BY SOFTWARE,
PLEASE READ THE FOLLOWING BEFORE OPENING THE PACKAGE.

This software contains files to help you utilize the models described in the
accompanying book. By opening the package, you are agreeing to be
bound by the following agreement:

This software product is protected by copyright and all rights are reserved by
the author and John Wiley & Sons, Inc. You are licensed to use this software
on a single computer. Copying the software to another medium or format for
use on a single computer does not violate the U.S. Copyright Law. Copying
the software for any other purpose is a violation of the U.S. Copyright Law.

This software product is sold as is without warranty of any kind, either
express or implied, including but not limited to the implied warranty of merchantability and fitness for a particular purpose. Neither
Wiley nor its dealers or distributors assumes any liability of any
alleged or actual damages arising from the use of or the inability
to use this software. (Some states do not allow the exclusion of
implied warranties, so the exclusion may not apply to you.)

WILEY